The Politics of Human Vulnerability to Climate Change

This book compares how the social consequences of climate change are similarly unevenly distributed within China and the United States, despite different political systems.

Focusing on the cases of Atlanta, USA, and Jinhua, China, Julia Teebken explores a set of path-dependent factors (lock-ins), which hamper the pursuit of climate adaptation by local governments to adequately address the root causes of vulnerability. Lock-ins help to explain why adaptation efforts in both locations are incremental and commonly focus on greening the environment. In both these political systems, vulnerability appears as a core component along with the reconstitution of a class-based society. This manifests in the way knowledge and political institutions operate. For this reason, Teebken challenges the argument that China's environmental authoritarian structures are better equipped in dealing with matters related to climate change. She also interrogates the proposition that certain aspects of the liberal democratic tradition of the United States are better suited in dealing with social justice issues in the context of adaptation. Overall, the book's findings contradict the widespread assumption that developed countries necessarily have higher adaptive capacity than developing or emerging economies.

This volume will be of great interest to students and scholars of climate justice and vulnerability, climate adaptation and environmental policy and governance.

Julia Teebken is a political and social scientist, and currently is a Postdoc in the Peking-Princeton Postdoctoral Program. She holds a Ph.D. degree from the Freie Universität Berlin, Germany. Her research focuses on the political responses and causes of (social) vulnerability to climate change. In this capacity, she researches inequality and climate change adaptation policy across different political systems.

Routledge Studies in Climate Justice

Series Editor: Tahseen Jafry, *Professor of Climate Justice and Director of The Centre for Climate Justice, Glasgow Caledonian University, UK*

Climate justice is a rapidly growing field of critical enquiry which concentrates on the social dimensions of climate change, including the unequal nature of its physical, socioeconomic and political impacts and humanity's responses to them. From the stark warnings of the Intergovernmental Panel on Climate Change (IPCC) to the direct action protests of Extinction Rebellion, there is growing interest in the study of the global inequalities of climate change.

Routledge Studies in Climate Justice will comprise monographs and edited collections addressing cutting-edge questions in the growing field of climate justice. The series will include a diverse range of topics, including climate justice and international development, intersectionality and climate inequality, climate governance and policy, gender and climate change, climate migration and displacement, health and well-being, climate justice activism, pedagogy and participation, and urban climate justice.

Climate Justice and Collective Action
Angela Kallhoff

The Politics of Human Vulnerability to Climate Change
Exploring Adaptation Lock-ins in China and the United States
Julia Teebken

For more information about this series, please visit: www.routledge.com/Routledge-Studies-in-Climate-Justice/book-series/RSCJ

The Politics of Human Vulnerability to Climate Change

Exploring Adaptation Lock-ins in China and the United States

Julia Teebken

Routledge
Taylor & Francis Group

LONDON AND NEW YORK

earthscan
from Routledge

First published 2022
by Routledge
4 Park Square, Milton Park, Abingdon, Oxon OX14 4RN

and by Routledge
605 Third Avenue, New York, NY 10158

Routledge is an imprint of the Taylor & Francis Group, an informa business

British Library Cataloguing-in-Publication Data
A catalogue record for this book is available from the British Library

Library of Congress Cataloging-in-Publication Data
Names: Teebken, Julia, author.
Title: The politics of human vulnerability to climate change : exploring adaptation lock-ins in China and the United States / Julia Teebken.
Description: Milton Park, Abingdon, Oxon ; New York, NY : Routledge, 2022. | Series: Routledge studies in climate justice | Includes bibliographical references and index.
Identifiers: LCCN 2021048872 (print) | LCCN 2021048873 (ebook)
Subjects: LCSH: Climatic changes—Government policy—China. | Climate change mitigation—China. | Environmental policy— China. | Climatic changes—Government policy—United States. | Climate change mitigation—United States. | Environmental policy— United States.
Classification: LCC QC903.2.C5 T44 2022 (print) | LCC QC903.2.C5 (ebook) | DDC 304.2/5—dc23/eng/20211223
LC record available at https://lccn.loc.gov/2021048872
LC ebook record available at https://lccn.loc.gov/2021048873

ISBN: 978-1-032-02403-5 (hbk)
ISBN: 978-1-032-02404-2 (pbk)
ISBN: 978-1-003-18325-9 (ebk)

DOI: 10.4324/9781003183259

Typeset in Goudy
by Apex CoVantage, LLC

Contents

List of figures   vii
List of tables   ix
Preface   x
Acknowledgments   xi
List of abbreviations   xii

1 Introduction: uneven human vulnerability to climate change   1

2 Vulnerability: core concepts   19

3 Vulnerability and adaptation lock-ins: theoretical foundations and main analytical framework   36

4 Methodological approach   61

5 Vulnerability and adaptation governance in China and the United States   82

6 Regional backgrounds and contextual lock-ins   114

7 Protracted vulnerability   136

8 Accidental adaptation policy   168

9 Lock-ins of political epistemology across different political systems   215

10 Adaptation policy and transformation?   232

11 Conclusion   255

    Appendices   268
    Index   278

Figures

2.1	IPCC's core concepts related to vulnerability	21
2.2	A conceptualization of vulnerability assessment approaches	26
3.1	Analytical framework of potential lock-in dimensions	44
3.2	Overarching explanatory framework of lock-in dynamics as anchored in the political ecology	45
4.1	Holistic multiple case study design	63
4.2	Case study region within the United States, location as outlined by the third National Climate Assessment	66
4.3	Case study region within China	67
4.4	Spatial distribution of SoVI in Georgia, 1980–2010, and Fulton County as persistently vulnerable	68
4.5	Spatial distribution of SoVI in the Yangtze River Delta region and Lanxi and Pan'an as most vulnerable within Jinhua prefecture	69
5.1	Adaptation progress in China	88
5.2	Adaptation progress in the United States (Hawaii and Alaska not included)	93
6.1	Increasing inequality in China and the United States	115
6.2	Location of Atlanta in Fulton County and location of Fulton County within Georgia	118
6.3	Income segregation along public transport routes and interstates in Atlanta	120
6.4	Location of case study unit, major cities and elevation river rich Zhejiang province	125
6.5	Administrative division of Jinhua	126
7.1	Composite vulnerability in Fulton County and Metro Atlanta	139
7.2	Heat hazard index across Zhejiang province with Jinhua at the center (summers 2008–2013)	141
7.3	Differentiation of external populations and causal links to policy factors that influence population vulnerability	152
7.4	Lock-ins related to knowledge and politics that explain protracted vulnerability	157
8.1	Atlanta abandoned train track 2009 and Beltline 2012	176
8.2	Old Fourth Ward flood park before	179

8.3   Old Fourth Ward after   180
8.4   Floodwater landscape in Jinhua, before and after   186
8.5   Flood-resilient green infrastructure landscape in central Jinhua   187
8.6   Flood-resilient green infrastructure landscape in central Jinhua,
      flooded   187
8.7   Lock-ins related to accidental and infrastructure-focused
      adaptation   197
8.8   Factors leading to the perception of overcomplexity   205
9.1   Political epistemological lock-ins   218
10.1 Overview of key vulnerability and adaption lock-ins preventing
      transformative adaptation   238
10.2 Ideal-type phases and substages of the adaptation planning
      process   245

Tables

3.1 Areas and types of adaptation 38
4.1 Comparative characterization of environmental and climate governance 65
4.2 Comparison of regions in which the cases are located 70
5.1 Climate change policy related adaptation efforts in China, the United States and at the global level 85
5.2 Core Executive Orders related to climate adaptation 91
5.3 Comparison of national CCA policy measures 94
5.4 Climate vulnerability priority sectors in China and the United States 95
5.5 Overview of U.S. National Climate Assessments 96
5.6 Overview of China's National Climate Assessments 97
5.7 Environmental consciousness, defining moments and regulations in China and the United States 106
6.1 Floating population in Zhejiang, China, 2000 and 2010 129
6.2 Number of rural-urban populations, population share (2000 and 2010 compared), area, and population density (2010) in Lanxi and Jinhua 129
7.1 Overview of social vulnerability variables 140
7.2 Summary of similarities and differences between Atlanta and Jinhua 144
7.3 Vulnerable groups through the eyes of local officials and policy advisors 154
8.1 Policy efforts addressing climate vulnerability – Atlanta 170
8.2 Atlanta's transition from shocks and stresses to vision-oriented targets 173
8.3 Adaptation priorities according to Zhejiang's 2010 Plan 184
8.4 Overview of relevant policy efforts – Jinhua 191
8.5 Differences in problem recognition, Atlanta and Jinhua 196

Preface

The political dimension of human vulnerability

This book looks at the political nature of uneven human affectedness to climate change. Although our lived environments have actively co-produced the way the climate is changing, we need to find ways how to deal with an uncertain future, which is signified by constantly evolving dynamics of vulnerability. In light of rapidly intensifying climate change, adaptation has become an expansive field of research and practice. In contrast to dominant adaptation practices, research has become more differentiated by also looking at how local populations are being affected differently and which characteristics correspond with greater human exposure. Related vulnerability discourses are manifold and stem from different processes of meaning-making. Although the omnipresence of the concept has been overwhelming, vulnerability remains a relevant concept, as Covid-19 has vividly demonstrated. The most recent debates point to the complex nature and entanglements of different forms of vulnerability. Yet, when approaching the issue in political practice, especially as it relates to climate change and social vulnerability, answers for how to deal with this complex challenge and formulating adequate political responses are not readily available.

This book explores a set of interrelated, path-dependent factors (lock-ins), which hamper substantive adaptation responses. The book also examines to what extent the political dimension has mattered in vulnerability co-creation and maintenance thereby hampering social vulnerability to be addressed. Despite a proliferation of vulnerability debates and adaptation practices, a majority of adaptation efforts continue to focus on green infrastructure and planning. This book looks at the political nature of this and why the transformative responses are not readily available.

Providing comprehensive insights on how inequality perpetuates vulnerability to climate change and which challenges scholars and practitioners face when approaching adaptation options (and defining vulnerability), this book will be of great interest to those studying climate change adaptation and the socio-political dimension of vulnerability across different political systems.

Acknowledgments

This study has been a long journey and helped me adjust my own ideas about vulnerability – especially what it means to call someone "vulnerable." I would like to thank my interviewees in China and the United States who gave me their time and provided me with the necessary insights to conduct this research.

Special thanks go to Miranda Schreurs and Genia Kostka, who both have pushed me to stay true to my own research interests and at times hinted me at the highlights of what I had found. Special thanks go to Terry Cannon, who helped me nail down my research through one tiny comment at a conference in Cardiff on the creation of vulnerability. From there, everything fell in line.

Thanks go to the many people living and working at the Teardown Community in Atlanta, the Vulnerability and the Human Condition Initiative at Emory University, the Ethics Institute at Georgetown University, the Shanghai Jiao Tong University, Graduate School of East Asian Studies and Environmental Policy Research Centre at Freie Universität Berlin. All have provided me with a safe harbor to expand upon my ideas and critically questioning them at the same time.

Finally, I would like to thank Anton – tremendously, Professor Pucci and all my Berliner Flitzpiepen for supporting me along the way. Lastly, danke liebe Eltern, dass ihr einfach nicht müde werdet. Ihr seid die Besten.

Abbreviations

ACCC	Adapting to Climate Change in China
AR	Assessment Report
CAP	Climate Action Plan
CASS	Chinese Academy of Social Sciences
CCA	Climate Change Adaptation
CDC	Centers for Disease Control and Prevention
CPC	Communist Party of China
CEQ	White House Council on Environmental Quality
CMA	Chinese Meteorological Administration
CVA	Climate Vulnerability Assessments
DEE	Department of Ecology and Environment
DRC	Department of Development and Reform Commission
DRR	Disaster Risk Reduction
EO	Executive Order
EPB	Environmental Protection Bureau
EPA	Environmental Protection Agency
FYP	Five-Year Plan
FWT	Five Water Treatment
GCC	Georgia Climate Conference
GEMA	Georgia Emergency Management Agency
GPCR	Great Proletarian Cultural Revolution
ICLEI	Local Governments for Sustainability (former International Council for Local Environmental Initiatives)
ICCATF	The Interagency Climate Change Adaptation Task Force
IPCC	Intergovernmental Panel on Climate Change
MARTA	Metropolitan Atlanta Rapid Transit Authority
MEE	Ministry of Ecology and Environment
MEM	Ministry of Emergency Management
MoA	Ministry of Agriculture
MoF	Ministry of Finance
MoHURD	Ministry of Housing and Urban-Rural Development
MOT	Ministry of Transport
MSA	Multiple Streams Approach

MTS	Mandatory Target System
MWR	Ministry of Water Resources
NAS	National Adaptation Strategy
NCA	National Climate Assessment
NCCCLSG	National Climate Change Coordinating Leading Small Group
NGO	Non-Governmental Organization
NLCCC	National Leading Committee on Climate Change
NLGACC	National Leading Group for Addressing Climate Change
NDRC	National Development and Reform Commission
NOAA	National Oceanic and Atmospheric Administration
OSTP	Office of Science and Technology Policy
PA	Paris Agreement
RCS	River Chief System
SDGs	Sustainable Development Goals
SCP	Sponge City Program
SFGA	State Forestry and Grassland Administration
SLR	Sea-Level Rise
SOA	State Oceanic Administration
SoVI	Social Vulnerability Index
SVAs	Social Vulnerability Assessments
UHI	Urban Heat Island Effect
UNFCCC	United Nations Framework Convention on Climate Change
USGCRP	United States Global Change Research Program
VAs	Vulnerability Assessments
VT	Vulnerability Theory
YRD	Yangtze River Delta
100RC	100 Resilient Cities Initiative

1   Introduction

Uneven human vulnerability to climate change

This book explores the persistence of incremental government responses in the context of uneven human vulnerability to climate change. Within studies on climate change adaptation, social vulnerability to climate change has been pointed out as an adaptive capacity deficit. Addressing the root causes of (social) vulnerability is considered a core tenet to achieve transformative adaptation. Meanwhile, adaptation practitioners are struggling to formulate adequate policy responses. This book examines why this is the case by reflecting upon the political dimension of "social" vulnerability and examining the persistence of interdependent factors ("lock-ins") that maintain, recreate and deteriorate inequal vulnerability to climate change. Lock-ins are conceptualized within a political ecology lens, which helps us to frame vulnerability and adaptation ideas not just as results of political institutions but likewise as an outcome of our lived, contextual environments and human-nature relations. The political ecology perspective enables a critical view on ongoing and path-dependent governance processes, how they interact with questions of power and influence the persistence of vulnerability across multiple scales and areas. These path-dependent processes structure the constraints of adaptation and looking at them helps to critically question the current parameters of adaptation. Ultimately, this angle may offer insights into the question of what role government can play in addressing questions of justice related to vulnerability and adaptation. Thereby, the book attempts to reconcile adaptation critical works nestled in political ecology studies with adaptation as a field of policy research and practice.

The planet has warmed by approximately 1.0 °C (1.8 °F) compared to the pre-industrial period with local variation (IPCC 2018). Limiting global warming to 1.5 °C could reduce the number of people exposed to climate-related risks by up to several hundred million by 2050 (IPCC 2018). At this moment in time, the world looks less likely to be able to stay within a 2.0 °C warming (Wise et al. 2014). Thereby, the IPCC has emphasized (2018), the world is shifting from a period of relative stability to a new era that is signified by increasing instability and unpredictability. No matter the degree of current climate mitigation progress and actual implementation of political commitments, adverse climate impacts have already intensified (IPCC 2018) and will continue to be unevenly distributed across and within nations, as well as generations. The IPCC has pointed out that

DOI: 10.4324/9781003183259-1

within society, disadvantaged and marginalized people are unevenly affected, as livelihoods, poverty and inequality interact with climate change in multifaceted ways (Olsson et al. 2014).

Political ecologists have critically questioned the idea of a changing external environment and stress that climatic change should not be viewed as an externality to the process and independent variable. Instead, it shall be considered "an enduring element – both cause and effect – of the active productions of humans as part of obdurately uneven lived environments" (Taylor 2015: 64). Rather than viewing the disadvantage of populations as a result of an externalized climate, this perspective emphasizes vulnerability as actively co-produced by humans and their lived environment, emphasizing that questions of adaptation cannot be separated from socio-ecological trajectories (Taylor 2015). Political ecologists are interested in factors that shape relations of power among human groups and that influence relations between these groups and diverse aspects of their environment (Paulson et al. 2003; Robins 2012; Perreault et al. 2015; Svarstadt et al. 2018). Works on the political ecology of adaptation have evolved as well (e.g., Taylor 2015; Sovacool and Linnér 2016). These studies are interested in the underlying processes that might affect the inequitable distribution of adaptation costs and benefits. Yet, the question of how social justice and equity concerns shall be realized, also as part of adaptation projects and (new) political pathways remains open.

The northern hemispheric heat wave of 2018 presents an example of intensified extremes, marked by concurrent heat events, which simultaneously affected large areas in Northern America, Europe and Asia (Vogel et al. 2019). Such heat waves have been deadly in the past and will become more deadly in the future (e.g., Stone et al. 2014; Kang and Eltahir 2018; USGCRP 2016). An array of studies focuses on different social impacts of heat waves: Aside from physiological indicators such as age, gender and pre-existent diseases, increasing death rates of heat waves coincide with gender, living alone, socioeconomic status, ethnicity, occupation and living conditions (McGregor et al. 2007; Windisch 2019; Alonso and Renard 2020). The political ecologist would contest that these externalized heat waves are of our own making and that these "social" indicators must be considered proxies of (past) human relations with the environment and likewise as an outcome of the political economy (see also Taylor 2015; Sovacool and Linnér 2016).

An established viewpoint in adaptation scholarship and political practice is that the magnitude of change depends not only on the future of emission pathways and mitigation efforts at limiting the temperature increase but likewise on the scale and choices of local adaptation. In this context, adaptation is presented as an indispensable policy response (see, e.g. Nadin et al. 2016; Moser et al. 2017; IPCC 2018). Although climate adaptation and mitigation are both considered complex political problems, governments are generally considered to have special obligations and ethical responsibilities to remedy harm based on the assumption that they have relative capacity for effectively addressing problems involving many hands (Miller 2001, 2008; De Campos 2017).

Against this background and because the intensity and frequency of climate extremes and slow-onset changes are already being felt across the globe, adaptation has evolved into a new distinct policy field (see also Massey and Huitema 2016). Governmental adaptation strategies, plans and policy activities have advanced significantly over the past decade. Although governments are comparatively late to the table, they have become increasingly aware that intensifying climate impacts, such as extreme heat waves, interrelate with drastic human and financial losses and should therefore receive more attention. But governments have been struggling to reconcile reactions to ad hoc and extreme events with their daily political business, competing policy priorities and their limited capacity for addressing complex problems that stretch across different policy sectors, political administrations and governance levels.

The devastating floods in Germany and Northwest Europe in the summer of 2021 have vividly demonstrated this: Still in the midst of a global pandemic, regions within western Germany recorded an amount of torrential rain within 24 hours that they usually receive in a month. This caused massive landslides, washing away streets, bridges, houses, and ultimately leading to human casualties. Aside from the destruction that has become immediately evident, the long-term effects are not visible right away, but are probably associated with changes to the social fabric of these places and further costs down the line. In the catastrophic heights of the events, politicians rushed to the spot, calling out the need to do much more for adapting to the unavoidable impacts of climate change, even at the EU level. Adaptation scholars have long researched the cognitive components such as environmental experience and close proximity to an event, as important factors for triggering change, not just with regard to political action. At the same time, it is far from clear what foresighted planning to uncertain conditions, but also strategies that include different interests of people look like.

Within this context, political ecologists demand caution on the proliferation of governmental rules, techniques and knowledge to manage social life. They point to the nature of governmentality and how the governmental rationality is signified by "the will to improve: the distance between what is discursively possible (a better society) and what is available for improvement (what actually exists)" (Valdivia 2015: 470). Improvement is about programming governmental interventions to achieve a desired outcome. Within this frame, experts and state administrators fulfill the task of voicing truths about populations and formulating governmental measures from "above." Too little time is spent on looking at how individuals shape their own behavior from "below" and listening to how people define their own experience and recognizing the multiple identities people may have (Valdivia 2015; Liverman 2015). They may be vulnerable, "they may be poor, but also leaders, networkers, and key sources of community knowledge" (Liverman 2015: 308; Kaijser and Kronsell 2014; Nightingale 2011).

Main theoretical framework

This book builds upon works in the political ecology and adaptation critical scholarship such as that of Taylor (2015), who actively questions climate adaptation as a self-evident concept. According to Taylor, the intrinsically technocratic politics of adaptation aims to contain perceived threats within existing institutions, marginalizing questions of power and production. Instead, caution about the seeming naturalness of adaptation is called for and shifting the question away from the governmentality of climate change to its inversion: "how do we make climate so powerful?" (Taylor 2015: 191). Thereby, Taylor moves the focus from the question of how to adapt to a changing externalized environment to the question of how we can produce ourselves differently.

Although these are important concerns, this book takes a step back in an effort to reconcile perspectives that are critical of adaptation and governance, with existing adaptation scholarship interested in questions such as: Which factors impact incremental government responses? Why do some people continue to be more affected than others? And what are enabling conditions to deal with uneven human vulnerability? Although the literature on adaptation policymaking and governance looks at the underlying issue of vulnerability from a distinctively different angle, that is effective governance, it nevertheless provides insights on the role state actors (can) play in this regard. The growing body of public policy research on climate adaptation finds that current institutions are at best an "incrementalist muddling through," signified by a "wait-and-see" approach (Huitema et al. 2016; Siebenhüner et al. 2017). Incremental adaptations can refer to reactive modifications such as vaccinating against newly emerging, climate-sensitive diseases (e.g., malaria, dengue fever, Lyme disease). Suppressing forest fires is another example of a routine response in reaction to an event (Kates et al. 2012). Transformative adaptation is clearly differentiated from incremental responses. Although there are different definitions, of what transformation in the adaptation context can mean (see also O'Brien 2012), transformative adaptation is largely understood as efforts, that involve shifts at a larger scale or intensity likely to occur at different levels and dimensions (Kates et al. 2012; Lonsdale et al. 2015). In that sense, it is thought of involving a paradigmatic shift and fundamental change.

Besides a growing number of recent government efforts, transformative responses are empirically rare and often focus on infrastructural modifications such as landscape readjustments and greening the environment. The social dimension of adaptation and vulnerability is slowly receiving more attention from policy practitioners but, with few exceptions, has not really been anchored in strategic responses to climate change.[1] In light of this empirical shortage, transformational responses are increasingly discussed and their need appears to be ever-growing. Moreover, adaptation actions are often considered to rely on weak political instruments, signified by little regulatory power (e.g., law-based frameworks setting performance standards, compliance or mandatory review mechanisms).

Main research questions

Accepting that governments are important actors, which have some capacity and responsibility to address the interrelation of uneven affectedness and a changing climate, this book examines why incrementalism in terms of uneven vulnerability persists even though climate-sensitive natural hazards are rapidly intensifying.

The main research question is: which lock-ins can explain this? The question is broken down into two main empirical chapters, which each examine lock-ins as they relate to vulnerability recognition (Chapter 7) and adaptation practices (Chapter 8). Although vulnerability and adaptation are often considered two sides of the same coin, examining each field for itself offers more in-depth answers and may even deliver different explanations. Governmental responses to vulnerability and adaptation do not necessarily carry the same logic. For the first empirical chapter on vulnerability lock-ins, two underlying research questions pave the way to understand how vulnerability is perceived and dealt with in political practice:

(1) What is the prevailing connotation of local human vulnerability to climate change in Jinhua and Atlanta as per local vulnerability assessments?
(2) How does this materialize with local decision-makers?

These two questions lead to the main research question which is: Why does it operationalize the way it does? (Which lock-ins explain protracted vulnerability?)

The following chapter (8) examines lock-ins related to the governance of adaptation. Beforehand, governmental efforts related to adaptation are analyzed, to get an understanding of the state of current adaptation efforts. The underlying research questions, which lead to the main question, are:

(1) What type of adaptation efforts exist?
(2) What is the problem recognition of local decision-makers regarding adaptation (and the uneven distribution of climate impacts) in the context of existing government efforts?

Why are adaptation efforts incremental, reactive and accidental? (Which lock-ins can explain it?)

Main concepts

Against the background of only minor governmental responses to complex problems, authors from different disciplines have called out adaptation as a formidable wicked policy challenge (e.g., see Nadin et al. 2016). The IPCC (2014) points to the opportunities and limits of adaptation and multiple scales at which adaptation can occur and reflects upon development pathways and opportunities, which exist to facilitate adaptation planning. Adaptation pathways are understood as a series of different adaptation choices involving trade-offs between incremental and transformative actions, goals and values (Pelling 2011; Wise et al. 2014). It becomes obvious that adaptation is a field that consists of manifold responses, actors and interests. The literature on barriers to adaptation seeks to explain the status of stalled adaptation efforts. With few exceptions, this research provides rather symptomatic explanations of problems, which are more deeply rooted and has largely failed to explain the protracted nature of political processes. Protracted political processes are reflected in the persistence of adaptation deficits and long-term maintenance of the uneven distribution of vulnerability to climate change (in the following often abbreviated by using the term "social vulnerability").

Main analytical framework: lock-ins

In contrast to earlier studies on adaptation barriers, this book finds that the concept of lock-ins is better equipped to explain stalled governance efforts and protracted political processes related to vulnerability and adaptation. Lock-ins can be understood as entrenched mechanisms that determine and are determined by behavior, thinking and values, dominant economic systems, cultural habits and political customs. Lock-ins are an agglomeration of factors, which act together and sustain certain pathways. The lock-in perspective aims to highlight the interaction of different aspects and their underlying mechanisms. In the adaptation context, lock-ins are defined as mutually reinforcing factors and an outcome of path-dependent processes, which constrain adaptation action (Siebenhüner et al. 2017). Lock-ins signify the institutional resistance to change and manifest in the inability to significantly reduce vulnerability and enable (equal) human and environmental well-being (see also Hölscher and Frantzeskaki 2020). Empirical research on this matter is only under development.[2] This emerging research examines lock-ins as one explanatory angle for sustained policy stability. In the context of this book, however, lock-ins are used as an angle to make sense of sustained uneven vulnerability to climate change and incremental adaptation responses in this regard.

Qualities of lock-ins

Lock-ins can have different qualities: they may enable as well as disable adaptation.[3] Adaptation enabling lock-ins, for instance, can include China's long historical tradition with managing water. The good treatment of water and natural hazards has always been related to political legitimacy concerns, thereby making it a top priority (see Mertha 2008). The right treatment of water has, to varying degrees and with deviations during the Mao period, always been a major political concern and may have ultimately resulted in China's adaptation responses being more progressed in the water sector than in others. This historically path-dependent tradition has resulted in high awareness and a different motivation for addressing issues related to water politically and advancing expertise in water governance. Due to scope, these adaptation enabling lock-ins unfortunately remain underconceptualized as part of this book, but present an interesting avenue for future research.

Lock-ins generally describe developments or paths that are difficult to change because of reinforcing feedback mechanisms. In this book, lock-ins are understood as a set of interrelated factors that are deeply embedded in existing structures of culture, environment, politics and the economy. Lock-ins have different dimensions, which contribute to both (social) vulnerability manifestation and adaptation inertia. These dimensions act in concert and build on the preliminary framework provided by Siebenhüner and colleagues (2017), focusing on four descriptive and entangled elements: infrastructure, actors, knowledge and institutions. This book examines how their interrelation plays out and impacts sustained vulnerability and incremental adaptation.

Lock-ins across two different political systems

Aside from often focusing on adaptation barriers to examine adaptation deficits, most adaptation studies have focused on analyzing adaptation pioneers or those already invested in progressive action (e.g., Preston et al. 2011; Moser and Boykoff 2013; Nadin et al. 2016). In order to examine why adaptation is occurring only incrementally, it is important to study supposed "non-pioneer regions" and apply the emerging framework of lock-ins empirically to deliver explanations that go beyond the description of barriers.

To better understand how different, path-dependent factors interact and shape, the predisposition of society and governments to deal with and be affected by climate change, cases of local governmental decision-making were examined in Atlanta, southeastern Georgia, the United States, and Jinhua, Zhejiang province in eastern China. Both cities have not taken deliberate adaptation action but are considered to be signified by strong adaptation pressure, especially at the social level. Summarizing the selection criteria, in both cities adaptation deficits manifest in (1) intensifying climate change (2) particularly high and uneven social vulnerability to climate change and (3) limited adaptation action despite supposedly different political circumstances.

In principle, the chosen local political contexts are equipped very differently in dealing with the governance of vulnerability to climate change. Localities within China and the United States were chosen, because they exemplify inaction at the local level and slow progress related to the governance of adaptation and social vulnerability to climate change. Although knowledge and information on the matters have diversified, social vulnerability and adaptation deficits have manifested in the two municipalities of Atlanta and Jinhua. To get a sense of how social vulnerability manifests differently but also what overarching patterns can be observed, these cities were chosen as case units for the examination of the research questions.

The political systems of China and the United States are very differently equipped in dealing with the related challenges. Whereas the Chinese government offers characteristics of fragmented authoritarianism and top-down decision-making, the United States is a highly decentralized liberal democracy with most action being initiated bottom-up and in local jurisdictions. Yet, both countries are exemplary regarding, what is considered "high adaptation pressure," which is met by continued inactivity that becomes especially evident at the local level. They not just have large populations at risk while being confronted by insufficient mitigation action, the incremental nature of adaptation action is also commonly detected in both countries (see, e.g., Preston et al. 2011; Park et al. 2012; Wise et al. 2014; Hart et al. 2014; Nadin et al. 2016). Comparing why high social vulnerability persists across two units, which are located in different political contexts is a gap this research aims to examine. High social vulnerability to climate change illustrates the extent of both units' adaptation challenges.

Examining highly uneven population vulnerability as major adaptation challenge

Vulnerability assessments are often the starting point for the facilitation of climate adaptation planning. They identify and characterize the extent to which different systems and public policy sectors are sensitive to climate change. Based on the assessments, targeted adaptation strategies are formulated with the objective of vulnerability abatement, enhanced awareness and increased adaptive capacity (Downing et al. 2005; Kelly and Adger 2009). Political ecologists have provided thorough critiques on universalized vulnerability approaches (also see chapter 2). Existing assessments disclose a core dilemma, which is the lack of an integrated perspective. In light of their distinctive policy expertise and the need to set political priorities, practitioners commonly adopt a sectoral lens in line with specific public policy sectors. Because the causes and effects of climate change are unevenly distributed, this book analyzes vulnerability from a non-sectoral perspective, by examining the transverse social and political aspects.

Vulnerability as a constant condition and relational concept

In line with the understanding of political ecology studies, vulnerability and adaptation are considered relational concepts and products of our cultural ecology and political economy. Political ecologists, feminist legal scholars and phenomenologists have pointed to vulnerability as a relational, intersectional and context-bound phenomenon. Accordingly, a person's position comes into being through relationships, which are ever changing and constituted at multiple scales (Neely and Nguse 2015). As a result different relationships intersect and also shape various forms of power as well as vulnerability. Human vulnerability to climate change is both a product of the cultural ecology – the way humans interact with the environment and are affected by it – and it is a source of the political economy and for that matter political institutions (Taylor 2015; Livermann 2015; Bergoffen 2016). Especially, the latter dimension sheds light onto the political nature of vulnerability, which is not just a product of our lived environments but fundamentally interacts with deeply political questions of power, stratification, production and environmental change (Taylor 2015). Vulnerability is also defined in relation to other people, how other people define themselves but also ascribe vulnerability onto others, thereby making it an embodied experience (also see, e.g., Livermann 2015; Bergoffen 2016). Vulnerability is constantly redesigned and a constant human condition (Fineman 2008; Bergoffen 2016).

Tension to reconcile different vulnerability angles

In this book, vulnerability is looked at from different angles including not only human vulnerability but also vulnerable political sectors and political institutions. Like all human institutions, political institutions are themselves vulnerable (to climate change) and impacted by different forms of lock-in. State vulnerability is best exemplified by the "external" conditions over which the state has only limited or no influence, such as private sector interests, terrorist attacks, the accidental collapse of a bridge or corruption (also see Fineman 2012, 2018). Thereby, vulnerability conceptions are expanded from human susceptibility to including the vulnerability of political institutions (also see Fineman 2008). One core tension that will become visible throughout the book is the difficulty of reconciling the different readings of vulnerability. The clash lies between the theorized ideal of vulnerability as a relational concept and embodied experience, also with respect to political institutions, the reliance on available methods to examine and compare local aspects of social vulnerability across different countries and ultimately, the way (social) vulnerability is perceived in political practice. The book attempts to make these tensions visible, but may fail in one place or the other.

Non-sectoral and integrative perspective on vulnerability and adaptation

The previous section already implied that vulnerability and adaptation are often examined per policy sector such as health, transportation or water. Human vulnerability is often subsumed by looking at what is done in the public health sector. This book adopts a non-sectoral lens by looking at population affectedness to climate change more broadly, and because population vulnerability intersects with many different public institutions. Because studies on specific disciplinary concentrations risk to provide an incomplete understanding of vulnerability (see, e.g., Ford et al. 2018), this book tries to avoid "siloed" thinking, to the extent that this is possible. Since vulnerability and adaptation mutually define each other, an integrative framework was adopted that looks at them in a combined approach.

Lock-ins and systemic shifts

Within the adaptation literature, the manifold debates on transformative adaptation have uncovered the urgent need for large-scale systemic shifts for achieving resilience.[4] Lock-ins are the main analytical framework applied in this book that aims to enable a systemic perspective of (path-dependent) and intertwined factors that together act as enablers and disablers for structural change. This systemic perspective also offers insights into why transformational shifts do not readily happen. Lock-ins are not necessarily historical but may likewise point to institutional environments and political factors that are protracted as part of different policy fields and overlapping systems. At the same time, lock-ins may follow broader patterns and be quite similar across different political systems, as the examined empirical cases of Atlanta in Georgia, the United States, and Jinhua in Zhejiang, China, will demonstrate.

Refocusing adaptation research

After years of criticism on the apolitical nature of adaptation (e.g., see Brunnengräber und Dietz 2013; Taylor 2015), the politics of adaptation is now slowly becoming an evolving item of research (e.g., see Bond and Barth 2020; Granberg and Glover 2020, 2021). While different forms of participation as part of public adaptation planning have increasingly been discussed in discourses on transformative adaptation, more and more works are pointing out the limitations of participation in adaptation politics (e.g., Prutsch et al. 2018; Oels 2019). In this context, related works have been shifting the attention to political capabilities of vulnerable populations to shape adaptation decisions (e.g., Holland 2017; Granberg and Glover 2020, 2021). Studies on adaptation justice have, for instance, focused on increasing the adaptive capacity by applying a Nussbaumian capabilities approach by combining recognition of the vulnerability of basic needs with a process of public involvement (Schlosberg 2012). Others explore the root causes of vulnerability and what factors generate the pre-existing vulnerabilities thereby shifting from approximating social variables such as poverty or lack of capacity to asking why capacity is lacking (Ribot 2014). Quite in line with capability approaches, vulnerability here is linked to the lack of freedom to influence the political economy that shapes certain entitlement (Ribot 2014). Despite the rapidly progressing literature, transformative responses in political practice that relate to the social dimension of vulnerability and to preventing the multiple origins and uneven nature of vulnerability are not remotely in sight. This book aims to explore as to why this is the case and thereby contributing to the increasing works on the politics of adaptation.

Research design and methods

This book uses a "holistic multiple-case study design" (Yin 2018: 84) and empirically examines political processes, which lock-in population vulnerability to climate change and adaptation deficits. The unit of analysis is holistic in that it looks at decision-making practices as main object of investigation across two cases (vulnerability, adaptation). The multiple cases correspond with different study areas of lock-ins, which help to uncover similar and different dimensions of lock-ins. The reasoning is similar across the multiple cases by looking at path-dependent and structurally embedded lock-ins. Thereby, a multicausal framework of analysis is applied.

The case studies are instrumental as opposed to intrinsic by aiming to understand broader phenomena (the maintenance of "adaptation deficits").[5] The cases are only secondary in that they are helpful to facilitate this broader understanding, but are not the main purpose of the study. Instrumental qualitative research refers to research that is rather theory-building, though it may consider theory engagement. Insights are generated from the examined cases to facilitate an understanding of broader phenomena (see Mills et al. 2010).

This book is grounded on a post-positivist approach of political and social sciences and follows an explorative research design. Because public entities and their policies play a fundamental role as the orchestrators of the political adaptation process (Knoepfel et al. 2007), the book focuses on local-governmental decision-making and political practices at the subnational level. Specifically, the following two case units were selected: Atlanta, in southeastern Georgia State, the United States, and Jinhua, in eastern Zhejiang province, China. Comprehending the action logic of governments across different contexts helps us to understand how to approach and deal with complex problems. The study of different policies and ideas increases our understanding of politics in terms of governmental plans, intentions and priorities.

The main sources of data are 53 semi-structured expert interviews, which were conducted through field research from 2016 to 2018 in China and the United States. The interviews were conducted in English and Chinese. The interviews were complemented with participant observation, an analysis of relevant policy documents, and secondary literature. The empirical data was processed by applying a Grounded Theory framework and has a strong iterative dimension. The iterative dimension implies that the framework of adaptation barriers, which partially informed the research and also guided the analysis of collected data, was reconsidered in the research process through applying the framework of lock-ins. The same applied for social vulnerability assessments, which were initially used as case selection criteria and baseline information to learn about some of the local vulnerability characteristics. However, throughout the research process, the vulnerability was reframed as a result of path-dependent, political processes that relate to different lock-ins and are theoretically grounded in the political ecology. The same holds true for adaptation deficits – which are commonly understood as the gap between an ideal state of a system that helps to minimize climate impacts on natural and human systems and an actual state. Rather than viewing these as adaptation deficits, the iterative approach implies that they can also be considered deficits of how political systems, cultural patterns and the global economy operate.

Structure of the book

Chapter 2 revisits some of the central concepts related to vulnerability thinking. It provides a brief background on the evolution of vulnerability conceptualizations and vulnerability assessments as (policy) instruments to guide adaptation research and practice. The different understandings of vulnerability often correspond with different prototypes of vulnerability assessments, which are used in political practice. These are briefly presented, as they will matter for the examination of the empirical cases. Chapter 3 reviews the relevant literature on adaptation and dominant explanations for adaptation deficits as represented in the research of adaptation barriers. It lays out why lock-ins were chosen as main analytical framework and conceptualized in the political ecology to explain adaptation deficits. The chapter also identifies core research gaps and demonstrates how the analytical framework corresponds with these gaps.

Chapter 4 explains the instrumental qualitative research framework and how the research questions were operationalized through the collection of empirical material and document analysis. Chapter 5 provides some background information about the current state of adaptation efforts and vulnerability governance at the national and subnational levels in China and the United States. It also shows how this corresponds with the current state of adaptation governance more broadly. Chapter 6 introduces the larger political trends of both countries and regions, in which local decision-making processes are to be located. The chapter finds that Atlanta and Jinhua are a reflection of inequality patterns at the national levels and have distinct local characteristics. The chapter already uncovers some empirical material and lock-in dimensions related to critical infrastructure. Due to scope, the biophysical aspects of Atlanta's and Jinhua's geographical environments cannot be laid out in greater detail. However, these are considered in Appendix 1.

The first main empirical Chapter 7 examines lock-in dimensions related to vulnerability governance. Because uneven social vulnerability is the primary starting point for the facilitation of adaptation and of main interest in this book, dominant discursive vulnerability conceptions and social vulnerability assessments are compared first. Next, it is examined how they align with perceptions of local political practitioners and how they operate in political practice. The chapter explores how local authorities in the policy process think about matters related to human vulnerability to climate change. Aside from a relatively low problem recognition about social vulnerability in the adaptation context, the chapter hints at the problematic aspects of vulnerability assessments that will be examined in closer detail in Chapter 9. On the one hand, they present "scientistic" constructions of vulnerability that are based on idealized assumptions and lack evidence for political implementation. On the other hand, some vulnerability categories are found to be rather unideal, because they stigmatize and manifest vulnerability further. Political practices of vulnerability acknowledgment are only selective and thereby further establish the vulnerability of some. The last part of the chapter analyzes the lock-ins, which help to explain the findings. The dominant lock-ins sit at the nexus of knowledge and politics. Past-dependent-knowledge patterns and processes of sense-making, coupled with the way political institutions operate can explain low-problem recognition about the social aspects of adaptation and vulnerability maintenance. These types of lock-ins become especially evident with regard to the limited access of shaping the dominant discourses on vulnerable populations. Further facets will be discussed in the chapter.

Chapter 8 proceeds by first examining existing adaptation efforts and policy rationales from the perspective of human vulnerability. This second empirical chapter is interested in how actual governmental efforts enable societies to adapt ("how does adaptation occur?"). The first part of this chapter analyzes the dynamics of local adaptation planning. The second part explains why the dynamics play out the way they do. Adaptation appears to be at a very early planning stage in both political contexts, often occurring accidentally and with a strong focus on green infrastructure. Adaptation related to social vulnerability can at best be considered incremental. The chapter also hints at the problem of maladaptation and different adaptation targets (floodwater management versus protecting vulnerable populations), which appear to conflict with each other. The main findings, accidental and green infrastructure-focused adaptation, were then explained with lock-ins that sit at the intersection of knowledge and politics. Three facets were discussed in greater detail: 1) the dominance of certain knowledge paths, 2) the lack of shared knowledge and 3) the perception of overcomplexity.

Chapter 9 discusses the empirical findings on vulnerability and adaptation lock-ins. First, highly unevenly distributed vulnerability to climate change as an inherent characteristic of two different political systems is pointed out. The subsequent section revisits the lock-in concept and summarizes the main findings. The way knowledge is produced, accessed and distributed, in combination with the way political institutions operate, are core elements of the inherent vulnerability of both countries. The chapter continues by conceptualizing them as political epistemological lock-ins by drawing from the current literature on political epistemology. Three facets of political epistemological lock-ins are discussed. The last part of the chapter argues that knowledge can be considered an important indicator for the reconstitution of a class-based society, which manifests in China and the United States. To place the argument, interviews and recent literature on the matter are drawn from.

Chapter 10 discusses the main empirical findings on political epistemological lock-ins against the background of dominant discourses on transformative adaptation. Before the chapter delves into the roles that policymaking and the state can play in addressing transformative adaptation, the first chapter section lays out the field why climate change adaptation is an inherently political concept and why greater attention needs to be devoted to questions of power and conflict in the context of uneven vulnerability. In the following, knowledge is presented as a fundamental aspect that matters for transformative adaptation. This section briefly revisits some of the dominant discourses to place the argument. This chapter also discusses some of the implications for dominant scientific practices and what science can do differently. The subsequent chapter then explores the role of the state in addressing transformative adaptation. It critically asks if the policy field of adaptation is at all equipped to deal with grand challenges, such as social justice or whether vulnerability governance in the sense of avoiding its roots causes is the golden limit of political decision-making. Is transformative adaptation a retain as a utopia? The chapter ends with a couple of low- and high-hanging policy fruits in an attempt to approach the identified lock-ins.

Chapter 11 provides a conclusion of this book and goes back to the main research questions, which guided the study: what lock-ins can explain adaptation deficits that are particularly apparent in high social vulnerability to climate change? What factors can explain why vulnerability is such a deeply rooted phenomenon across two different political systems and adaptation occurring only incrementally? The chapter revisits the main findings and contributions. Core contributions lie in the newly emerging field of research on adaptation lock-ins, which examines the interrelation of path-dependent factors that significantly impede adaptation action. This research drew from political ecology studies to make sense of lock-ins. Further, this book makes a contribution to critical adaptation studies by exploring the parameters against which adaptation action operates and taking a closer look at factors that determine, recreate and maintain vulnerability. Critical adaptation studies examine the way adaptation measures are interpreted, transformed and implemented and how these measures interfere with power relations. After years of criticism on the apolitical nature of adaptation, the politics of adaptation is now slowly becoming an evolving item of research. The argument of this book is that lock-ins to adaptation in the sense of addressing social vulnerability are deeply connected with questions of social justice. In the examined cases, issue related to knowledge production, the (lack of) power to frame dominant discourses on vulnerability and little awareness about the social dimension due to path-dependent knowledge patterns focused on engineering solutions and green infrastructure were detected. The chapter ends with allies for future research.

Notes

1   Whenever the book refers to "social vulnerability" it does not mean to imply that vulnerability is of "social" origin. Instead, the term is used synonymously with the uneven distribution of climate change impacts. In this book, vulnerability is conceptualized as an embodied experience and as an intersectectional phenomenon in terms of being the result of different, interdependent societal stratification processes. In line with the political ecology perspective, these vulnerability dynamics are understood as actively coproduced by nature and society, which are influenced by power relations that affect access to natural resources and the uneven distribution costs and benefits of climate change.

2   See the website of the cross-disciplinary study and interdisciplinary research project, "Climate adaptation policy lock-ins: a 3×3 approach", available at: https://uol.de/en/adapt-lockin/, accessed August 26, 2019.

3   Lock-in qualities are not meant in a normative sense in terms of implying that adaptation is the desired end-state.

4   Due to scope, this book was unable to critically reflect upon the debate about resilience vis-à-vis transformation as started by Pelling (2011) and others. In this discussion, resilience is considered an early stage on the way to transformation, which corresponds to stability. The next stage, "transition", refers to incremental societal change. Transformation is the ultimate stage, which corresponds with the change in political regimes. Pelling (2011) refers to this three-stage model as resilience–transition–transformation framework.

5   Whenever the term "adaptation deficit" is used, the book does not mean to imply that adaptation policies are necessarily the normative goal, but instead the term aims to point out continued deficiencies of adjusting to climate change.

References

Alonso, Lucille, and Florent Renard. "A Comparative Study of the Physiological and Socio-Economic Vulnerabilities to Heat Waves of the Population of the Metropolis of Lyon (France) in a Climate Change Context." *International Journal of Environmental Research and Public Health* 5, no. 17 (2020): 1004. https://doi.org/10.3390/ijerph17031004.

Bergoffen, Debra. "The Flight from Vulnerability." In *Dem Erleben auf der Spur. Feminismus und die Philosophie des Leibes*, edited by Hilge Landweer and Isabella Marcinski, 137–152. Bielefeld: Transcript Verlag/De Gruyter, 2016.

Bond, Sophie, and J. Barth. "Care-Full and Just: Making a Difference through Climate Change Adaptation." *Cities* 102 (2020): 102734.

Brunnengräber, Achim, and Kristina Dietz. "Transformativ, politisch und normativ: für eine Re-Politisierung der Anpassungsforschung." *GAIA* 22, no. 4 (2013): 224–227.

Campos, Thana C. de. *The Global Health Crisis: Ethical Responsibilities*. Cambridge: Cambridge University Press, 2017.

Downing, Thomas E., Anand Patwardhan, Richard J. T. Klein, Elija Mukhala, Linda Stephen, Manuel Winograd, and Gina Ziervogel. "Assessing Vulnerability for Climate Adaptation." In *Adaptation Policy Frameworks for Climate Change: Developing Strategies, Policies and Measures*, edited by Bo Lim et al., 61–90. Cambridge: Cambridge University Press, 2005.

Fineman, Martha Albertson. "The Vulnerable Subject: Anchoring Equality in the Human Condition." *Yale Journal of Law and Feminism* 20, no. 1 (2008).

Fineman, Martha Albertson. "'Elderly' as Vulnerable: Rethinking the Nature of Individual and Societal Responsibility." *Emory Legal Studies Research Paper* no. 12–224, 2012.

Fineman, Martha Albertson. "Injury in the Unresponsive State: Writing the Vulnerable Subject into Neo-Liberal Legal Culture." In *Emory Legal Studies Research Paper Forthcoming in Injury and Injustice: The Cultural Politics of Harm and Redress*, edited by Anne Bloom, David M. Engel, and Michael McCann, 50–75. Cambridge: Cambridge University Press, Cambridge Studies in Law and Society, 2018.

Ford, James D., Tristan Pearce, Graham McDowell, Lea Berrang-Ford, Jesse S. Sayles, and Ella Belfer. "Vulnerability and its Discontents: The Past, Present, and Future of Climate Change Vulnerability Research." *Climatic Change* 151 (2018): 189–203. Doi: 10.1007/s10584-018-2304-1.

Granberg, Mikael, and Leigh Glover. *The Politics of Adapting to Climate Change*. Cham, Switzerland: Palgrave MacMillan, 2020.

Granberg, Mikael, and Leigh Glover. "The Climate Just City." *Sustainability* 13(3), no. 1201 (2021). https://doi.org/10.3390/su13031201.

Hart, Craig, Zhu Jiayan, Ying Jiahui, Cyril Cassisa, and Hugh Kater. "Mapping China's Policy Formation Process." *Beijing: Report Development Technologies International for the China Carbon Forum*, November (2014): 1–27.

Holland, Breena. "Procedural Justice in Local Climate Adaptation: Political Capabilities and Transformational Change." *Environmental Politics* 26, no. 3 (2017): 391–412. https://doi.org/10.1080/09644016.2017.1287625.

Hölscher, Katharina, and Niki Frantzeskaki. "A Transformative Perspective on Climate Change and Climate Governance." In *Transformative Climate Governance. A Capacities Perspective to Systematise, Evaluate and Guide Climate Action*, edited by Katharina Hölscher and Niki Frantzeskaki, 3–48. Cham, Switzerland: Palgrave MacMillan, 2020.

Huitema, Dave, W. Neil Adger, Frans Berkhout, Eric Massey, Daniel Mazmanian, Stefania Munaretto, Ryan Plummer, and Katrien Termeer. "The Governance of Adaptation: Choices, Reasons, and Effects. Introduction to the Special Feature." *Ecology and Society* 21, no. 3(37) (2016).

IPCC. "Summary for policymakers. In: Climate Change 2014: Impacts, Adaptation, and Vulnerability. Part A: Global and Sectoral Aspects." In *Contribution of Working Group II to the Fifth Assessment Report of the Intergovernmental Panel on Climate Change*, edited by C. B. Field, V. R. Barros, D. J. Dokken et al., 1–32. New York, NY and Cambridge: Cambridge University Press, 2014.

IPCC. *Global Warming of 1.5°C, an IPCC Special Report on the Impacts of Global Warming of 1.5°C Above Pre-Industrial Levels and Related Global Greenhouse Gas Emission Pathways, in the Context of Strengthening the Global Response to the Threat of Climate Change, Sustainable Development, and Efforts to Eradicate Poverty.* Edited by V. Masson-Delmotte, P. Zhai, H. O. Pörtner, D. Roberts, J. Skea, P. R. Shukla, A. Pirani, W. Moufouma-Okia, C. Péan, R. Pidcock, S. Connors, J. B. R. Matthews, Y. Chen, X. Zhou, M. I. Gomis, E. Lonnoy, T. Maycock, M. Tignor, and T. Waterfield, 2018. www.ipcc.ch/report/sr15/.

Kaijser, Anna, and Annica Kronsell. "Climate Change Through the Lens of Intersectionality." *Environmental Politics* 23, no. 3 (2014): 417–433. https://doi.org/10.1080/09644016.2013.835203.

Kang, Suchul, and Elfatih A. B. Eltahir. "North China Plain Threatened by Deadly Heatwaves Due to Climate Change and Irrigation." *Nature Communications* 9, no. 2894 (2018). https://doi.org/10.1038/s41467-018-05252-y.

Kates, Robert W., William R. Travis, and Thomas J. Wilbanks. "Transformational Adaptation When Incremental Adaptations to Climate Change Are insufficient." *Proceedings of the National Academy of Sciences* 109, no. 19 (2012): 7156–7161. https://doi.org/10.1073/pnas.1115521109.

Kelly, Mick P., and W. Neil Adger. "Theory and Practice in Assessing Vulnerability to Climate Change and Facilitating Adaptation." In *The Earthscan Reader on Adaptation to Climate Change*, edited by E. Lisa, F. Schipper, and Ian Burton. London: Earthscan, 2009.

Knoepfel, Peter, Corinne Larrue, Frederic Varone, and Michael Hill. *Public Policy Analysis.* Bristol: The Policy Press, 2007.

Livermann, Diana. "23 Reading Climate Change and Climate Governance As Political Ecologies." In *The Routledge Handbook of Political Ecology*, edited by Perreault, Tom, Gavin Bridge, and James McCarthy, 303–319. Abingdon: Routledge, 2015.

Lonsdale, Kate, Patrick Pringle, and Briony Turner. *Transformative Adaptation: What It Is, Why It Matters & What Is Needed. UK Climate Impacts Programme (UKCIP).* Oxford: University of Oxford, 2015.

Massey, Eric, and Dave Huitema. "The Emergence of Climate Change Adaptation as a New Field of Public Policy in Europe." *Regional Environmental Change* 16 (2016): 553–564. https://doi.org/10.1007/s10113-015-0771-8.

McGregor, Glenn, Mark Pelling, Tanja Wolf, and Simon Wolf. "The Social Impacts of Heat Waves." *Environment Agency*, Science Report, SC2006, 2007.

Mertha, Andrew C. *China's Water Warriors: Citizen Action and Policy Change.* Ithaca and London: Cornell University Press, 2008.

Miller, David. "Distributing Responsibilities." *The Journal of Political Philosophy* 9, no. 4 (2001): 453–471. Doi: 10.1111/1467-9760.00136.

Miller, David. "Global Justice and Climate Change: How Should Responsibilities Be Distributed?" *The Tanner Lectures on Human Value.* Speech delivered at Tsinghua University, Beijing, March 24–25, 2008. https://tannerlectures.utah.edu/_documents/a-to-z/m/Miller_08.pdf

Mills, Albert J., Gabrielle Durepos, and Elden Wiebe. *Encyclopedia of Case Study Research.* Thousand Oaks, CA: Sage Publications, 2010. Doi: 10.4135/9781412957397.

Moser, Susanne, and Maxwell Boykoff. *Successful Adaptation to Climate Change: Linking Science and Practice in a Rapidly Changing World.* London: Routledge, 2013.

Moser, Susanne, Joyce Coffee, and Aleka Seville. "Rising to the Challenge, Together. A Review and Critical Assessment of the State of the US Climate Adaptation Field." A report prepared for the Kresge Foundation, December 2017.

Nadin, Rebecca, Sarah Opitz-Stapleton, and Xu Yinlong, eds. *Climate Risk and Resilience in China*. London: Routledge, 2016.

Neely, Abigail, and Thokozile Nguse. "10 Relationships and Research Methods. Entanglements, intra-actions, and diffraction." In *The Routledge Handbook of Political Ecology*, edited by Tom Perreault, Gavin Bridge, and James McCarthy. Abingdon: Routledge 2015, Routledge Handbooks Online. 10.4324/9781315759289.ch36.

Nightingale, Andrea J. "Bounding Difference: Intersectionality and the Material Production of Gender, Caste, Class and Environment in Nepal." *Geoforum* 42, no. 2 (2011): 153–162. https://doi.org/10.1016/j.geoforum.2010.03.004.

O'Brien, Karen. "Global Environmental Change II: From Adaptation to Deliberate Transformation." *Progress in Human Geography* 36, no. 5 (2012): 667–676.

Oels, Angela. "The Promise and Limits of Participation in Adaptation Governance: Moving Beyond Participation Towards Disruption." In *Research Handbook on Climate Change Adaptation Policy*, edited by E. Carina, H. Keskitalo, and Benjamin L. Preston. Cheltenham: Edward Elgar Press, 2019.

Olsson, L., M. Opondo, P. Tschakert, A. Agrawal, S. H. Eriksen, S. Ma, L. N. Perch, and S. A. Zakieldeen. "Livelihoods and Poverty." Chapter. In *Climate Change 2014 – Impacts, Adaptation and Vulnerability: Part A: Global and Sectoral Aspects: Working Group II Contribution to the IPCC Fifth Assessment Report*, Vol. 1, 793–832. Cambridge: Cambridge University Press, 2014. Doi: 10.1017/CBO9781107415379.018.

Park, Sarah E., Nadine A. Marshall, Emma Jakku, Anne M. Dowd, S. Mark Howden, Emily Mendham, and Aysha Fleminge. "Informing Adaptation Responses to Climate Change through Theories of Transformation." *Global Environmental Change* 22 (2012): 115–126.

Paulson, Susan, Lisa L. Gezon, and Michael Watts. "Locating the Political in Political Ecology: An Introduction." *Human Organization* 62, no. 3 (2003): 205–217. https://doi.org/10.17730/humo.62.3.e5xcjnd6y8v09n6b.

Pelling, Mark. *Adaptation to Climate Change. From Resilience to Transformation*. New York: Routledge, 2011.

Perreault, Tom, Gavin Bridge, and James McCarthy. *The Routledge Handbook of Political Ecology*. Abingdon: Routledge, 2015.

Preston, Benjamin L., Richard M. Westaway, and Emma J. Yuen. "Climate Adaptation Planning in Practice: An Evaluation of Adaptation Plans from Three Developed Nations." *Mitigation and Adaptation Strategies for Global Change* 16 (2011): 407–438.

Prutsch, Andrea, Reinhard Steurer, and Theresa Stickler. "Is the Participatory Formulation of Policy Strategies Worth the Effort? The Case of Climate Change Adaptation in Austria." *Regional Environmental Change* 18, (2018): 271–285. https://doi.org/10.1007/s10113-017-1204-7.

Ribot, Jesse. "Cause and Response: Vulnerability and Climate in the Anthropocene." *The Journal of Peasant Studies* 41, no. 5 (2014): 667–705. Doi: 10.1080/03066150.2014.894911.

Robins, Paul. *Political Ecology: A Critical Introduction*. Chichester and Malden: J. Wiley & Sons, 2012.

Schlosberg, David. "Climate Justice and Capabilities: A Framework for Adaptation Policy." *Ethics & International Affairs* 26, no. 4 (2012): 445–461. Doi: 10.1017/S0892679412000615.

Siebenhüner, Bernd, Torsten Grothmann, Dave Huitema, Angela Oels, Tim Rayner, and John Turnpenny. "Lock-ins in Climate Adaptation Governance. Conceptual and Empirical Approaches." Conference Paper 2017, unpublished.

Sovacool, Benjamin, and Björn-Ola Linnér. *The Political Economy of Climate Change Adaptation*. London: Palgrave Macmillan 2016.

Stone, Brian Jr., Jason Vargo, Peng Liu, Dana Habeeb, Anthony DeLucia, Marcus Trail, Yongtao Hu, and Armistead Russell. "Avoided Heat-Related Mortality through Climate Adaptation Strategies in Three US Cities." *PLoS One* 9, no. 6 (2014). Doi: 10.1371/journal.pone.0100852.

Svarstad, Hanne, Tor A. Benjaminsen, and Ragnhild Overå. "Power Theories in Political Ecology." *Journal of Political Ecology* 25, no. 1 (2018): 350–363. https://doi.org/10.2458/v25i1.23044.

Taylor, Marcus. *The Political Ecology of Climate Change Adaptation: Livelihoods, Agrarian Change and the Conflict of Development*. London and New York: Routledge Explorations in Development Studies, 2015.

U.S. Global Change Research Program (USGCRP). *The Impacts of Climate Change on Human Health in the United States: A Scientific Assessment*. Edited by A. Crimmins, J. Balbus, J. L. Gamble, C. B. Beard, J. E. Bell, D. Dodgen, R. J. Eisen et al. Washington, DC: U.S. Global Change Research Program, 2016. http://dx.doi.org/10.7930/J0R49NQX

Valdivia, Gabriela. "Eco-governmentality." In *The Routledge Handbook of Political Ecology*, edited by Tom Perreault, Gavin Bridge, and James McCarthy. Abingdon: Routledge 2015, Routledge Handbooks Online. 10.4324/9781315759289.ch36.

Vogel, Martha M., Jakob Zscheischler, Richard Wartenburger, Dick Dee, and Sonia I. Seneviratne. "Concurrent 2018 Hot Extremes Across Northern Hemisphere due to Human-Induced Climate Change." *Advancing Earth and Space Science* 7, no. 7 (2019): 692–703. https://doi.org/10.1029/2019EF001189.

Windisch, Margareta. "Denaturalising Heatwaves: Gendered Social Vulnerability in Urban Heatwaves, a Review." *Australian Journal of Emergency Management Monograph* 4 (2019): 146–153.

Wise, Russell, Ioan Fazey, Mark Stafford Smith, Sarah E. Park, Hallie Eakin, Emma Archer Van Garderen, and Bruce Campbell. "Reconceptualising Adaptation to Climate Change as Part of Pathways of Change and Response." *Global Environmental Change* 28 (2014): 325–336.

Yin, Robert K. *Case Study Research Design and Methods* (6th ed.). Thousand Oaks: Sage Publishing, 2018.

2   Vulnerability

Core concepts

This chapter revisits some of the central concepts and the evolution of vulnerability conceptualizations related to climate change and human exposure. There is an array of academic work with different vulnerability underpinnings, which crucially inform (adaptation) policy responses and (governmental) interventions (also see Taylor 2015: 49 ff.). Despite the multiplicity of vulnerability understandings, the standard definition used in political practice is that of the IPCC, which positions vulnerability as the centerpiece of the adaptation paradigm. Taylor (2015) questions the self-evidence of related concepts, and together with others, the fallacies of unified vulnerability approaches have been sufficiently examined (also see Preston et al. 2011; Liverman 2015). Others have looked at the question why climate vulnerability research is such a contested process more broadly by identifying, characterizing and evaluating concerns over the use of vulnerability approaches (Ford et al. 2018a; Sherman et al. 2018). The authors derive the conclusion that besides the fact that vulnerability research may have its limits in covering the complex climate-society dynamics, revitalized efforts for vulnerability are needed, including a focus on the social drivers, a more substantial exploration of institutions and politics to counter the limited influence that vulnerability research has had for decision-making and foster a next generation of empirical studies.

This book does not aim to repeat what has already been written elsewhere. At the same time, it tries to deal with the core tension of different, but oftentimes overlapping vulnerability understandings: The main conceptualization of vulnerability is derived from works in the political ecology, viewing vulnerability as a relational concept and embodied experience. At the same time, and when interviewing local decision-makers on matters related to vulnerability, it became visible that vulnerability definitions are themselves a relational boundary object. Definitions and understandings of vulnerability appeared to be construed in relation to understandings at the global, national and/or regional level. Getting a shared understanding with interviewees about the point of reference proved to be especially challenging. Additionally, a standardized vulnerability understanding was used to inform the case selection of the study by relying on social vulnerability indices. Despite their severe shortcomings, social vulnerability indices were used as a baseline to unpack one interpretation of vulnerable groups and some of the local characteristics of vulnerability.

DOI: 10.4324/9781003183259-2

Unveiling these different vulnerability understandings is important, as they are representative of local processes of sense-making. This was a difficult task throughout the research process. These tensions could unfortunately not be omitted and will become visible throughout the book. Because of that, this chapter provides a brief context on vulnerability thinking nevertheless. Different interpretations of vulnerability are part of different epistemic traditions, with knowledge deriving from various origins, rationalities, beliefs as well as academic practices. Therefore, this chapter provides a brief background on vulnerability thinking, its evolution (section "Evolution of vulnerability thinking") and related policy instruments in the form of different assessments (section "Vulnerability assessments as policy instrument to guide adaptation"). This chapter briefly reflects on the origin of these types of assessments and how they have informed adaptation policymaking before exploring later in the book how they function in practice (chapter 7) and what their fallacies are (chapter 9). Based upon the review of existing vulnerability research in the context of climate change, several syntheses are derived, which can be considered gaps of existing research and provided guidance for the book.

Evolution of vulnerability thinking

Since vulnerability began to play a bigger role in climate politics in 1992, much has changed but the outcome remains the same: there is little consensus about what vulnerability can mean (Dow 1992; Downing et al. 2005; O'Brien et al. 2004; Füssel and Klein 2006). Broadly defined, "vulnerability" can refer to "the quality or state of being exposed to the possibility of being attacked or harmed, either physically or emotionally."[1] Vulnerability has become a central analytical framework for examining environmental change to help prioritize actions needed for vulnerability management (Luers 2005).

The social science perspective in climate policy (research)

The dominant perspective put forth by the IPCC defines vulnerability as "the propensity or predisposition to be adversely affected" (IPCC 2014: 5). It is commonly considered a combined function of exposure, sensitivity and (the lack of) adaptive capacity of a given system (e.g., IPCC 2007, 2014). The IPCC's Fifth Assessment Report (2014) (AR5) refers to vulnerability in both human and natural systems. According to this understanding, the climate is a result of both natural and anthropogenic causes and is impacted and simultaneously impacts socioeconomic processes (see Figure 2.1). This mutual dependence is considered to affect the climate impact risk. The widespread understanding is that the risks can be reduced through changed socioeconomic pathways, governance mechanisms and two types of policy action: adaptation and mitigation.

The literature further distinguishes between outcome and contextual understandings of vulnerability (O'Brien et al. 2007; Füssel 2010). Whereas the latter is considered a starting-point interpretation of vulnerability, by explaining the roots

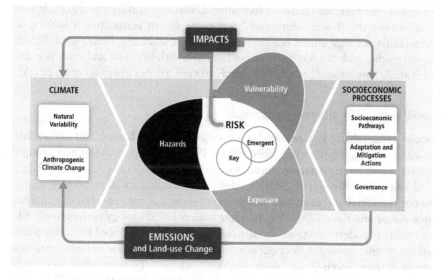

Figure 2.1 IPCC's core concepts related to vulnerability.
Source: IPCC (2014: 3)

of vulnerability as part of the political economy and the pre-existent social vulnerability, endpoint-based interpretations examine climate change as a root problem that leads to vulnerability. The policy context of outcome vulnerability is focused on mitigating climate change and articulating technical adaptation responses, whereas policy measures under a contextual vulnerability understanding seek to engage in social adaptation and sustainable development (Füssel 2010). Though the recent literature has engaged in different interpretations of vulnerability based on a contextual understanding of the sociopolitical and economic factors that impact the sensitivity of a system, earlier IPCC definitions of vulnerability corresponded to an outcome-based understanding (Füssel 2010).

The human dimension of vulnerability research

In line with the IPCC's development of embracing the contextual factors, the lack of considering the "human dimension" of climatic exposure was criticized in the early 2000s (see, e.g., Agrawala 2001; IPCC 2007). In the subsequent years, studies on the human dimension flourished. As part of these developments, some systematized the human dimension by narrowing down a research agenda to three aspects: 1) how humans have altered the environment, 2) how they are affected by the consequences of environmental change and 3) how they are responding to the experience or the expectation of global change (also see NRC 1999). Other researchers studied the human dimension by looking at factors that mediate human perceptions of environmental change and behavioral responses at the local level and how this interrelates with processes of production and dissemination of global and national climate information (Brondizio and Moran 2015). When emphasizing the "human dimension" a popular, recent perspective is looking at how climate change affects human health (Ford et al. 2018a, Pearce et al. 2018). Fewer studies look outside the realm of human health to define and study population affectedness more broadly (also see Ford et al. 2018b; Sherman et al. 2018).

Social vulnerability as bottom-up approach

Social vulnerability has received a lot of attention by hazard and geographic researchers (e.g., Blaikie et al. 1994; Cutter 1996; Cutter et al. 2003) and strongly overlaps with more political economy focused works (e.g., Bohle et al. 1994; Adger and Kelly 1999; Ribot et al. 2005). In this understanding, social vulnerability aims to put the social and economic well-being of society at the center of the analysis (see Adger and Kelly 1999). Much in line with political ecology perspectives, vulnerability is understood to be socially constructed and influenced by institutional and economic dynamics. Socioeconomic and biophysical characteristics are manifest at different levels (local, national, global), whereas the vulnerability is considered to be population-specific (Adger and Kelly 1999). Since the early 2000s, the frameworks and methods to define social vulnerability have constantly evolved.

Much of the concept originated in human geography, where one assumption is that certain social and geographic characteristics correspond with the potential for loss (Cutter 1996). Recognizing the variability of vulnerable populations and places is understood to improve management plans (Cutter and Finch 2008).[2] The aim is to design policy interventions that address the multilevel nature of vulnerability. This marks a conceptual advance from earlier vulnerability definitions (Adger 2006). Vulnerability assessments have gradually shifted from only considering natural hazards and climate variability as primary contributors to vulnerability (see earlier), to a more differentiated understanding of social inequality and pre-existent conditions that result in greater exposure to climate change for certain groups. The main research questions are: "Who is vulnerable to climate change and why?" and "How can vulnerability be reduced?" (O'Brien et al. 2004: 3). Because vulnerability to climate change is shaped by the local conditions and specific cultural characteristics, it is especially important to study the local level, such as municipalities. Adaptation challenges and vulnerability materialize on the basis of specific local conditions as well as sociopolitical environments.

Avoiding the essentialism trap by accepting different perspectives

A recent collection on the matter of the human dimension was criticized for their "scientized" understanding of society, emphasizing that there is more than one way to comprehend the human dimension (Castree 2016). Although the critique was referring to a specific collection of articles, it hit a kernel that relates to scientific essentialism by pointing out how the environmental social science perspectives often fail to acknowledge the plurality by implying there is "right" concepts and models which "can derive a fairly objective understanding of the climate-energy-nexus" (p. 731). Castree has criticized this tandem of epistemic realism and ontological monism, as it implies that one may know what "works" (p. 731). This knowledge essentialism trap is also visible in the many voices that critique the lack of consensus and clarity on the concept of human vulnerability (and as we will later see transformative adaptation as well). Maneuvering through these ambivalences is not an easy task. O'Brien and colleagues (2007) also challenge related critiques, who call for uniform concepts, by exploring in greater detail why different interpretations of vulnerability matter (also see the following section). Accordingly, different types of frames produce different types of knowledge and thereby emphasize different types of policy responses. They argue that different interpretations cannot necessarily be integrated into a common framework and should therefore be recognized as complementary approaches.

Existing works in the political ecology have also summarized some of the main approaches with a focus on how vulnerability relates to questions of power and entitlements (e.g., see Liverman 2015). Especially in the context of empirical analyses of vulnerability, the use of proxies to determine and map local vulnerability has been a highly contested practice confronted by methodological challenges connected with quantifying or categorizing vulnerability. Vulnerability indices and mapping often provide only a snapshot of vulnerability, overlook contextual factors and create a false sense of confidence and accuracy (see Liverman 2015: 309; Preston 2011). When attempting to assess vulnerability and use the knowledge for political decisions and setting the adaptation agenda, care must be taken to not erase agency (also see Liverman 2015).

Political ecology vis-à-vis climate policy research

Vulnerability is the conceptual centerpiece of today's climate politics and has become the "core nucleus" of the adaptation paradigm (Taylor 2015; Sovacool and Linnér 2016). As part of climate policy discourses and practitioner-oriented perspectives, vulnerability and resilience are presented as two opposing positions with adaptation strategies often aiming at vulnerability reduction. In this context, resilience is commonly defined as a robust or system, and/or a safe and secure environment that can prevent, withstand and/or recover from shocks (e.g., Vogel et al. 2007; RIC 2012; IPCC 2014; Joakim et al. 2015). Despite the enormous recent popularity of resilience research in diverse academic disciplines, the disillusion of the concept followed rapidly (also see Zebrowski 2009). Political ecologists criticize the narrow-minded construction of resilience as an achievable end-state, calling for a more dynamic understanding of a constantly changing society (Beymer-Farris et al. 2012). Further, they signal two faulty resilience assumptions commonly found in the literature on socio-ecological systems: resilience as an end or an outcome of action assumes that there is consensus on the "desired state" or that a desired state even exists. Second, they criticize that resilience, understood as a process, often overlooks conflicts over resources and the importance of power asymmetries.

Political ecologists have been keen on examining the dynamics of social, cultural or contextual factors and how they interrelate with questions of power and access. Accordingly, vulnerability is understood as a relational and context-bound phenomenon. It is considered both a product of the cultural ecology – the way humans interact with the environment and are affected by it – and it is a source of the political economy and for that matter political institutions (Taylor 2015; Livermann 2015; Bergoffen 2016). Especially, the latter dimension sheds light onto the political nature of vulnerability, which is not just a product of our lived environments but fundamentally interacts with deeply political questions of power, production and environmental change (Taylor 2015). Vulnerability is also defined in relation to other people, how other people define themselves but also ascribe vulnerability onto others, thereby making it an embodied experience (also see, e.g., Livermann 2015; Bergoffen 2016).

Vulnerability assessments as policy instrument to guide adaptation

Vulnerability assessments (VAs) deliver the baseline information needed for adaptation planning. The assessment reports (ARs) by the IPCC provide a science-based understanding of the latest climate impact developments, vulnerability, adaptation opportunities and limits. Their goal is to support policymaking through "policy-relevant information that creates opportunities for decision-making that can lead to climate-resilient development pathways" (Burkett et al. 2014). VAs function as a major policy instrument that seeks to identify and prioritize vulnerabilities to build matching adaptation strategies with the aim of improving the resilience of a system (such as climatic, ecological, economic or human systems). The appropriate assessment approach depends on the research or policy questions it aims to address (Füssel and Klein 2006).

Aside from the aforementioned gap of looking at the social dimension as part of earlier vulnerability research, exploring the social dimension of adaptation responses has been slowly gaining traction in political practice as of late. More and more governments are referring to the social aspects of vulnerability and point out that they should be considered when assessing vulnerability (see, e.g., EEA 2018; Breil et al. 2018). This chapter will show that this is not a new demand and that concepts to cover the social dimension – even in political practice have been around for a while. These have even moved to a next generation of validation studies and examining evidence of social vulnerability (e.g., Otto et al. 2017).

In contrast to social vulnerability assessments (SVAs), the use of climate vulnerability assessments (CVAs) to inform political decision-making is already common practice. The revived interest in the social and human dimensions of adaptation and vulnerability may lead to a proliferation of SVAs to inform policymaking in the future. Yet, answers are needed as to why they are not as readily used by political decision-makers. Aside from these fairly political developments and research demands, SVAs were used as case selection criteria to compare local units of decision-making on questions of vulnerability and adaptation in China and the United States.

Different prototypes of vulnerability assessments

Füssel and Klein (2006) differentiate four prototypes of vulnerability assessments: impact assessments, first- and second-generation vulnerability assessments, and adaptation policy assessments. Earlier attempts to assess vulnerability were criticized for their narrow focus on biophysical and environmental aspects or geographical processes, and their neglect of social aspects (Saarinen et al. 1973; Blaikie et al. 1994; Cutter 1996; Cutter et al. 2003). The nature of VAs shifted toward evaluating climate impacts on society.[3] In the IPCC context, this is reflected in the differentiation between outcome and contextual understandings of vulnerability (also see above; O'Brien et al. 2007; Füssel 2010). Adaptation and justice researchers have pushed toward considering starting-point interpretations of vulnerability and engaging with the social aspects inside policy planning.

Social vulnerability assessments as a bottom-up approach

Determining the local social vulnerability beforehand is considered a bottom-up approach to inform climate adaptation policy (see Figure 2.1). Economic resources, infrastructure, institutions, technology, equity and information determine the adaptive capacity of populations as well as their social vulnerability (Street et al. 2016). The top-down approach is characterized by select global climate projections (e.g., in relation to world developments, global greenhouse gases), which are downscaled to regional levels and into biophysical assessments to identify sectoral impacts (Street et al. 2016). Climate adaptation policy can occur across the spectrum of these two different vulnerability conceptions (physical vis-à-vis social) (see Figure 2.1).

Within political practice, regional or national government adaptation responses such as the European Union's Adaptation Strategy (EC 2013) are considered top-down approaches, which are complemented by bottom-up efforts at the local level (Keskitalo and Preston 2019). Corresponding vulnerability assessments argue for the improved consideration of social vulnerability factors for a better understanding of the social justice implications of climate change (Breil et al. 2018; EEA 2018). The renewed emphasis on the inclusion of social vulnerability assessments as part of top-down and bottom-up strategies marks a growing political momentum of the concept lately.

Social vulnerability assessments (SVA) are understood to represent a starting point approach that has become increasingly popular based on the widespread agreement that a more holistic understanding of the multifaceted nature of vulnerability is needed (Cutter 1996; IPCC 2012; Yang et al. 2014; Zhou et al. 2014; Hou et al. 2016). This perspective redefines vulnerability as a social condition,

'Top-down' and 'bottom-up' approaches used to inform climate adaptation policy

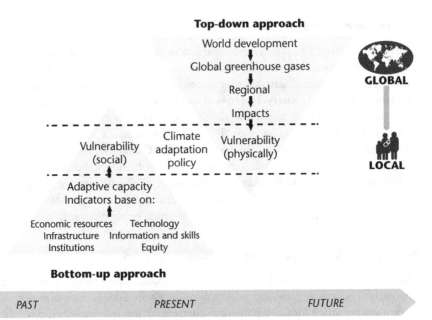

Figure 2.2 A conceptualization of vulnerability assessment approaches.

Sources: Nadin et al. (2016: 21), adapted from Dessai and Hulme (2004)

which varies over time and is affected by geographic and place-based aspects (Blaikie et al. 1994; Cutter 1996).

Synthesis 1: Exploring vulnerability as a relational concept

The first gap is located at the intersection of social vulnerability and the role of political institutions. There is a paucity of research that explores the multidimensional and political drivers of (social) vulnerability. Within social vulnerability literature, two of the most common assumptions are that society has become increasingly vulnerable and that environmental and social drivers have a significant impact on human well-being at a time when weather patterns and global average temperatures are changing (see, e.g., Otto et al. 2017). The political nature of uneven population vulnerability is frenetically avoided. There is only little emphasis on investigating the political causes of social susceptibility, as the acknowledgment that there are vulnerable groups in society, such as ethnic minorities, women, or homeless people, might be difficult for politicians (Lewis and Lewis 2014). As O'Brien et al. (2007: 83) argue: "'vulnerability reduction' as policy objective may be rhetorically non-controversial, but what this means in practice depends on the particular interpretation of vulnerability." This book therefore views vulnerability as a dynamic phenomenon that is relational and context-specific as well as constantly redesigned and a constant human condition (Fineman 2008; Bergoffen 2016). It explores how vulnerability intersects with political institutions.

Social vulnerability research compared

Social vulnerability research in China

Although the number of studies on natural hazards and social vulnerability has increased, in China there has been little attention given to the social factors tied to vulnerability (Chen et al. 2013; Zhou et al. 2014; Yang et al. 2014). The reasons for this gap are manifold. Chinese scholars have criticized the strong and continued emphasis on physical vulnerability, which they explain with the dominance of the natural science sector and disaster risk exposure frameworks (see, e.g., Chen et al. 2013; Su et al. 2015). There is a prevailing view of "disasters as acts of God, not acts of people, dominates both popular and academic understanding" (Chen et al. 2013: 170). Vulnerability assessments to inform adaptation planning are considered a fairly new approach in China (Street et al. 2016). Sectoral vulnerability assessments are likewise criticized for being too narrowly focused on biophysical impacts, for failing to perform holistic analyses and for neglecting non-climatic factors such as social, economic and technological determinants (Li et al. 2007; Street et al. 2016). This said, in recent years, some scholars have noted an increasing research interest into the social factors that determine pre-existent vulnerabilities (see, e.g., Nadin et al. 2016; Ge et al. 2017a).

Research on social vulnerability has become more prominent for emergency management given recent devastating disaster events and the large number of affected people (see, e.g., Zhou et al. 2014; Yang et al. 2014; Su et al. 2015; Hou et al. 2016). There are a handful of comprehensive assessments that assess social vulnerability from different perspectives in China, including provincial social vulnerability to disasters, urban social vulnerability and social vulnerability to climate change (see, e.g., Zhou et al. 2014; Ge et al. 2017a, 2017b, 2017c). Studies on social inequality are gaining attention because of widening social inequality (also see, e.g., Dillon 2015). Due to rapid urbanization coupled with severe environmental degradation; growing media attention and long-standing environmental protests, (social) vulnerability studies with a particular focus on air pollution and the health impacts of particulate matter have taken on significance as well (see, e.g., Kan et al. 2012; Ge et al. 2017b).[4]

The vulnerability of people working outdoors, people living in cubicle housing and people who have little knowledge about climate change is often emphasized in China (see, e.g., Chen et al. 2013; Ge et al. 2013). The same research outlines the particular vulnerability of rural Chinese migrant workers, due to their limited access to social welfare and healthcare based on their *hukou* status. *Hukou* refers to the Chinese household registration system, which determines what type of access a person has to certain public goods. Rural-to-urban migrants, for instance, cannot access education, healthcare or social services outside their registered place of birth. It was used as an initial instrument to control population mobility and to regulate population distribution (Chan and Zhang 1999). Its functions, however, have been going far beyond.[5]

For instance, Zheng et al. (2016) specifically look at rural communities in central and northern parts of Ningxia province that are affected by heat stress, droughts and serious water shortages. Their social vulnerability is closely related to their reliance on agriculture and livestock raising (Zheng et al. 2016). Ethnic background is another factor that shapes vulnerability: the response to climatic stress might be handled differently, due to the cultural circumstances. In Ningxia province, 34.8 percent of the people belong to the Islamic Hui population. Due to the higher birthrates and limited access to healthcare, childhood morbidity is significantly higher in Hui communities. The same factors, such as limited public health care, chronic diseases and increased elderly morbidity also play a role within non-Hui farmer communities in Ningxia province (Zheng et al. 2016).

Du et al. (2016), on the other hand, look at social vulnerability within urban communities in Guangdong province. Here, the *hukou* system is taken as an important factor, which explains the disproportionate vulnerability of migrants, particularly women migrants that face severe labor conditions without adequate access to social welfare and healthcare services. Population growth and urban density are key determinants that make Guangdong province and megacities therein particularly vulnerable despite the high spatial exposure to coastal flooding and extreme weather events. These intersectional perspectives already point to the political nature of uneven human vulnerability, which is not limited to climate change.

Social vulnerability studies in the United States

The social vulnerability paradigm gained prominence in the United States in the early 1990s and has been a popular research area ever since. The reasons for its popularity may lie in the long tradition of focusing academic debates on justice-related measures in different fields, such as climate ethics, distributive justice, egalitarian justice, environmental and social justice in light of the country's persistent and omnipresent problem with social inequality. One major strand of research frames existent social vulnerability studies in the context of environmental justice, as this has grown into a dominant discourse since the early 1970s (see, e.g., Huang and London 2012; Cooley et al. 2012; Nutters 2012). In China, the predominant social vulnerability studies are largely detached from climate justice or environmental justice-related discourses. As the previous chapter indicated, this may be explained by the lack of social science studies, an argument, which also applies to the very weak status of public health-focused adaptation research and practice (see Chapters 6 and 7).

Afro-American communities; Hispanics; single parents; the elderly; people with lower socioeconomic status, lower educational attainment and health constraints are considered to be particularly exposed in the United States (Lynn et al. 2011; KC et al. 2015). But why these factors correspond with higher vulnerability remains insufficiently explored and poses a major research gap.

Synthesis 2: Systemic study of social vulnerability origins across contexts

There is a need to uncover the underlying drivers of sustained vulnerability across two different political systems. The general lack of and need for understanding different vulnerabilities among groups of people rather than economic sectors is emphasized (Nadin et al. 2016). In light of "no understanding" about human health related adaptation measures (and human vulnerability more broadly), a course of action is particularly needed in this field in addition to addressing the knowledge gaps persistent in other impacted sectors such as agriculture, water resources and forestry (Nadin et al. 2016). Local decision-makers are largely inexperienced with vulnerability and risk assessments (Nadin et al. 2016). Little is known as to why this is the case. Research emphasizes the urgency for using (social) vulnerability assessments for improved risk management and sustainable development in China (Ge et al. 2013; Chen et al. 2013; Yi et al. 2014) and the United States (Turner et al. 2003, 2010). However, to what extent vulnerability conceptions match those of the practitioners is not well-researched.

Synthesis 3: Pairing dominant social vulnerability conceptions with views of local decision-makers

Researchers emphasize the urgency for using (social) vulnerability assessments for improved risk management and sustainable development in China (Ge et al. 2013; Chen et al. 2013; Yi et al. 2014) and the United States (Turner et al. 2003; Turner 2010). However, to what extent vulnerability conceptions match those of the practitioners is not well-researched. As has been pointed out by Ford and colleagues (2018a) previously, empirical vulnerability studies rarely examine the actual implementation process. No known research examines how political practitioners view population vulnerability to climate change and how dominant social vulnerability conceptions work at local political levels. However, understanding how local practitioners perceive of vulnerability and what policy options they consider viable is essential for understanding the dynamics of and opportunities for adaptation efforts and policy change. This research gap corresponds with the two underlying research questions of the first empirical chapter, which attempts to understand the characteristics that determine uneven human vulnerability and how it is understood and dealt with in political practice. The main research question on vulnerability lock-ins is grounded upon this preliminary assessment.

Preliminary conclusion

This chapter introduced some of the basic concepts and terminologies of climate change-related science and policy. First, the dominant understanding of vulnerability, as put forth by the IPCC, was laid out. The IPCC views vulnerability as an interplay of climate impacts, which are the result of both natural and anthropogenic causes, and are simultaneously affected by socioeconomic pathways and governance mechanisms (adaptation and mitigation, also see Chapter 3).

Throughout the years, the IPCC's understanding shifted from an end-point-based interpretation, which perceived of climate change as the root problem that results in vulnerability, to a starting point interpretation that engages in a contextual understanding of the sociopolitical and economic factors which predetermine the adaptive capacity of populations. The growing interest in the human dimension was briefly reflected upon by pointing to the perspective of social vulnerability and also briefly reflecting upon different prototypes of vulnerability assessments used in political practice. Social vulnerability assessments are one approach that engages in a starting-point interpretation of vulnerability, by looking at the contextual conditions that predetermine population vulnerability. Two localized social vulnerability assessments were used for the case selection, as will be laid out in Chapter 4. Further, they critically informed the empirical research process (see Chapter 7).

The chapter also pointed to the tension of many different and at a time overlapping frames. There are, for instance, significant overlaps between the way political ecologists define vulnerability and the way climate policy scholarship does. Yet, within climate policy, the target of vulnerability reduction is rarely questioned. Political ecology offers insightful critiques on resilience understandings and uniform concepts by also pointing to the interconnection of other contextual factors such as questions of power and access. What has become visible throughout this chapter is not just the observation of vulnerability as a relational concept and embodied experience of local populations themselves, but also that the process of defining vulnerability is as much a relational one, striving of dominant frames from the global governance landscape.

From this brief review of existing literature, the chapter synthesized core gaps that this book aims to address. These are:

1   There is a paucity of research that explores the multidimensional and political drivers of (social) vulnerability. This book aims to address this gap by exploring the intersection of social vulnerability and the role of political institutions.
2   A systemic study of social vulnerability origins across different political contexts is needed. Most commonly, developed countries are being compared, lacking insights on similar vulnerability drivers across different political systems.
3   There is a need to pair dominant social vulnerability conceptions with views of local decision-makers.

Notes

1   See *Oxford Dictionaries*, s.v. "vulnerability," available at: https://en.oxforddictionaries. com/definition/vulnerability, accessed July 10, 2018.
2   For a better understanding of the multidimensional concept of social vulnerability, also see the "hazards-of-place-model" (Cutter et al. 2003: 244).
3   For a more detailed discussion, see Füssel and Klein (2006).
4   Chai Jiang's *Under the Dome* documentary (2015) is commonly compared to Al Gore's *An Inconvenient Truth*. The film resonated with many Chinese people in light of growing

concerns over deteriorating air pollution and public health regrind the daily exposure to 2.5 particulate matter. The documentary was initially supported by then environmental minister Chen Jining and later on removed from online website and later on banned.

5 On the history of the hukou and its recent developments, see K. Chan and L. Zhang, "The Hukou System and Rural-Urban Migration in China: Processes and Changes," *The China Quarterly* 160 (1999): 818–855. Doi: 10.1017/S0305741000001351.

References

Adger, W. Neil " 'Vulnerability'." *Global Environmental Change-Human and Policy Dimensions* 16 (2006): 268–281. https://doi.org/10.1016/j.gloenvcha.2006.02.006.

Adger, W Neil, and P. Mick Kelly. "Social Vulnerability to Climate Change and the Architecture of Entitlements." *Mitigation and Adaptation Strategies for Global Change* 4 (1999): 253–266. https://doi.org/10.1023/A:1009601904210.

Agrawala, Shardul. "Integration of Human Dimensions in Climate Change Assessments." Open Meeting of the International Human Dimensions of Global Change Community Rio de Janeiro, Brazil, October 6–8, 2001.

Bergoffen, Debra. "The Flight from Vulnerability." In *Dem Erleben auf der Spur. Feminismus und die Philosophie des Leibes*, edited by Hilge Landweer and Isabella Marcinski, 137–152. Bielefeld: Transcript Verlag/De Gruyter, 2016.

Beymer-Farris, Betsy, Thomas J. Bassett, and Ian Bryceson. "Chapter 15: Promises and Pitfalls of Adaptive Management in Resilience Thinking: The Lens of Political Ecology." In *Resilience and the Cultural Landscape: Understanding and Managing Change in Human-Shaped Environments*, edited by T. Plieninger and C. Bieling. Cambridge, UK: Cambridge University Press, 2012.

Blaikie, Piers M., Terry Cannon, Ian Davis, and Ben Wisner. *At Risk: Natural Hazards, People's Vulnerability, and Disasters*. London: Routledge, 1994.

Bohle, Hans G., Thomas E. Downing, and Michael J. Watts. "Climate Change and Social Vulnerability. Toward a Sociology and Geography of Food Insecurity." *Global Environmental Change* 4, no. 1 (1994): 37–48. https://doi.org/10.1016/0959-3780(94)90020-5.

Breil, Margaretha, Clare Downing, Aleksandra Kazmierczak, Kirsi Mäkinen, and Linda Romanovska. "Social Vulnerability to Climate Change in European Cities – State of Play in Policy and Practice." European Topic Centre on Climate Change Impacts, Vulnerability and Adaptation (ETC/CCA), Technical Paper, 2018.

Brondizio, Eduardo S., and Emilio F. Moran. "Human Dimensions of Climate Change: The Vulnerability of Small Farmers in the Amazon." *Royal Society* 363, no. 1498 (2015). https://doi.org/10.1098/rstb.2007.0025

Burkett, V. R., A. G. Suarez, M. Bindi, C. Conde, R. Mukerji, M. J. Prather, A. L. St. Clair, et al. "Point of Departure." In *Climate Change 2014: Impacts, Adaptation, and Vulnerability. Part A: Global and Sectoral Aspects. Contribution of Working Group II to the Fifth Assessment Report of the Intergovernmental Panel on Climate Change*, edited by C. B. Field, V. R. Barros, D. J. Dokken, K. J. Mach, M. D. Mastrandrea, T. E. Bilir, M. Chatterjee, et al. 169–194. Cambridge, UK and New York, NY: Cambridge University Press, 2014.

Castree, Noel. "Broaden Research on the Human Dimensions of Climate Change." *Nature Climate Change* 6, no. 731 (2016). https://doi.org/10.1038/nclimate3078.

Chan, Kam W., and Li Zhang. "The Hukou System and Rural-Urban Migration in China: Processes and Changes." *The China Quarterly* 160 (1999): 818–855. Doi: 10.1017/S0305741000001351.

Chen, Wenfang, Susan L. Cutter, Christopher T. Emrich, and Peijun Shi. "Measuring Social Vulnerability to Natural Hazards in the Yangtze River Delta Region, China." *International Journal of Disaster Risk Science* 4, no. 4 (2013): 169–181. Doi: 10.1007/s13753-013-0018-6.

Cooley, Heather, Eli Moore, Matthew Heberger, and Lucy Allen. "Social Vulnerability to Climate Change in California." California Energy Commission, Pacific Institute, White paper, 2012.

Cutter, Susan L. "Vulnerability to Environmental Hazards." *Progress in Human Geography* 20, no. 4 (1996): 529–539. https://doi.org/10.1177%2F030913259602000407.

Cutter, Susan L., Bryan J. Boruff, and W. Lynn Shirley. "Social Vulnerability to Environmental Hazards." *Social Science Quarterly* 84, no. 2 (2003): 242–261. Doi: 10.1111/1540-6237.8402002.

Cutter, Susan L., and Christina Finch. "Temporal and Spatial Changes in Social Vulnerability to Natural Hazards." *Proceedings of the National Academy of Sciences* 105, no. 7 (2008): 2301–2306. https://doi.org/10.1073/pnas.0710375105.

Dessai, Suraje, and Mike Hulme. "Does Climate Adaptation Policy Need Probabilities?" *Climate Policy* 4, no. 2 (2004): 107–128. Doi: 10.1080/14693062.2004.9685515.

Dillon, Nara. *Radical Inequalities. China's Revolutionary Welfare State in Comparative Perspective.* Cambridge, MA and London: Harvard University Press, 2015.

Dow, Kirsten. "Exploring Differences in our Common Future(s): The Meaning of Vulnerability to Global Environmental Change." *Geoforum* 23, no. 3 (1992): 417–436. https://doi.org/10.1016/0016-7185(92)90052-6.

Downing, Thomas E., Anand Patwardhan, Richard J. T. Klein, Elija Mukhala, Linda Stephen, Manuel Winograd, and Gina Ziervogel. "Assessing Vulnerability for Climate Adaptation." In *Adaptation Policy Frameworks for Climate Change: Developing Strategies, Policies and Measures,* edited by Bo Lim et al., 61–90. Cambridge, UK: Cambridge University Press, 2005.

Du Yaodong, Zeng Yunmin, Ma Wenjun, Chen Xiaohong, Sarah Opitz-Stapleton, Rebecca Nadin, and Samantha Kierath. "Guangdong." In *Climate Risk and Resilience in China,* edited by Rebecca Nadin et al., 272–297. New York: Routledge, 2016.

European Commission (EC). *The EU Strategy on Adaptation to Climate Change.* Brussels: European Commission, 2013.

European Environmental Agency (EEA). *National Climate Change Vulnerability and Risk Assessments in Europe, 2018.* EEA Report no 1, 2018. Luxembourg: Publications Office of the European Union, 2018.

Fineman, Martha Albertson. "The Vulnerable Subject: Anchoring Equality in the Human Condition." *Yale Journal of Law and Feminism* 20, no. 1 (2008).

Ford, James D., Tristan Pearce, Graham McDowell, Lea Berrang-Ford, Jesse S. Sayles, and Ella Belfer. "Vulnerability and its Discontents: The Past, Present, and Future of Climate Change Vulnerability Research." *Climatic Change* 151 (2018a): 189–203. Doi: 10.1007/s10584-018-2304-1.

Ford, James D., Mya Sherman, Lea Berrang-Ford, Alejandro Llanos, Cesar Carcamo, Sherilee Harper, Shuaib Lwasa et al. "Preparing for the Health Impacts of Climate Change in Indigenous Communities: The Role of Community-Based Adaptation." *Global Environmental Change* 49 (2018b). https://doi.org/10.1016/j.gloenvcha.2018.02.006.

Füssel, Hans-Martin. *Review and Quantitative Analysis of Indices of Climate Change Exposure, Adaptive Capacity, Sensitivity, and Impacts.* Washington, DC: World Bank, 2010.

Füssel, Hans-Martin, and Richard J. T. Klein. "Climate Change Vulnerability Assessments: An Evolution of Conceptual Thinking." *Climatic Change* 75 (2006): 301–329. https://doi.org/10.1007/s10584-006-0329-3.

Ge, Yi, Wen Dou, and Jianping Dai. "A New Approach to Identify Social Vulnerability to Climate Change in the Yangtze River Delta." *Sustainability* 9, no. 12 (2017a): 2236. Doi: 10.3390/su9122236.

Ge, Yi, Wen Dou, Zhihui Gu, Xin Qian, Jinfei Wang, Wei Xu, Peijun Shi et al. "Assessment of Social Vulnerability to Natural Hazards in the Yangtze River Delta, China." *Stochastic Environmental Research and Risk Assessment* 27, no. 8 (2013): 1899–1908. https://doi. org/10.1007/s00477-013-0725-y.

Ge, Yi, Wen Dou, and Haibo Zhang. "A New Framework for Understanding Urban Social Vulnerability from a Network Perspective." *Sustainability* 9, no. 10 (2017b): 1723. Doi: 10.3390/su9101723.

Ge, Yi, Haibo Zhang, Wen Dou, Wenfang Chen, Ning Liu, Yuan Wang, Yulin Shi, and Wenxin Rao. "Mapping Social Vulnerability to Air Pollution: A Case Study of the Yangtze River Delta Region, China." *Sustainability* 9, no. 1 (2017c): 109. https://doi. org/10.3390/su9010109.

Hou, Jundong, Jun Lv, Xin Chen, and Shiwei Yu. "China's Regional Social Vulnerability to Geological Disasters: Evaluation and Spatial Characteristics Analysis." *Natural Hazards* 84 (2016): 97–111. https://doi.org/10.1007/s11069-015-1931-3.

Huang, Ganlin Huang, and Jonathan K. London. "Cumulative Environmental Vulnerability and Environmental Justice in California's San Joaquin Valley." *International Journal of Environmental Research and Public Health* 9, no. 5 (2012): 1593–1608. https://doi. org/10.3390/ijerph9051593.

IPCC. *Climate Change 2007: Synthesis Report. Contribution of Working Groups I, II and III to the Fourth Assessment Report of the Intergovernmental Panel on Climate Change*. Edited by Core Writing Team, R. K. Pachauri, and A. Reisinger. Geneva, Switzerland: IPCC, 2007.

IPCC. *Special Report for the Intergovernmental Panel on Climate Change Managing the Risks of Extreme Events and Disasters to Advance Climate Change Adaptation*. Edited by Field, C. B., V. Barros, T. F. Stocker, D. Qin, D. J. Dokken, K. L. Ebi, M. D. Mastrandrea, K. J. Mach, G.-K. Plattner, S. K. Allen, M. Tignor, and P. M. Midgley. Cambridge, UK: Cambridge University Press, 2012.

IPCC. "Summary for policymakers. In: Climate Change 2014: Impacts, Adaptation, and Vulnerability. Part A: Global and Sectoral Aspects." In *Contribution of Working Group II to the Fifth Assessment Report of the Intergovernmental Panel on Climate Change*, edited by Christopher B. Field, Vincente R. Barros, David J. Dokken, Katharine J. Mach, Michael D. Mastrandrea et al., 1–32. New York, NY: Cambridge University Press, Cambridge, 2014.

Joakim, Erin P., Linda Mortsch, and Greg Oulahen. "Using Vulnerability and Resilience Concepts to Advance Climate Change Adaptation." *Environmental Hazards* 14, no. 2 (2015): 137–155. https://doi.org/10.1080/17477891.2014.1003777.

Kan, Haidong, Renjie Chena, and Shilu Tong. "Ambient Air Pollution, Climate Change, and Population Health in China." In *Emerging Environmental Health Issues in Modern China*, edited by Yong-Guan Zhu and Ming H. Wong, *Environment International* 42, (2012): 10–19. https://doi.org/10.1016/j.envint.2011.03.003

KC, Binita, J. Marshall Shepherd, and Cassandra Johnson Gaither. "Climate change vulnerability assessment in Georgia." *Applied Geography* 62 (2015): 62–74. https://doi. org/10.1016/j.apgeog.2015.04.007.

Keskitalo, E. Carina H., and Benjamin L. Preston. *Research Handbook on Climate Change Adaptation Policy*. Cheltenham, UK, and Northampton, MA: Edward Elgar Press, 2019.

Lewis, James, and Sarah A. V. Lewis. "Processes of Vulnerability in England? Place, Poverty and Susceptibility." *Disaster Prevention and Management* 23, no. 5 (2014): 586–609. https://doi.org/10.1108/DPM-03-2014-0044.

Li, Yue, Wei Xiong, and Yanjuan Wu. "Climate Change Impacts, Vulnerability and Adaptation in China September 2007." Report published online by the BASIC Project, 2007.

Livermann, Diana. "Chapter 23 Reading Climate Change and Climate Governance As Political Ecologies." In *The Routledge Handbook of Political Ecology*, edited by Tom Perreault, Gavin Bridge, and James McCarthy. London, UK and New York: Routledge, 2015.

Luers, Amy L. "The Surface of Vulnerability: An Analytical Framework for Examining Environmental Change." *Global Environmental Change* 15, no. 3 (2005): 214–223. https://doi.org/10.1016/j.gloenvcha.2005.04.003.

Lynn, Kathy, Katharine MacKendrick, and Ellen M. Donoghue. *Social Vulnerability and Climate Change: Synthesis of Literature*. General Technical Report PNW-GTR-838. United States Department of Agriculture (Portland: USDA), 2011.

Nadin, Rebecca, Sarah Opitz-Stapleton, and Xu Yinlong, eds. *Climate Risk and Resilience in China*. London: Routledge, 2016.

National Academy Press (NAP). "Chapter: 7 Human Dimensions of Global Environmental Change." In *Global Environmental Change: Research Pathways for the Next Decade*, edited by the Committee on Global Change Research, 293–376. Washington, DC: National Academy Press, 1999.

Nutters, Heidi. "Addressing Social Vulnerability and Equity in Climate Change Adaptation Planning." White Paper, Adapting to Rising Tides Project, San Francisco Bay Conservation and Development Commission, 2012. Doi: 10.13140/RG.2.1.3442.4163.

O'Brien, Karen, Siri Eriksen, Lynn P. Nygaard, and Ane Schjolden. "Why Different Interpretations of Vulnerability Matter in Climate Change Discourses." *Climate Policy* 7, no. 1 (2007): 73–88.

O'Brien, Karen, Siri Eriksen, Ane Schjolden, and Lynn Nygaard, "What's in a Word? Conflicting Interpretations of Vulnerability in Climate Change Research." CICERO Working Paper no. 4, 2004.

Otto, Ilona M., Diana Reckien, Christopher P. O. Reyer, Rachel Marcus, Virginie Le Masson, Lindsey Jones et al. "Social Vulnerability to Climate Change: A Review of Concepts and Evidence." *Regional Environmental Change* 17, no. 6 (2017): 1651–1662. Doi: 10.1007/s10113-017-1105-9

Preston, Benjamin L. "Putting Vulnerability to Climate Change on the Map: A Review of Approaches, Benefits and Risks." *Sustainability Science* 6 (2011): 177–202.

Ribot, Jesse C., Antonio Rocha Magalhães, and Stahis Panagides. *Climate Variability, Climate Change and Social Vulnerability in the Semi-Arid Tropics*. Cambridge: Cambridge University Press, 2005.

Rotterdam Climate Initiative (RCI). "Rotterdam Climate Change Adaptation Strategy." De Urbanisten in cooperation with Management team of Rotterdam Climate Proof (RCP), 2012.

Saarinen, Thomas, Kenneth Hewitt, and Ian Burton. "The Hazardousness of a Place: A Regional Ecology of Damaging Events." *Geographical Review* 63, no. 1 (1973): 134–136. Doi: 10.2307/213252.

Sovacool, Benjamin, and Björn-Ola Linnér. *The Political Economy of Climate Change Adaptation*. London: Palgrave Macmillan, 2016.

Street, Roger, Sarah Opitz-Stapleton, Rebecca Nadin, Cordia Chu, Scott Baum, and Declan Conway. "Climate Change Adaptation Planning to Policy: Critical Considerations and

Challenges." In *Climate Risk and Resilience in China*, edited by Rebecca Nadin Sarah Opitz-Stapleton, and Xu Yinlong, 11–37. New York: Routledge, 2016.

Su Shiliang, Jianhua Pi, Chen Wan, Huilei Li, Rui Xiao, and Binbin Li. "Categorizing Social Vulnerability Patterns in Chinese Coastal Cities." *Ocean & Coastal Management* 116 (2015): 1–8. http://dx.doi.org/10.1016/j.ocecoaman.2015.06.026.

Taylor, Marcus. *The Political Ecology of Climate Change Adaptation. Livelihoods, Agrarian Change and the Conflict of Development*. London and New York: Routledge, 2015.

Turner II, Billie L. "Vulnerability and Resilience: Coalescing or Paralleling Approaches for Sustainability Science?" *Global Environmental Change* 20, no. 4 (2010): 570–576. https://doi.org/10.1016/j.gloenvcha.2010.07.003.

Turner II, Billie L., Roger E. Kasperson, Pamela A. Matson, James J. McCarthy, Robert W. Corell, Lindsey Christensen, Noelle Eckley et al. "A Framework for Vulnerability Analysis in Sustainability Science." *Proceedings of the National Academy of Sciences of the United States of America* 100, no. 14 (2003): 8074–8079. https://doi.org/10.1073/pnas.1231335100.

Vogel, Coleen, Susanne C. Moser, Roger E. Kasperson, and Geoffrey D. Dabelko. "Linking Vulnerability, Adaptation, and Resilience Science to Practice: Pathways, Players, and Partnerships." *Global Environmental Change* 17 (2007): 349–364. Doi: 10.1016/j.gloenvcha.2007.05.002.

Yang, Xuchao, Wenze Yue, Honghui Xu, Jingsheng Wu, and Yue He. "Environmental Consequences of Rapid Urbanization in Zhejiang Province, East China." *International Journal of Environmental Research and Public Health* 11, no. 7 (2014): 7045–7059. Doi: 10.3390/ijerph110707045.

Yi Lixin, Zhang Xi, Ge Lingling, and Zhao Dong. "Analysis of Social Vulnerability to Hazards in China." *Environmental Earth Science* 71, no. 7 (2014): 3109–3117. https://doi.org/10.1007/s12665-013-2689-0.

Zebrowski, Chris. "Governing the Network Society: A Biopolitical Critique of Resilience." *Political Perspectives* 3, no. 1 (2009): 1–38.

Zheng Yan, Meng Huixin, Zhang Xiaoyu, Zhu Furong, Wang Zhanjun, Fang Shuxing et al. "Ningxia." In *Climate Risk and Resilience in China*, edited by Rebecca Nadin, Sarah Opitz-Stapleton, and Xu Yinlong, 213–262. New York: Routledge, 2016.

Zhou, Yang, Ning Li, Wenxiang Wu, and Jidong Wu. "Assessment of Provincial Social Vulnerability to Natural Disasters in China." *Natural Hazards* 71 (2014): 2165–2186. Doi: 10.1007/s11069-013-1003-5.

3 Vulnerability and adaptation lock-ins

Theoretical foundations and main analytical framework

This chapter presents the main analytical framework of lock-ins. The main goals of this research are to 1) understand the specific targeting of vulnerable populations as part of local policymaking processes and 2) explain vulnerability maintenance and incremental adaptation from this perspective. This is realized through applying an integrative framework of lock-ins to vulnerability and adaptation. Incremental adaptation is signified by differential exposure of specific population groups such as ethnic minorities in Atlanta or rural migrant communities in Jinhua to climate impacts such as extreme heat waves and flooding. Their social vulnerability has manifested over long periods.

This section reviews prevailing research attempts on adaptation barriers, which remain the dominant research perspective to explain incremental adaptation efforts and adaptation deficits. The section continues by introducing lock-ins as a set of inter- and path-dependent processes, which can help to explain some of the root causes of vulnerability production and maintenance. Lock-ins are self-reinforcing factors that prevent larger changes. The chapter argues that these dynamics are anchored in the political ecology.

This chapter finds that the concept of lock-ins is better equipped to explain stalled governance efforts regarding adaptation. The concept of lock-ins better grasps the long-term maintenance of adaptation deficits and the persistence (if not deterioration) of uneven vulnerability to climate change. Because social vulnerability and adaptation are relatively new concepts in the local governance of climate impacts in the chosen regions, lock-ins can act as a framework to identify more deeply rooted structures, which prevent policy action from occurring. The chapter ends with a brief literature review of adaptation research in China and the United States and a summary of the key research gaps this book addresses.

Setting the adaptation research agenda

Adaptation scholars have outlined the difficulty of defining the adaptation policy field due to the problem of overlapping policy levels and arenas. The term adaptation lacks conceptual and methodological clarity (Dupuis and Biesbroek 2013; Purdon 2014). Different interpretations of adaptation and how to operationalize it within policymaking further complicate how outcomes of adaptation should and

DOI: 10.4324/9781003183259-3

can be measured. Dupuis and Biesbroek (2013) call this the "dependent variable problem in adaptation research" (p. 8). It is not just difficult measuring policy outcomes, but it is also challenging to define the research field and policy planning.

Due to its complex nature, governments have broken adaptation policymaking down into different policy domains such as education, health, transportation, technology, urban planning, despite increasing calls for the need for cross-sectoral frameworks (Heltberg et al. 2010; Perry and Ciscar 2014; Charan et al. 2018). There is growing recognition of the need to integrate the different sectors, as a multi-sectoral approach is considered a precondition to increase society's capacity to manage both indirect and direct climate risks (see, e.g., Heltberg et al. 2009). There are also increasing research interests and academic acknowledgment of the need to link multi-scale policy arenas such as disaster risk reduction, climate change adaptation and sustainable development. However, the difficulty of combining these three large policy fields is likewise outlined (Schipper et al. 2016; Kelman et al. 2017). The theoretically integrated frameworks, which outline multi-sectoral approaches or the integration of multi-scale policy arenas, continue to be disconnected from the policymaking structures and practical approaches of public institutions.

Frameworks for assessing adaptation

When analyzing the state of adaptation action, the anatomy of adaptation to climate change and variability developed by Smit et al. (2000), helps to set the research agenda, which can be broken down into four distinctive steps and research questions: (i) adapt to what? (climate-related stimuli across time and space, coupled with non-climate factors) (ii) who or what adapts? (system definition and characteristics) and (iii) how does adaptation occur? (processes and outcomes). Lastly, (iv) evaluating adaptation based on preset criteria and principles. Here, the observed and projected climate-related stimuli of the chosen regions are laid out based on a literature review of climate vulnerability assessments.

Adaptation deficits are "the gap between the current state of a system and a state that would minimize adverse impacts and from existing climate conditions and variability, that is essentially inadequate adaptation to the current climate conditions" (Noble et al. 2014: 839). In this book, the adaptation deficits are marked by high adaptation pressure in climatological and social terms.

Aspects of time of adaptation measures can be distinguished between short- (less than 10 years), medium- (10 to 30 years) and long-term approaches (more than 30 to 100 years) (CACCA 2010). Depending on the managing level, researchers also differentiate between ad hoc and incremental, transformative, just, equitable, pro-active or anticipatory adaptation. The literature itself is rich in concepts related to reactive and incremental vis-à-vis anticipatory and proactive adaptation (see, e.g., Smit et al. 2000; Wolf et al. 2009; Bierbaum et al. 2013; Wise et al. 2014). Although some agreement exists that both forms of adaptation (reactive/incremental and anticipatory/proactive) are needed, reactive forms of adaptation

Table 3.1 Areas and types of adaptation

Areas	Type of adaptation				
Space	Local (municipal & county)	Metro/ Regional	National	Global	
Actor	Public (Government)		Private (NGO, community-based, business)		
Timing/ Stimulus	Reactive		Proactive/Anticipatory		
Intent	Planned	Implemented	Autonomous		
Timespan	Short-term	Medium	Long-term		
Sector	Agriculture	Education	Health/ Human	Transport	Water
Approach	Behavioral	Ecosystem-based	Financial	Structural	Technological
Degree of change	Maintenance	Incremental	Transformation		

Source: The author, based on: Pelling (2011), O'Brien (2012), IPCC (2014), Lonsdale et al. (2015) and Garrelts et al. (2018) (grey = aspect in focus of this book)

are oftentimes judged to be too incremental; nondeliberate and non-transformational in nature (e.g., O'Brien 2012; Bierbaum et al. 2013).

Social scientific understanding of adaptation

Social scientific responses to a natural science-dominated field have been scarce. Although research on the social aspects of adaptation has been flourishing, the incorporation of social science into decision-making structures remains understudied. SVAs constitute one form of social scientific practice and shall be comparatively studied regarding related political practices at the local level, context-specific applications and perceptions of human vulnerability to climate change. Adaptation practitioners and experts consider the use of SVAs an important element for enabling transformative adaptation. How they are being considered by decision-makers offers insights to offers about their political practicability and where processes can potentially be improved.

Areas of adaptation

Areas and types of adaptation are manifold. They occur at different levels of space, actor, timing, intent, considering different timespans, sectors, approaches and ultimately varying degrees of change (see Table 3.1). The IPCC (2014) outlines three broad types of adaptation intent: Autonomous, which can be spontaneous without explicit previous planning, incremental with the central aim system maintenance and transformational with more fundamental changes. This study focuses on public adaptation planning at local political levels. As deliberate adaptation concerns have not been mainstreamed into policy process yet, the areas: timespan, sector and approach are studied from a broader perspective. The degree of change is of particular research interest, as it includes an overriding timespan, different degrees of change and a multi-sectoral focus.

Explaining adaptation deficits

Adaptation deficits are oftentimes explained by referring to climate change adaptation (CCA) as a particularly wicked policy problem (e.g., Moser and Boykoff 2013; Perry 2015; Nadin et al. 2016). Others have framed climate adaptation as a collective action problem (Sovacool and Linnér 2016). Like collective action problems, wicked problems are defined by their complex, temporally dispersed and highly localized nature, which involves a variety of actors across different levels of government, society as well as non-state sectors. The wicked nature of adaptation is characterized as a problem for which no clear-cut solutions exist. Drastic uncertainty involved in the planning, the magnitude of change underway, large disagreement on values and norms and strongly sectoral efforts make this a particularly challenging policy field. The need to plan for longer time horizons and related economic costs of longer distance adaptation routes make CCA a particularly high-hanging policy fruit.

Biesbroek et al. (2013) have referred to this as a "dependent variable problem." Wicked problems cannot be definitively formulated and can be considered symptoms of other problems (Hartmann et al. 2012).[1] The term was first introduced in planning theory and emphasized that "planning problems are inherently wicked" (Rittel and Webber 1973: 160). Turnbull and Hoppe (2018) reject the notion of "wicked problems" put forward by Rittel and Webber to analyze public problems. Categorizing some public policy problems as particularly resistant public policy problems has resulted in a frustrating experience, which often stretched the concept too far (Turnbull and Hoppe 2018). As a consequence, scholars failed to locate the wicked problem concept within the larger questioning of the capacities and limitations of governmental policymaking" (Turnbull and Hoppe 2018: 2).

Adaptation barriers

Research on barriers initially intended to respond to the question of limited capacity by looking at different dimensions which hinder action from occurring. In light of major and continued adaptation deficits, research on adaptation barriers, limits and thresholds has been growing (see, e.g., Adger et al. 2009; Moser and Ekstrom 2010; Jones and Boyd 2011; Prutsch et al. 2014). Barriers are defined as obstacles to adaptation mainstreaming; implementation and/or transformation. With concerted efforts and under a variety of different conditions (e.g., change of thinking, prioritization, shifts in resources) barriers can be overcome (Moser and Ekstrom 2010). Identifying barriers to adaptation is said to increase the understanding of the political dynamics of a given place and the nature of existent adaptation efforts and to help in the prioritization of adaptation action. Barriers to adaptation are conceptualized in many different ways but attempted definitions and differentiations are elusive.

Most frameworks consist of different sources of barriers, which interact in both influencing and prescribing adaptation (Jones and Boyd 2011). For instance, "structural barriers" often refer to constraining and interconnected contextual factors such as actors, the governance context and socioeconomic conditions (e.g., Moser and Ekstrom 2010). Other structural barriers commonly outlined are the lack of human and financial resources, lack of best practices, knowledge and information, which persist in science–policy gaps (Adger et al. 2009; Prutsch et al. 2014; Nadin et al. 2016). Parallel to these barrier studies, research began to investigate how to strengthen the science–policy interface (Horton and Brown 2018; also in UNEP 2017). Other barriers include the lack of political leadership or lack of political support from non-state actors and the public.

The structural lens was criticized for not focusing on the underlying norms and social institutions through which decisions are made. As a consequence, the emphasis shifted toward exploring the social barriers to adaptation (see, e.g., Adger et al. 2009; Jones 2010; Jones and Boyd 2011; Bierbaum et al. 2013; Biesbroek et al. 2013, 2014; Dow et al. 2013). Social barriers explore the cognitive, normative and institutional factors that influence adaptation (see, e.g., Adger et al. 2009; Jones and Boyd 2011; Evans et al. 2016). Jones and Boyd (2011), for instance, distinguish between cognitive barriers such as psychological factors, normative barriers which persist in behavior as well as values and institutional barriers, which relate to institutional inequities and governance structures. These different dimensions are all considered social barriers, which act in concert (Jones and Boyd 2011). Social barriers exist in addition to natural barriers (e.g., physical and environmental components) and human as well as informational factors (knowledge, technology and economic barriers) (Jones and Boyd 2011).

The social perspective of barriers intended to offer an alternative conception of limits that are endogenous, through "emerging from 'inside' society" (Adger et al. 2009). According to this understanding, ethical components are coupled with cultural valuation, societal risk perception, and structural deficits such as the lack of precise knowledge. At the same time, they are influenced by cognitive components and psychosocial uncertainties (see, e.g., Wolf et al. 2009; Dow et al. 2013). In this line of thinking, the direct experience with climate impacts, such as extreme heat or flooding and close proximity to an event is said to structure adaptation responses, ultimately leading to greater engagement with the issue. Other angles have looked at a mix of "external" factors in terms of complex institutional environments in which adapation decisions are made that are characterized by, unprecedented environmental change and an increasing demand on limited resources (scarcity and overpopulation), as barriers to transformative change and factors that negatively influence people's adaptive capacity (e.g., Eakin at al. 2016). Some of these angles have begun to hint at other reasons, such as dominant political practices of water resources allocation or the dominance of urban interests, which impact local choices and conditions to adapt to a changing climate.

Critique: Descriptive nature of prevailing explanatory attempts

Overall, however, research on adaptation barriers is criticized for its focus on developed countries and lack of comparative studies across different contexts (Biesbroek et al. 2013). Additionally, the descriptive nature of barrier thinking and barrier classification as modus operandi has come into question (e.g., Dupuis and Biesbroek 2013; Oels 2019). The explanatory value is limited, as most studies do not touch upon the underlying reasons. Criticisms recommend going beyond "asking the questions 'if' and 'which' barriers to adaptation exist and begin asking 'how' and 'why' barriers emerge" (Biesbroek et al. 2013: 1127).

One exemption is the study of Fieldman (2011), who provides a different explanatory angle by looking at systemic barriers to climate adaptation outside of the climate policy literature. Here, neoliberalism and "the global institutionalization of laissez-faire economics" are considered a systemic barrier to climate adaptation due to diminished social welfare functions of the state (Fieldman 2011). Besides few exemptions, the barrier literature largely fails to explain, why vulnerability to climate change continues to be so high, and in the language of policymakers: adaptation remaining in a nascent stage, or why implementation of planned action continues to occur with extreme limitations. The underlying causal mechanisms are rarely examined:

> The vast majority of policy-oriented and scholarly publications on climate-related vulnerability and adaptation attend to response rather than causality (Bassett and Fogelman 2013, 47). They seek to identify who is vulnerable rather than why, indicators rather than explanation, fixes rather than causes – as if cause were not part of redressing vulnerability and its production. . . . Many stop with convenient proximate explanations such as assets or poverty – without asking how these are produced. Others, from adaptation and resilience schools, cordon off causality in capacities – like adaptive capacity or the capacity to bounce back (Manyena 2006). These approaches focus attention on 'innate' characteristics of the individual, household or group – the unit at risk (Gaillard 2010: 220). Capacity is now a common explanatory factor in most risk and vulnerability frameworks (Yohe and Tol 2002; Manyena 2006; Folke et al. 2010; Cardona et al. 2012: 72). But capacity as cause is not enough – it begs us to ask what shapes capacity.
>
> (Ribot 2014: 669f.)

The majority of criticisms of barrier frameworks essentially criticize the palliative nature of explanations, which at first appear to clarify which barriers need addressing but fall short of analyzing how they manifest as part of deeper structures. They are a phenomenon of more deeply rooted issues. Ribot points to a range of problematic research practices, lots of which follow a similar pattern by only identifying the proximate causes. Thereby, the causalities are neglected. Going a step further, one would have to ask, what the identified proximate problems are a greater phenomenon of. For example: Why is the risk perception so critically influenced by direct affectedness? Why does limited access to common goods such as education, health and transport persist for certain parts of the population? This would ultimately shift the discussion of "social" vulnerability and barriers to an examination of the role political institutions (can) play in breaking with past dependencies.

Going beyond barriers

Adaptation limits

In earlier publications, limits, barriers and thresholds were often used interchangeably. Recently, agreement seems to have developed differentiating between limits as absolute obstacles and barriers as obstacles that can be overcome (Adger et al. 2007; Moser and Ekstrom 2010; Evans et al. 2016). Limits are considered a state, which is positively correlated with unacceptable consequences in the form of drastic ecological, human and/or financial losses. The permissible risk threshold, or risk tolerance, is viewed as the limit beyond which systems can no longer adapt.

> Where barriers block or divert the adaptation process, limits often present more absolute thresholds. . . . We define limits as the point at which adaptation is not just *inefficient* (as influenced by barriers) but *ineffective at reducing vulnerability of those undertaking adaptation*.
>
> (Evans et al. 2016: 2)

The understanding of limits as absolute thresholds that are more complex to overcome enables a more systemic view of the protracted, mutually reinforcing and complex nature of inbuilt dependencies. The IPCC (2014) points to the opportunities and limits of adaptation and multiple scales at which adaptation can occur and reflects upon development pathways and opportunities, which exist to facilitate adaptation planning (Klein et al. 2014). Examples of constraining factors that the IPCC mentions include a range of biophysical, institutional, financial, social, and cultural factors, which can "constrain the planning and implementation of adaptation options and potentially reduce their effectiveness" (Klein et al. 2014: 902). The IPCC also points to the ethical and justice dimensions of opportunities and of differential vulnerability. For example, gender is given as a factor leading to lowered adaptive capacity, or individuals without sufficient access to resources and access to transportation (p. 916). Yet, the circumstances that maintain the patriarchal conventions or limited access are not investigated. Further, access to resources, which are mediated by public and private institutions are largely looked at from a policy perspective to initiate adaptation action and access to resources as part of decision-making. The insights remain vague to the extent that it is not examined how public actors influence entitlements to resources and how patterns of power structure adaptation limits. The report on the limits to adaptation nevertheless provides important background information on the multiplicity of factors that set the parameters of the adaptation policy field. In this book, these factors are conceptualized in the political ecology. The political ecology has insights to offer on power and agency in questions related to the environment and governance. These perspectives will be briefly reflected upon in the subsequent paragraphs.

Adaptation and vulnerability lock-ins

Although barriers might be helpful to describe policy processes related to incremental decision-making, they do not explain why barriers emerge and persist (Biesbroek et al. 2013) and how to deal with complex policy challenges. Recent research has approached factors, which hinder adaptation as path-dependent lock-ins to go beyond providing barrier taxonomies (Huitema et al. 2016; Siebenhüner et al. 2017). Lock-ins are a "particular conceptual approach to understand path-dependencies and rigidities in policy processes" (Siebenhüner et al. 2017: 2). This emerging framework builds on criticisms of the concepts of barriers and aims to offer a more nuanced understanding of the nature of barriers, how they relate to one another, how they emerged, and might be addressed (Siebenhüner et al. 2017). In line with the criticized atheoretical delineation of "barriers" and "opportunities" (Keskitalo and Preston 2019), lock-ins are not per se historical or negative but also describe inbuilt trajectories that can be positive (Siebenhüner et al. 2017). Lock-ins generally describe developments or paths that are difficult to change because of reinforcing feedback mechanisms. They are neither negative nor positive by definition, but call for an empirical assessment of the effect lock-ins can have. The German energy transition is provided as an example of positive discursive lock-in, based on the fact that the energy transition only changed its form (recentralization of political control) but not its commitment to transition away from nuclear-fossils (Siebenhüner et al. 2017: 6).

Siebenhüner et al. (2017) identify four descriptive dimensions that in combination can lead to adaptation lock-ins (see Figure 3.1). These are 1) lock-ins of knowledge, information and expertise such as discursive, epistemological and ontological dimensions in climate adaptation research, which revolve around the study of what constitutes legitimate knowledge on adaptation. Another lock-in dimension is related to 2) physical infrastructure such as planning design, which connects places and communities but often has limited room for variation. Water infrastructure is provided as an area where these dynamics can be demonstrated due to infrastructure construction. 3) Institutional lock-ins refer to the political dimension of formal procedures and rules as well as norms. Policy tools are taken as one particular element for the examination of lock-ins, which refer to the aides "policymakers use to prepare, implement and evaluate policy, including cost-benefit analysis, scenarios, computer models or participatory methods" (Siebenhüner et al. 2017: 8). Lastly, 4) actors in combination with cognitive, psychological and cultural factors such as conservatism are considered another dimension of lock-in.

Siebenhüner and colleagues frame these as adaptation lock-ins. As part of this book, they are however understood as adaptation and vulnerability lock-ins alike. Aside from seeking responses for the incremental nature of political adaptation action, we have to likewise ask: Why is vulnerability to climate change continuously

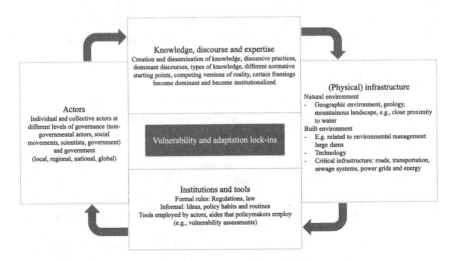

Figure 3.1 Analytical framework of potential lock-in dimensions.

Source: The author, adjusted from Siebenhüner et al. (2017, 2019)

high and so unevenly distributed? The assumption is that lock-ins may manifest differently regarding vulnerability and adaptation. For instance, aspects of knowledge related to vulnerability may focus on different aspects than types of knowledge related to adaptation. There may be strong overlap, but to take another example, discursive practices related to adaptation may have different normative starting points and logics than lock-ins in the context of (social) vulnerability. Further, and in line with the previous insights on the limits to adaptation, there may be certain aspects that lie outside the field of adaptation policy. This is not to say that they do not matter for addressing lock-ins from an adaptation perspective, but thinking of them as vulnerability lock-ins alike opens the perspective to factors that may need to be addressed outside the realm of adaptation.

Characteristics of lock-ins

Aside from their path-dependent and at times mutually reinforcing nature (Siebenhüner et al. 2017), lock-ins are matters of coevolution (e.g., Seto et al. 2016). Lock-ins stem from an interrelated set of factors that have their origins at multiple scales (Siebenhüner et al. 2019).[2] In relation to adaptation, path dependencies can signify institutional inertia, imply growing inflexibility in form of established political and cultural practices. In line with Corvellec et al. (2013), the innovative potential for the notion of lock-in lies in the question "are we in a lock-in?" Providing a systematic analysis of lock-in mechanisms and different lock-in components can provide new explanatory avenues (Siebenhüner et al. 2019).

The notion of lock-ins can be used as a practical starting point for locating critical junctures, where constraining lock-ins may be overcome. This would enable

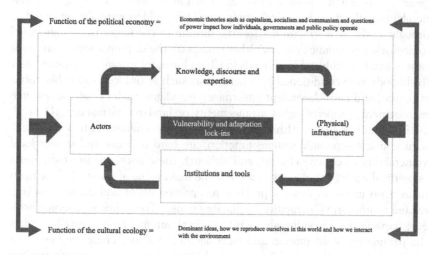

Figure 3.2 Overarching explanatory framework of lock-in dynamics as anchored in the political ecology.

Source: The author

current trajectories to break with former reproductions and co-creations of vulnerability through political institutions and potentially result in transformative adaptation. The lock-in concept has been marginalized in adaptation research (Hetz and Bruns 2014; Siebenhüner et al. 2017). Although the conceptual development of lock-in effects needs further research, this research looks at lock-ins as a new model for explaining policy processes related to vulnerability and adaptation governance.

The nature of specific lock-in dimensions, their interrelation and path-dependencies will be explored in this research. Building upon the preliminary research framework provided by Siebenhüner et al. (2017), the conceptualization of adaptation lock-ins is expanded to also include vulnerability.

Anchoring lock-ins in the political ecology

In this book, lock-ins are considered the analytical framework, which is conceptualized in the political ecology. Thus, they are, on the one hand, a function of the cultural ecology, which relates to the dominant ideas, cultural routines and practices as well as how we choose to reproduce ourselves and structure our relations with nature. On the other hand, lock-ins are considered products of the political economy, interacting with questions of power against dominant economic theories, and path-dependent processes of, for instance, capitalism or socialism. The descriptive dimensions of knowledge, physical infrastructure, institutions and actors, how they interact and maintain vulnerability are to be seen in this light.

According to this understanding, the pattern of locking in differential exposure to climate impacts is divided along class lines. In the context of critical geography research, Taylor (2015) has hinted at an uneven topography of vulnerability to flooding. In Taylor's study, small landholders were also impacted by drought under new irrigation systems, pointing to the larger factors anchored in dominant economic systems. Other studies on the disproportionate affectedness of local agricultural livelihoods in China have, for instance, examined how livelihoods were conditioned by other factors such as (in)access to public infrastructure, and path-dependent governance paradigms such as natural resource management, property rights arrangements or land rental markets (Li et al. 2019; Qi and Li 2021). Although it is important to understand how the dominant political economic systems function, are hard to break and impact local vulnerability to climate change, and although these questions sit at the heart of political ecology studies, they are not the main focus of the study. The main focus is on instances of local problem recognition about and decision-making related to uneven vulnerability to climate change. The political economy however impacts the way vulnerability is brought about, and fundamentally renders the parameters of adaptation and local decision-making. These are important context factors to consider when approaching the issue of uneven vulnerability to climatic changes.

Synthesis 1: Lock-ins as they relate to political institutions

Although some works explore the adaptive capacity of governments (e.g., Grothmann et al. 2013), the role formal political institutions play in enabling adaptive capacity of climate-vulnerable populations is largely under-researched. In line with vulnerability understanding put forth in Chapter 2, vulnerability is relational and thus a function of certain societal practices and society-nature relations. Yet, the role of politics in reproducing and maintaining vulnerability is not well-researched and thus is a focal point of this study. In the early 2000s, the debates on climate vulnerability concentrated on the biophysical and geography-specific impacts of climate change and less on the pre-existent social conditions that make populations particularly vulnerable (Cutter et al. 2003). Although adaptation planning still misses a holistic approach that considers social vulnerability also as a result of political institutions. Politically, the neglect of governmental institutions in formulating and implementing risk-reducing policies has been present on an international as well as at the domestic level. In order to address these deficiencies, this research looks at how governmental bodies engage in increasing the resilience of climate-vulnerable population groups and political institutions. Although this research gap is an underlying theme throughout the chapters, it is specifically addressed in the last empirical chapter and corresponds with the last research question, which lock-ins can explain adaptation deficits.

The growing field of research on how to govern big transformations needs to be seen in this context. Lock-ins help to understand, why the larger shifts are not occurring, and transformative change is so difficult. Adaptation barrier research has been criticized for being only descriptive. This book applies an evolving analytical framework of adaptation and vulnerability lock-ins. This lens enables a better understanding of the systemic nature of protracted vulnerability and thereby the absence of (transformative) adaptation. At the same time, it appears to be a better tool to grasp policy processes and dynamics at local government levels in China and the United States by examining different forms of path dependencies and their intertwinement.

Adaptation studies in China and the United States

Adaptation research foci in China

Climate adaptation opportunities have commonly been studied per policy sector that exhibit specific vulnerability and are considered of important political significance. In China, CCA research has put a particular emphasis on aspects related to agricultural and food security, as well as water management in drought-affected provinces in the northwestern part of the country. This may be explained with rapid environmental degradation and a particular regional climate vulnerability which is more drastically coupled with poverty in this region. Further, the agriculture-based economy region is considered representative of attempts to adapt to climate change in semi-arid regions (see, e.g., Zhao et al. 2014; Yang et al. 2015; Zheng et al. 2016).

Another topic of great research interest is how to adapt to SLR in the southern part of the country, with a particular focus on Guangzhou province (see, e.g., Yang et al. 2015; Meng and Dabrowski 2016; Du et al. 2016). Guangzhou province has been considered a pioneer regarding early adaptation planning and leadership. The province has a long-standing historical background regarding its coexistence with water, nature-based solutions in minimizing flood risk, besides being strategically located in the heart of a major economic zone, the Pearl River Delta (PRD). Broader studies look at how to deal with China's total exposure to floods (see, e.g., Wang et al. 2014). Most of the adaptation research on flood and drought risk management is simultaneously placed in the contexts of urbanization. Research on urban climate adaptation governance flourishes but is still considered far from comprehensive (Li 2013; Fu et al. 2017).

In light of the continued implementation gaps and only fragmented policy application, some studies investigate the awareness of local politicians about the need to adapt to a changing climate across different provinces (see, e.g., Jin and Francisco 2013; Liu et al. 2016). Liu et al. (2016), for example, conducted a survey among 85 governmental departments responsible for adaptation planning in five different provinces. They found that the understanding of adaptation issues differs widely, but that over half of survey respondents have knowledge about climate change adaptation measures (Liu et al. 2016). The study also touches upon the limited awareness of adaptation measures related to human health. Here, ecological security and resources security were considered to be the primary goals of adaptation planning. Although opinions on best practices regarding adaptation approaches differed among the surveyed decision-makers, all considered developing an adaptation plan necessary, with 60 percent ranking it "highly necessary and absolutely urgent" (Liu et al. 2016: 48).

Barriers: China

The lack of political awareness of local politicians about intensifying impacts and adaptation opportunities has been considered a dominant barrier, which, coupled with other factors, can collectively constrain adaptation action (Jin and Francisco 2013; Wang et al. 2020). Perception studies have been undertaken to discuss barrier types related to matters of awareness. For instance, Kibue et al. (2016) examine adaptation barriers in Anhui and Jiangsu province in the context of climate variability, agriculture and rural livelihoods. Accordingly, adaptation to climate variability is hampered by socioeconomic and sociocultural factors such as the lack of information, attachment to traditional crops, lack of access to credit and water shortage (Kibue et al. 2016: 981).

Zhai et al. (2018) also study a behavioral perspective that is a barrier to adaptation from the perspective of factors that impact farmers' awareness. The authors find that lack of money and lack of information are two important barriers. However, the authors inadequately explore why richer farmers should have greater adaptive capacity to address climate change, or why it is that only certain farmers, who have access to knowledge. These causal links are not explored and the underlying patterns, which can be detected ignored.

Wang and colleagues (2020) applied the barrier research framework by Jones and Boyd (2011) to peasant households in the Qinghai-Tibetan Plateau. They find that multiple barriers such as cognitive, technology, normative and institutional barriers prevent farmers from taking adaptation intention to specific adaptation action: "Barriers are manifested not only in the internal limitations of farming household (e.g., lack of knowledge, low household income, outdated traditional beliefs or values), but also in constraints from the external environment (e.g., lack of infrastructure, uneven distribution of resources, harsh natural environment)" (Wang et al. 2020). Their framework goes a little further than most barrier studies by identifying factors, which influence these barriers such as the lack of policy guidance, limited policy incentives or differences in adaptation requirements.

Barrier studies in China are only developing but equally stressing the need to systematically identify the interaction of barriers and their dynamics (e.g., see Wang et al. 2020). Although the more recent studies appear to be interested in the origins of factors, which act as barriers, it becomes obvious that the barrier perspective has its limitations when it comes to exploring the interrelations of different barriers and path-dependence. Barrier studies in China also appear to be focusing on distinct policy sectors such as agriculture and farming or water. An integrative lens of barriers that examines vulnerability and adaptation challenges from a broader perspective seems to be lacking.

The United States: Adaptation research

In light of the highly decentralized and bi-partisan nature of the U.S. political system, much of the research concentrates on demonstrating political difficulties related to public climate governance. Due to widespread climate skepticism; weak national guidance and limited political maneuvering space, research on local governments as central actors in addressing climate hazard management and urban adaptation continues to flourish (see, e.g., Hansen et al. 2013; Gerber 2015). The emergence of multilateral and non-state actors has become a primary research interest. Examples include the Rockefeller Foundation, the Kresge Foundation, ICLEI or the UN Habitat's Cities and Climate Change Initiative and how they are exemplary of non-state support structures in urban adaptation and resilience planning in light of a deadlocked federal government (see, e.g., Stone 2012; Bulkeley and Tuts 2013).

This is in line with international discourses on revived multilateralism and polycentric governance, which have emphasized the importance of the local scale (see, e.g., Bulkeley and Kern 2006; Ford 2018; Morchain 2018). In China, the evaluation and role of distinct actors has not been a research priority. Here, procedural elements of adaptation governance have played only a subordinate role. Besides a diversification of non-state actors in the policy process, nongovernmental actors have only limited influence in climate policymaking and continue to lack expertise besides climate change still being a relatively new topic (see, e.g., Liu 2009; Liu et al. 2017).

Preston et al. (2011) conclude that the majority of adaptation plans are largely underdeveloped. The critique of persistently inadequate adaptation efforts echoes earlier research that explains the adaptation deficit with institutional neglect regarding anticipatory risk management (see, e.g., Schipper and Burton 2009).

Little research examines the absence of adaptation action relative to the adaptation need (the adaptation gap) (see, e.g., Preston et al. 2011 for the national level and Chen et al. 2016 for municipal levels). Most of these studies investigate climate adaptation plans and implementation deficits at subnational levels and across sectors (Poyar and Beller-Sims 2010; Hansen et al. 2013; Bierbaum et al. 2013; Moser and Boykoff 2013; Ray and Grannis 2015; Lysák and Bugge-Henriksen 2016). Corresponding with the political developments and a range of Executive Orders issued under the Obama administration, researchers also began to investigate CCA planning by the federal agencies after Obama's first and second terms of office in 2009 and 2013, respectively (see, e.g., Leggett 2015; Conners et al. 2015). The findings of initial adaptation gap studies differ widely and range from detecting only minor gaps in planning targets to finding major deficits (Chen et al. 2016).

Barriers

In the United States, the literature on adaptation barriers is expansive and has been constantly developing. Bierbaum et al. (2013) conclude that lack of funding, policy and institutional constraints and difficulty in anticipating climate change based on the state of information negatively affect adaptation efforts and policy mainstreaming into existing policies and plans. Becker and Kretsch (2019) look at the aspect of adaptation leadership and perception thereof, which results in a leadership void, serving as a significant adaptation barrier.

A literature review finds that the understanding of barriers is limited and highly fragmented within the academic community (Biesbroek et al. 2013). According to this analysis, climate and non-climatic factors condition adaptation barriers. Biesbroek et al. (2013) also find that barriers are commonly studied in developed countries and conclude that most barriers are not specific to climate change adaptation but also a result of the short-term dynamics of politics and point to the importance of perceiving barriers as not just climate-related but broader phenomena. The barrier research in the United States recently appears to have put greater emphasis on the broader patterns and historical path dependencies. McGuire (2018), for instance, refers to the causes for the disconnect between information and action, which is considered an adaptation barrier, as a historical path dependence, which can be explained through key policy instruments. These "cumulatively act as barriers to adopting climate change assessment recommendations in coastal regions" (McGuire 2018: 1). This is one example of research attempts, which seek to go beyond describing and mapping barriers and a good starting point for the discussion of lock-ins.

Synthesis 2: Need to explain local adaptation deficits in both political contexts and non-pioneer regions

The body of adaptation scholarship is eclectic, characterized by a wealth of scientific understanding (Schipper and Burton 2009; Preston et al. 2011), which is, with only few exceptions, largely unmet by the political realities and governmental adaptation efforts on the ground (Vij et al. 2017). Although the need to adapt to climate impacts is now widely recognized and the field has received increased attention and although formal climate adaptation planning is rapidly emerging, there continues to be a strong disconnect between the priority placed on adaptation at the national and local levels (Preston et al. 2011; Siebenhüner et al. 2017). The scientific call for transformational adaptation is likewise unmet by the actual nature of adaptation actions, which continue to lack substance, are incremental, and also ad hoc in nature (Preston et al. 2011; Bierbaum et al. 2013; Biesbroek et al. 2013; Wise et al. 2014; Munck af Rosenschöld and Rozema 2019).

Although the adaptation field has been gaining tremendous attention since the Chinese government published the National Adaptation Strategy (NAS) in 2013, mitigation continues to be at the forefront of scholarly and political concerns. Comparative works on explaining policy processes related to climate change in China and the United States focus almost exclusively on mitigation (see, e.g., Gallagher et al. 2018) or the status of efforts in tackling problems of global environmental and energy governance (Kalatzakos 2017). Most of these works either illuminate the international relations perspective and big power rivalries or focus on (national) mitigation politics. Very little research focuses on explaining local policy processes related to adaptation from a comparative perspective, policy formation and local policy choices related to climate adaptation and persisting deficits.

Although a range of activities are taking place at different levels of government, most research argues that the United States is not doing enough to address climate impacts and the field of CCA (Moser 2009; Hansen et al. 2013; Moser et al. 2017). Some consensus seems to exist too that although the central government in China merged adaptation into domestic policies, more efforts are needed to enhance national and local CCA efforts (e.g., Chao et al. 2014; He et al. 2017), particularly in nontraditional areas such as the impact of climate change on human health (Nadin et al. 2016). In both cases, the effectiveness of climate change adaptation has rarely been evaluated because a majority of explicit public adaptation actions have only recently been initiated (Bierbaum et al. 2014; Moser et al. 2017; Nadin et al. 2016).

At a broader scale, climate adaptation approaches are compared with other developed countries (Gagnon-Lebrun and Agrawala 2006; Preston et al. 2011; Ford et al. 2011; Prutsch et al. 2014). For instance, in their 2009 study of adaptation plans from three developed countries (Australia, UK and the United States).

Preliminary conclusion: The political dimension of lock-ins across different systems and adaptation "non-pioneers"

The aim of this chapter was to review prevailing research attempts on adaptation deficits. Adaptation barriers were briefly discussed as the main explanatory attempt to make sense of the existing lack of adaptation action or limits of adaptation. Most research on social vulnerability and adaptation deficits is symptomatic and largely descriptive, failing to examine the underlying factors, which protract vulnerability and adaptation deficits. In line with Turnbull and Hoppe's critique of wicked problems (2019), existing angles fail to locate vulnerability and adaptation within a systemic perspective by not looking at the capacities and limitations of governmental decision-making. These unfold, the chapter argues against the political ecology. Only few research attempts undertake an integrative assessment of both vulnerability and adaptation. In addition to the research gaps of the foregone chapter, two major research gaps were identified that guided the empirical research of this book:

It was found that the concept of path-dependent lock-ins is better equipped in explaining matters of maintained vulnerability and adaptation deficits. Therefore, the main analytical framework of lock-ins was introduced and conceptualized within the political ecology lens. Besides their interdependent quality, they are considered outcomes of human-nature relations (cultural ecology) and likewise a function of dominant economic theories and questions of power, which determine how individuals, governments and public policy operate.

Because adaptation deficits are especially apparent at the social level, it is crucial to study how vulnerable groups are addressed and considered in local decision-making and as part of governmental frameworks. Because adaptation deficits and social vulnerability appear to be deeply manifested, it is important to look at the origins and intersections of path-dependencies – lock-ins – to explain this. Lock-ins will be examined with regard to two facets, i.e., studying supposed "adaptation non-pioneers" and looking at lock-ins as they related to the political aspects of social vulnerability.

(1) **Need to address emerging questions of political vulnerability and capacity:** Social vulnerability commonly emphasizes limited adaptive capacity of society; low resource entitlements as well as access as fundamental problems. The conceptualizations that exist in this context, occur only superficially and fail to look at power as an underlying determinant of protracted inequality and differential access. Further, the classification of vulnerable groups is problematic at different levels (also see Ribot 2014). There is a need for a critical assessment of the underlying causal mechanisms that protract population vulnerability and incremental decision-making with only reactive and unintentional adaptation efforts. Shifting the focus from looking at the adaptive capacity of society to analyzing the adaptive capacity in relation to political institutions is an important starting point for the facilitation of a different discussion. To what extent social vulnerability is a lock-in of our political institutions and an inherent characteristic of different political systems will be examined.

(2) **Comparative research of local policy processes of non-pioneers:** So far, a lot of studies focus on CCA of developed nations (e.g. Ford et al. 2011) or North America (e.g. Filho and Keenan 2017) or countries where climate adaptation is advanced, such as the UK or Australia. Little research compares vulnerability and adaptation as a combined approach across different countries and local contexts that are not considered "pioneers." At the same time, little research investigates the aspect of human susceptibility across different political systems. In order to understand the manifested nature of lock-ins, studying adaptation deficits in supposedly "non-pioneer" regions across different political systems provides deeper insights on pathways that prevent change from occurring and meaningful adaptation from taking place. Identifying similarities and differences across two supposedly different contexts can offer different explanatory angles of the more deeply rooted causes of deficits. This research gap corresponds to the underlying research question of the second case study, which analyzes adaptation policymaking in Atlanta and Jinhua. The main research question is based upon the analysis of local policy efforts and investigates, which lock-ins explain the accidental and incremental nature of adaptation measures.

Notes

1 Turnbull and Hoppe (2018) have articulated a compelling critique of the concept of wickedness that is unfortunately only inadequately considered in this book. See: "Problematizing 'Wickedness': A Critique of the Wicked Problems Concept, from Philosophy to Practice." *Policy and Society* 38, no. 2 (2018): 315–337. Doi: 10.1080/14494035.2018.1488796.

2 For a complete review of the lock-in concept and its theoretical foundations, see Siebenhüner et al. 2017/18 and Siebenhüner et al. 2017.

References

Adger, Neil W., Pramod Aggarwal, Shardul Agrawala, Joseph Alcamo, Abdelkader Allali, Oleg Anisimov, Nigel Arnell et al. "Climate Change 2007: Impacts, Adaptation and Vulnerability. Contribution of Working Group II to the Fourth Assessment Report of the Intergovernmental Panel on Climate Change. Summary for Policymakers." In *Working Group II Contribution to the Intergovernmental Panel on Climate Change Fourth Assessment Report*, edited by Martin L. Parry, Oswaldo F. Canziani, Jean P. Palutikof, P. J. van der Linden, and Clair E. Hanson. Cambridge, UK and New York, NY: Cambridge University Press, 2007.

Adger, Neil W., Irene Lorenzoni, and Karen L. O'Brien. *Adapting to Climate Change. Thresholds, Values, Governance.* Cambridge, UK: Cambridge University Press, 2009.

Bassett, Thomas J., and Charles Fogelman. "Déjà vu or Something New? The Adaptation Concept in the Climate Change Literature." *Geoforum* 48, no. 0 (2013): 42–53.

Becker, Austin, and Eric Kretsch. "The Leadership Void for Climate Adaptation Planning: Case Study of the Port of Providence (Rhode Island, United States)." *Frontiers in Earth Science* 7, no. 29 (2019): 1–13. Doi: 10.3389/feart.2019.00029.

Bierbaum, Rosina, Arthur Lee, Joel Smith, Maria Blair, Lynne M. Carter, F. Stuart Chapin, III, Paul Fleming et al. "Chapter 28: Adaptation." In *Climate Change Impacts in the United States: The Third National Climate Assessment*, edited by Jerry M. Melillo, Terese (T.C.) Richmond, and Gary Yohe. 670–706. Washington, DC: U.S. Global Change Research Program, 2014. Doi: 10.7930/J07H1GGT.

Bierbaum, Rosina, Joel B. Smith, Arthur Lee, Maria Blair, Lynne M. Carter, F. Stuart Chapin III et al. "A Comprehensive Review of Climate Adaptation in the United States: More than Before, but Less than Needed." *Mitigation and Adaptation Strategies for Global Change* 18, no. 3 (2013): 361–406. https://doi.org/10.1007/s11027-012-9423-1.

Biesbroek, G. Robbert, Judith E. M. Klostermann, Katrien C. J. A. M. Termeer, and Pavel Kabat. "On the Nature of Barriers to Climate Change Adaptation." *Regional Environmental Change* 13, no. 5 (2013): 1119–1129. Doi: 10.1007/s10113-013-0421-y.

Biesbroek, G. Robbert, Katrien C. J. A. M. Termeer, Judith E. M. Klostermann, and Pavel Kabat. "Rethinking Barriers to Adaptation: Mechanism-Based Explanation of Impasses in the Governance of an Innovative Adaptation Measure." *Global Environmental Change* 26, (2014): 108–118. http://dx.doi.org/10.1016/j.gloenvcha.2014.04.004.

Bulkeley, Harriet, and Kristine Kern. "Local Government and the Governing of Climate Change in Germany and the UK." *Urban Studies* 43, no. 12 (2006): 2237–2259. https://doi.org/10.1080/00420980600936491.

Bulkeley, Harriet, and Rafael Tuts. "Understanding Urban Vulnerability, Adaptation and Resilience in the Context of Climate Change." *Local Environment* 18, no. 6 (2013): 646–662. http://dx.doi.org/10.1080/13549839.2013.788479.

CACCA. "Climate Change Adaptation: Approaches for National and Local Governments." Report, *Committee on Approaches to Climate Change Adaptation (CACCA)*, published Nov. 2010.

Cardona, Omar-Dario, Maarten K. van Aalst, Jörn Birkmann, Maureen Fordham, Glenn McGregor, Rosa Perez, Roger S. Pulwarty, E. Lisa F. Schipper, and Bach Tan Sinh. "Determinants of Risk: Exposure and Vulnerability." In *Managing the Risks of Extreme Events and Disasters to Advance Climate Change Adaptation: Special Report of the Intergovernmental Panel on Climate Change*, edited by Christopher B. Field, Vicente Barros, Thomas F. Stocker, and Qin Dahe, 65–108. Cambridge: Cambridge University Press, 2012. Doi: 10.1017/CBO9781139177245.005.

Chao, Qingchen, Liu Changyi, and Yuan Jiashuang. "The Evolvement of Impact and Adaptation on Climate Change and Their Implications on Climate Policies." (Original: 气候变化影响和适应认知的演进及对气候政策的影响) *Climate Change Research* 10, no. 3 (2014): 167–174. Doi: 10.3969/j.issn.1673-1719.2014.03.002.

Charan, Dhrishna, Manpreet Kaur, and Priyatma Singh. "Chapter 19 Customary Land and Climate Change Induced Relocation: A Case Study of Vunidogoloa Village, Vanua Levu, Fiji." In *Limits to Climate Change Adaptation*, edited by Walter Leal Filho and Johanna Nalau, 345–359. Cham: Springer, 2018.

Chen, Chen, Meghan Doherty, Joyce Coffee, Theodore Wong, and Jessica Hellmann. "Measuring the Adaptation Gap: A Framework for Evaluating Climate Hazards and Opportunities in Urban Areas." *Environmental Science and Policy* 66 (2016): 403–419. Doi: 10.1016/j.envsci.2016.05.007.

Conners, Hanna M., Kathleen D. White, and Jeffrey R. Arnold. "Report Providing Comparison of Adaptation Plans Submitted to the White House in 2014." *US Army Corps of Engineers*: Washington, DC., 2015. https://usace.contentdm.oclc.org/digital/collection/p266001coll1/id/5230/.

Corvellec, Hervé, María José Zapata Camposa, and Patrik Zapata. "Infrastructures, Lock-In, and Sustainable Urban Development: The Case of Waste Incineration in the Göteborg Metropolitan Area." *Journal of Cleaner Production* 50 (2013): 32–39. https://doi.org/10.1016/j.jclepro.2012.12.009.

Cutter, Susan L., Bryan J. Boruff, and W. Lynn Shirley. "Social Vulnerability to Environmental Hazards." *Social Science Quarterly* 84, no. 2 (2003): 242–261. Doi: 10.1111/1540-6237.8402002.

Dow, Kirstin, Frans Berkhout, and Benjamin L. Preston. "Limits to Adaptation to Climate Change: A Risk Approach." *Current Opinion in Environmental Sustainability* 5, no. 3–4 (2013): 384–391. https://doi.org/10.1016/j.cosust.2013.07.005.

Du, Yaodong, Yunmin Zeng, Wenjun Ma, Xiaohong Chen, Sarah Opitz-Stapleton, Rebecca Nadin, and Samantha Kierath. "Guangdong." In *Climate Risk and Resilience in China*, edited by Rebecca Nadin, Sarah Opitz- Stapleton, and Xu Yinlong, 272–297. New York: Routledge, 2016.

Dupuis, Johann, and Robbert Biesbroek. "Comparing Apples and Oranges: The Dependent Variable Problem in Comparing and Evaluating Climate Change Adaptation Policies." *Global Environmental Change* 23, no. 6 (2013): 1476–1487. http://dx.doi.org/10.1016/j.gloenvcha.2013.07.022.

Eakin, Hallie, Abigail York, Rimjhim Aggarwal, Summer Waters, Jessica Welch, Cathy Rubiños, Skaidra Smith-Heisters et al. "Cognitive and Institutional Influences on Farmers' Adaptive Capacity: Insights into Barriers and Opportunities for Transformative Change in Central Arizona." *Regional Environmental Change* 16, no. 3 (2016): 801–814. https://doi.org/10.1007/s10113-015-0789-y.

Evans, Louisa S., Christina C. Hicks, W. Neil Adger, Jon Barnett, Allison L. Perry, Pedro Fidelman, and Renae Tobin. "Structural and Psycho-Social Limits to Climate Change Adaptation in the Great Barrier Reef Region." *PLoS One* 11, no. 3 (2016). Doi: 10.1371/journal.pone.0150575.

Fieldman, Glenn. "Neoliberalism, the Production of Vulnerability and the Hobbled State: Systemic Barriers to Climate Adaptation." *Climate and Development* 3, no. 2 (2011): 159–174. https://doi.org/10.1080/17565529.2011.582278.

Filho, Walter Leal, and Jesse M. Keenan eds. *Climate Change Adaptation in North America. Fostering Resilience and the Regional Capacity to Adapt.* Cham: Springer, Climate Change Management Series, 2017.

Folke, Carl, Stephen R. Carpenter, Brian Walker, Martin Scheffer, Terry Chapin, and Johan Rockström. "Resilience Thinking: Integrating Resilience, Adaptability, and Transformability." *Ecology and Society* 15, no. 4, 20 (2010): 1–9. http://www.ecologyandsociety.org/vol15/iss4/art20/.

Ford, James D., Lea Berrang-Ford, and J. Paterson. "A Systematic Review of Observed Climate Change Adaptation in Developed Nations." *Climatic Change* 106, no. 2 (2011): 106, 327. https://doi.org/10.1007/s10584-011-0045-5

Ford, James D., Mya Sherman, Lea Berrang-Ford, Alejandro Llanos, Cesar Carcamo, Sherilee Harper, Shuaib Lwasa et al. "Preparing for the Health Impacts of Climate Change in Indigenous Communities: The Role of Community-Based Adaptation." *Global Environmental Change* 49 (2018): 129–139. https://doi.org/10.1016/j.gloenvcha.2018.02.006.

Fu, Lin, Xiu Yang, and Xiaoya Feng. "Crisis and Countermeasure Analysis on Climate Change Adaptation of Urban Lifeline System." (Original: 城市生命线系统适应气候变化危机及其对策). *Environmental Economic Research* (环境经济研究) 1 (2017): 118–127. Doi: 10.19511/j.cnki.jee.2017.01.010.

Gagnon-Lebrun, Frédéric, and Shardul Agrawala. *Progress on Adaptation to Climate Change in Developed Countries: An Analysis of Broad Trends.* Paris: OECD, 2006.

Gaillard, J. C. "Vulnerability, capacity and resilience: perspectives for climate and development policy." *Journal of International Development* 22, no. 2 (2010): 218–232.

Gallagher, Kelly Sims, Xiaowei Xuan, John P. Holdren, Junkuo Zhang, Michael E. Kraft, and Sheldon Kamieniecki. *Titans of the Climate: Explaining Policy Process in the United States and China.* Cambridge, MA: MIT Press, 2018.

Garrelts, Heiki, Johannes Herbeck, and Michael Flitner. "Leaving the Comfort Zone. Regional Governance in a German Climate Adaptation Project." In *A Critical Approach to Climate Change Adaptation. Discourses, Policies, and Practices,* edited by Silja Klepp and Libertad Chavey-Rodriguez. Oxon, New York: Routledge, 2018.

Gerber, Brian J. "Local Governments and Climate Change in the United States Assessing Administrators' Perspectives on Hazard Management Challenges and Responses." *State and Local Government Review* 47, no. 1 (2015): 48–56. https://doi.org/10.1177/0160323X15575077.

Grothmann, Thorsten, Kevin Grecksch, Maik Winges, and Bernd Siebenhüner. "Assessing Institutional Capacities to Adapt to Climate Change – Integrating Psychological Dimensions in the Adaptive Capacity Wheel." *Natural Hazards Earth Systems Science* 13 (2013): 3369–3384. https://doi.org/10.5194/nhess-13-3369-2013.

Hansen, Lara, Rachel M. Gregg, Vicki Arroyo, Susan Ellsworth, Louise Jackson, and Amy Snover. "The State of Adaptation in the United States: An Overview." Report, *John D. and Catherine T. MacArthur Foundation,* EcoAdapt, 2013.

Hartmann, Thomas. "Wicked Problems and Clumsy Solutions: Planning as Expectation Management." *Planning Theory* 11, no. 3 (2012): 242–256. https://doi.org/10.1177/1473095212440427.

He, Xiaojia, Xueyan Zhang, and Xin Ma. "The Development of Climate Change Adaptation on Institution and Policies in China." (Original: 中国适应气候变化制度建设与政策发展方向研究). *Climate Change Research Letters* 6, no. 1 (2017): 40–45. Doi: 10.12677/CCRL.2017.61005.

Heltberg, Rasmus, Paul B. Siegel, and Steen L. Jorgensen. "Addressing Human Vulnerability to Climate Change: Toward a 'No-regrets' Approach." *Global Environmental Change* 19, no. 1 (2009): 89–99. Doi: 10.1016/j.gloenvcha.2008.11.003.

Heltberg, Rasmus, Paul B. Siegel, and Steen L. Jorgensen. "Chapter 10 Social Policies for Adaptation to Climate Change." In *Social Dimensions of Climate Change: Equity and Vulnerability in a Warming World*, edited by Robin Mearns and Andrew Norton, 259–275. Washington, DC: The World Bank, 2010.

Hetz, Karen, and Antje Bruns. "Urban Planning Lock-In: Implications for the Realization of Adaptive Options Towards Climate Change Risks." *Water International* 39 (2014): 884–900. https://doi.org/10.1080/02508060.2014.962679.

Horton, Peter, and Garrett W. Brown. "Integrating Evidence, Politics and Society: A Methodology for the Science – Policy Interface." *Palgrave Communications* 4, no. 42 (2018). https://doi.org/10.1057/s41599-018-0099-3.

Huitema, Dave, W. Neil Adger, Frans Berkhout, Eric Massey, Daniel Mazmanian, Stefania Munaretto, Ryan Plummer, and Katrien Termeer. "The Governance of Adaptation: Choices, Reasons, and Effects. Introduction to the Special Feature." *Ecology and Society* 21, no. 3 (2016): 1–15. Article 37. http://dx.doi.org/10.5751/ES-08797-210337.

IPCC. "Summary for Policymakers." In *Climate Change 2014: Impacts, Adaptation, and Vulnerability. Part A: Global and Sectoral Aspects*, edited by Christopher B. Field, Vincente R. Barros, David J. Dokken, Katharine J. Mach, Michael D. Mastrandrea et al., 1–32. Cambridge, UK and New York, NY: Cambridge University Press, 2014.

Jin, Jianjun, and Hermi Francisco. "Sea-Level Rise Adaptation Measures in Local Communities of Zhejiang Province, China." *Ocean and Coastal Management* 71 (2013): 187–194. Doi: 10.1016/j.ocecoaman.2012.10.020.

Jones, Lindsey. "Overcoming Social Barriers to Adaptation." *Overseas Development Institute*, Background Note, July 2010. http://dx.doi.org/10.2139/ssrn.2646812.

Jones, Lindsey, and Emily Boyd. "Exploring Social Barriers to Adaptation: Insights from Western Nepal." *Global Environmental Change* 21, no. 4 (2011): 1262–1274. https://doi.org/10.1016/j.gloenvcha.2011.06.002.

Kalatzakos, Sophia. *The EU, US, and China Tackling Climate Change: Policies and Alliances for the Anthropocene*. London: Routledge, 2017.

Kelman, Ilan, Jessica Mercer, and J. C. Gaillard, eds. *The Routledge Handbook of Disaster Risk Reduction Including Climate Change Adaptation*. London: Routledge, 2017.

Keskitalo, E. Carina H., and Benjamin L. Preston. *Research Handbook on Climate Change Adaptation Policy*. Cheltenham, UK, and Northampton, MA: Edward Elgar Press, 2019.

Kibue, Grace W., Xiaoyu Liu, Jufeng Zheng, Xuhui Zhang, Genxing Pan, Lianqing Li, and Xiaojun Han. "Farmers' Perceptions of Climate Variability and Factors Influencing Adaptation: Evidence from Anhui and Jiangsu, China." *Environmental Management* 57, no. 5 (2016): 976–986. Doi: 10.1007/s00267-016-0661-y.

Klein, R. J. T., G. F. Midgley, B. L. Preston, M. Alam, F. G. H. Berkhout, K. Dow, and M. R. Shaw. "Adaptation Opportunities, Constraints, and Limits." In *Climate Change 2014: Impacts, Adaptation, and Vulnerability. Part A: Global and Sectoral Aspects. Contribution of*

Working Group II to the Fifth Assessment Report of the Intergovernmental Panel on Climate Change, edited by C. B. Field et al., 899–943. Cambridge and New York, NY: Cambridge University Press, 2014.

Leggett, Jane A. "Climate Change Adaptation by Federal Agencies: An Analysis of Plans and Issues for Congress." CRS Report, Congressional Research Service, February 2015.

Li, Bingqin. "Governing Urban Climate Change Adaptation in China." *Environment and Urbanization* 25, no. 2 (2013): 413–427. https://doi.org/10.1177/0956247813490907.

Li, Dan, Tracy Hruska, Shalima Talinbayi, and Wenjun Li. "Changing Agro-Pastoral Livelihoods under Collective and Private Land Use in Xinjiang, China." *Sustainability* 11, no. 1(166) (2019). https://doi.org/10.3390/su11010166.

Liu, Haiying. "The Impact of Climate Change on China." In *The China Environment Yearbook: Crises and Opportunities*, edited by Dongping Yang, 79–97. Leiden: Brill, 2009.

Liu, Lei, Pu Wang, and Tong Wu. "Advanced Review. The Role of Nongovernmental Organizations in China's Climate Change Governance." *WIREs Climate Change* 8, no. 6 (2017): e483. Doi: 10.1002/wcc.483.

Liu, Tingting, Zhongyu Ma, Ted Huffman, Like Ma, Hongqiang Jiang, and Haiyan Xie. "Gaps in Provincial Decision-Maker's Perception and Knowledge of Climate Change Adaptation in China." *Environmental Science and Policy* 58 (2016): 41–51. https://doi.org/10.1016/j.envsci.2016.01.002.

Lonsdale, Kate, Patrick Pringle, and Briony Turner. *Transformative Adaptation: What It Is, Why It Matters & What Is Needed*. UK Climate Impacts Programme (UKCIP). Oxford: University of Oxford, 2015.

Lysák, Marin, and Christian Bugge-Henriksen. "Current Status of Climate Change Adaptation Plans across the United States." *Mitigation and Adaptation Strategies for Global Change* 21, no. 3 (2016): 323–342. https://doi.org/10.1007/s11027-014-9601-4.

Manyena, Siambabala Bernard. "The Concept of Resilience Revisited." *Disasters* 30, no. 4 (2006): 434–450. https://doi.org/10.1111/j.0361-3666.2006.00331.x.

McGuire, Chad J. "Examining Legal and Regulatory Barriers to Climate Change Adaptation in the Coastal Zone of the United States." *Cogent Environmental Science* 4, no. 1 (2018): 1–11. Doi: 10.1080/23311843.2018.1491096.

Meng, Meng, and Marcin Dabrowski. "The Governance of Flood Risk Planning in Guangzhou, China: Using the Past to Study the Present." Conference Paper, 17th IPHS Conference, Delft 2016.

Morchain, Daniel. "Multi Scalar Transformation to Enable Desirable Adaptation." Paper presented at the Annual International Conference of Geographical Landscapes, Royal Geographical Society, Cardiff, Wales, August 2018.

Moser, Susanne C. "Good Morning, America! The Explosive U.S. Awakening to the need for adaptation." Report, California Energy Commission and the NOAA Coastal Services Center, May 2009.

Moser, Susanne C., and Maxwell Boykoff. *Successful Adaptation to Climate Change: Linking Science and Practice in a Rapidly Changing World*. London: Routledge, 2013.

Moser, Susanne C., Joyce Coffee, and Aleka Seville. "Rising to the Challenge, Together. A Review and Critical Assessment of the State of the US Climate Adaptation Field." Report, Kresge Foundation, December 2017.

Moser, Susanne C., and Julia A. Ekstrom. "A Framework to Diagnose Barriers to Climate Change Adaptation." *Proceedings in the National Academies of Sciences* 107, no. 51 (2010): 22026–22031. Doi: 10.1073/pnas.1007887107.

Munck af Rosenschöld, Johan, and Jaap G. Rozema. "Moving from Incremental to Transformational Change in Climate Adaptation Policy? An Institutionalist perspective." In *Research*

Handbook on Climate Change Adaptation Policy, edited by E. Carina H. Keskitalo and Benjamin L. Preston, 91–107. Cheltenham, UK, and Northampton, MA: Edward Elgar Press, 2019.

Nadin, Rebecca, Sarah Opitz-Stapleton, and Xu Yinlong, eds. *Climate Risk and Resilience in China*. London: Routledge, 2016.

Noble, Ian R., Saleemul Huq, Yuri A. Anokhin, JoAnn Carmin, Dieudonne Goudou, Felino P. Lansigan, Balgis Osman-Elasha et al. "Adaptation Needs and Options. In *Climate Change 2014: Impacts, Adaptation, and Vulnerability. Part A: Global and Sectoral Aspects*, edited by C. B. Field et al., 833–868. Cambridge and NewYork, NY: Cambridge University Press, 2014.

O'Brien, Karen. "Global Environmental Change II: From Adaptation to Deliberate Transformation." *Progress in Human Geography* 36, no. 5 (2012): 667–676. Doi: 10.1177/0309132511425767.

Oels, Angela. "The Promise and Limits of Participation in Adaptation Governance: Moving Beyond Participation Towards Disruption." In *Research Handbook on Climate Change Adaptation Policy*, edited by E. Carina H. Keskitalo and Benjamin L. Preston, 138–156. Cheltenham: Edward Elgar Press, 2019.

Pelling, Mark. *Adaptation to Climate Change: From Resilience to Transformation*. New York: Routledge, 2011.

Perry, Jim. "Climate Change Adaptation in the World's Best Places: A Wicked Problem in Need of Immediate Attention." *Landscape and Urban Planning* 133 (2015): 1–11. Doi: 10.1016/j.landurbplan.2014.08.013.

Perry, Miles, and Juan Carlos Ciscar. "Chapter 15 Multi- Sectoral Perspective in Modelling of Climate Impacts and Adaptation." In *Routledge Handbook of the Economics of Climate Change Adaptation*, edited by Anil Markandya, Ibon Galarraga, and Elisa Sainz de Murieta, 301–317. London, Oxon, UK and New York: Routledge, 2014.

Poyar, Kyle A., and Nancy Beller-Simms. "Early Responses to Climate Change: An Analysis of Seven U.S. State and Local Climate Adaptation Planning Initiatives." *Weather, Climate and Society* 2, no. 6 (2010): 237–248. https://doi.org/10.1175/2010WCAS1047.1.

Preston, Benjamin L., Richard M. Westaway, and Emma J. Yuen. "Climate Adaptation Planning in Practice: An Evaluation of Adaptation Plans from Three Developed Nations." *Mitigation and Adaptation Strategies for Global Change* 16, no. 4 (2011): 407–438. https://doi.org/10.1007/s11027-010-9270-x.

Prutsch, Andrea, Thorsten Grothmann, Sabine McCallum, Inke Schauer, and Rob Swart, eds. *Climate Change Adaptation Manual, Lessons Learned from Europe and other Industrialized Countries*. London: Routledge, 2014.

Purdon, Mark. *The Comparative Turn in Climate Change Adaptation and Food Security Governance Research*. CCAFS Working Paper no. 92. Copenhagen, Denmark: CGIAR Research Program on Climate Change, Agriculture and Food Security (CCAFS), 2014.

Qi, Yingjun, and Wenjun Li. "A Nested Property Right System of the Commons: Perspective of Resource System-Units." *Environmental Science & Policy* 115 (2021): 1–7.

Ray, Aaron D., and Jessica Grannis. "From Planning to Action: Implementation of State Climate Change Adaptation Plans." *Michigan Journal of Sustainability* 3 (2015): 5–28. http://dx.doi.org/10.3998/mjs.12333712.0003.001.

Ribot, Jesse. "Cause and Response: Vulnerability and Climate in the Anthropocene." *The Journal of Peasant Studies* 41, no. 5 (2014): 667–705. Doi: 10.1080/03066150.2014.894911.

Rittel, H. W., and M. A. Webber. "Dilemmas in a General Theory of Planning." *Policy Sciences* 4, no. 2 (1973): 155–169. http://dx.doi.org/10.1007/BF01405730.

Schipper, E. Lisa F., and Ian Burton. *The Earthscan Reader on Adaptation to Climate Change.* London and Sterling, VA: Earthscan, 2009.

Schipper, E. Lisa F., Frank Thomalla, Gregor Vulturius, Marion Davis, and Karlee Johnson. "Linking Disaster Risk Reduction, Climate Change and Development." *International Journal of Disaster Resilience in the Built Environment* 7, no. 2 (2016): 216–228. https://doi.org/10.1108/IJDRBE-03-2015-0014.

Seto, Karen C., Steven J. Davis, Ronald B. Mitchell, Eleanor C. Stokes, Gregory Unruh, and Diana Ürge-Vorsatz. "Carbon Lock-In: Types, Causes, and Policy Implications." *Annual Review of Environment and Resources* 41, no. 1 (2016): 425–452. https://doi.org/10.1146/annurev-environ-110615-085934.

Siebenhüner, Bernd, Torsten Grothmann, Dave Huitema, Angela Oels, Tim Rayner, and John Turnpenny. "Lock-Ins in Climate Adaptation Governance. Conceptual and Empirical Approaches." Conference Paper 2017, unpublished.

Siebenhüner, Bernd et al. "Understanding Climate Adaptation Policy Lock-ins: A 3 x 3 Approach." Adapt-lock-in, project proposal, project call 2017/2018, published online 2019. https://uol.de/f/2/dept/wire/fachgebiete/ecoeco/Dokumente/ADAPT-LOCK-IN_full_proposal_public_.pdf.

Smit, Barry, Ian Burton, Richard J. T. Klein, and Johanna Wandel. "An Anatomy of Adaptation to Climate Change and Variability." *Climate Change* 45, no. 1 (2000): 223–251.

Sovacool, Benjamin, and Björn-Ola Linnér. *The Political Economy of Climate Change Adaptation.* London: Palgrave Macmillan 2016. https://doi.org/10.1057/9781137496737.

Stone, Brian. *The City and the Coming Climate: Climate Change in the Places We Live.* Cambridge and New York: Cambridge University Press, 2012.

Taylor, Marcus. *The Political Ecology of Climate Change Adaptation. Livelihoods, Agrarian Change and the Conflict of Development.* London and New York: Routledge, 2015.

Turnbull, Nick, and Robert Hoppe. "Problematizing 'Wickedness': A Critique of the Wicked Problems Concept, From Philosophy to Practice." *Policy and Society* 38, no. 2 (2019): 315–337. Doi: 10.1080/14494035.2018.1488796.

UNEP. "The Emissions Gap Report 2017." United Nations Environment Programme (UNEP), Nairobi, November 2017.

Vij, Sumit, Eddy Moors, Bashir Ahmad, Md. Arfanuzzaman, Suruchi Bhadwal, Robbert Biesbroek, Giovanna Gioli et al. "Climate Adaptation Approaches and Key Policy Characteristics: Cases from South Asia." *Environmental Science and Policy* 78 (2017): 58–65. https://doi.org/10.1016/j.envsci.2017.09.007.

Wang, Weijun, Xueyan Zhao, Jianjun Cao, Hua Li, and Qin Zhang. "Barriers and Requirements to Climate Change Adaptation of Mountainous Rural Communities in Developing Countries: The Case of the Eastern Qinghai-Tibetan Plateau of China." *Land Use Policy* 95 (2020): 104354. https://doi.org/10.1016/j.landusepol.2019.104354.

Wang, Yan-Jun, Chao Gao, Jian-Qing Zhai, Xiu-Cang Li, Bu-da Su, and Heike Hartmann. "Spatio-Temporal Changes of Exposure and Vulnerability to Floods in China." *Advances in Climate Change Research* 5, no. 4 (2014): 197–205. https://doi.org/10.1016/j.accre.2015.03.002.

Wise, Russell M., Ioan Fazey, Mark Stafford Smith, Sarah E. Park, Hallie C. Eakin, E. R. M. Archer Van Garderen, and Bruce Campbell. "Reconceptualising Adaptation to Climate Change as Part of Pathways of Change and Response." *Global Environmental Change* 28 (2014): 325–336. https://doi.org/10.1016/j.gloenvcha.2013.12.002.

Wolf, Johanna, Irene Lorenzoni, Roger Few, Vanessa Abrahamson, and Rosalind Raine. "Chapter 11, Conceptual and Practical Barriers to Adaptation: Vulnerability and Response to Heat Waves in the UK." In *Adapting to Climate Change*, edited by W. Neil

Adger, Irene Lorenzoni, and Karen L. O'Brien, 181–196. New York: Cambridge University Press, 2009.

Yang, Saini, Shuai He, Juan Du, and Xiaohua Sun. "Screening of Social Vulnerability to Natural Hazards in China." *Natural Hazards* 76, no. 1 (2015): 1–18. https://doi.org/10.1007/s11069-014-1225-1.

Yohe, Gary, and Richard S. J. Tol. "Indicators for Social and Economic Coping Capacity—Moving Toward a Working Definition of Adaptive Capacity." *Global Environmental Change* 12, no. 1 (2002): 25–40. https://doi.org/10.1016/S0959-3780(01)00026-7.

Zhai, Shi-yan, Gen-xin Song, Yao-chen Qing, Xin-yue Ye, and Mark Leipnik. "Climate Change and Chinese Farmers: Perceptions and Determinants of Adaptive Strategies." *Journal of Integrative Agriculture* 17, no. 4 (2018): 949–963. Doi: 10.1016/S2095-3119(17)61753-2.

Zhao, Hong-Yan, Jun-Qin Guo, Cun-Jie Zhang, Lan-Dong Sun, Xu-Dong Zhang, Jing-Jing Lin et al. "Climate Change Impacts and Adaptation Strategies in Northwest China." *Advanced Climate Change Research* 5, no. 1 (2014): 7–16. https://doi.org/10.3724/SP.J.1248.2014.007.

Zheng Yan, Meng Huixin, Zhang Xiaoyu, Zhu Furong, Wang Zhanjun, Fang Shuxing, Sarah Opitz- Stapleton et al. "Ningxia." In *Climate Risk and Resilience in China*, edited by Rebecca Nadin, Sarah Opitz-Stapleton, and Xu Yinlong, 213–262. New York: Routledge, 2016.

4 Methodological approach

The chapter describes the methods that were used to answer the research questions.[1]

As stated in the introduction, the core research endeavor is to understand why adaptation deficits persist across two different political contexts. The heightened uneven vulnerability to climate change in both cities is conceptualized as an instance of adaptation deficits. The aim is to better understand the political origins of these deficits and which factors are locked-in at deeper levels of path-dependent, interrelated processes. These are conceptualized within the political ecology, which provides valuable insights into the political economy of uneven affectedness and the ways in which human vulnerability to climate change also intersects with cultural ecological aspects such as understandings and attitudes to climate and vulnerable populations. This includes aspects such as climate skepticism as well as dominant vulnerability discourses. Why and how certain governmental choices are (not) made regarding climate-vulnerable populations and what role governments (can) take needs to be elaborated, meaning: what role do local governments play in the maintenance of population vulnerability?

Comparative qualitative case study research design

This book is based in the empirical social and political sciences and follows an explorative research design, which uses qualitative data. One core characteristic of qualitative social research is the openness toward generating theories from empirical material rather than testing hypotheses from pre-articulated theories (Gläser and Laudel 2004). This is considered a major benefit of qualitative studies: the ability to explain locally specific aspects of social life through an in-depth view rather than generalizing for the purpose of greater inference. This work is by nature inclined to be instrumental, theory-building, but considers theory engagement. Instrumental research facilitates a better understanding of something else as opposed to intrinsic research, which focuses on the cases themselves (Grandy 2012). Accordingly, the case is secondary to understanding a particular phenomenon but helps to offer a thick description and generate insights of broader patterns.

In line with Diekmann (2007), the social arena under study is largely unexplored. Therefore, hypotheses on how human vulnerability to climate change

DOI: 10.4324/9781003183259-4

are conceptualized locally and how they operate in political practice were not specified. Semi-structured expert interviews and document analysis paired with a review of relevant literature constitute the main data.

Why a comparative case study?

The case study method is defined through its reliance upon evidence drawn from a single case or multiple cases (Gerring 2007). Case studies are helpful as they provide a deep understanding of contemporary phenomena and are predestined for "how" and "why" questions (Yin 2018). Like many authors have emphasized, comparison lies at the heart of social science (see, e.g., Burnham et al. 2008; Harles 2017). We compare to contextualize knowledge, explain and highlight variations, as well as similarities across cases that may appear to be different at first (Burnham et al. 2008). According to Newton and Van Deth (2005), theorizing about the state is roughly categorized into two major strands: normative political theories and empirical political theories. Whereas the latter asks how the state *actually operates* and tries to explain why it *operates that way*, the former looks at normative issues related to the functioning of the state and what governments *ought to do*.

The comparative case study method in this book helps to connect normative political theories with empirical political theories by comparing how governments actually operate at the local level (incremental, unintentional and reactive) and how they ought to (transformative). This is done by providing a framework, which, on the one hand, uncovers underlying explanatory angles that explain how and why agendas are being set a certain way, and on the other hand identifies more ideal adaptation options. The coupling of local empirical circumstances, political practices and existing theories is useful for questioning existing forms of status quo reproduction, which may not be conducive to avoid vulnerability reproduction and maladaptation. The avoidance thereof is an essential precondition for a transformative turn.

A multiple case-study design

The understanding of a case study is based on Yin's twofold definition (2018) that views case study as an empirical method (*scope*) and other methodological characteristics as relevant *features* of a case study. Features correspond to the contextual conditions of a complex case such as geographic locations, social groups and organizations. The boundaries of the case study are by nature fuzzy and may become relevant characteristics when comprising an all-encompassing mode of inquiry (Yin 2018). The "case" is defined as a contemporary phenomenon within a real-world context. It is the concrete manifestation of an abstraction. Concrete examples include individuals, organizations, and projects compared with less concrete cases such as relationships, decisions or partnerships (Yin 2018: 66). Based on the application of specific selection criteria, the latter can become concrete cases. Yin (2018) offers different types of research designs, which each provide an

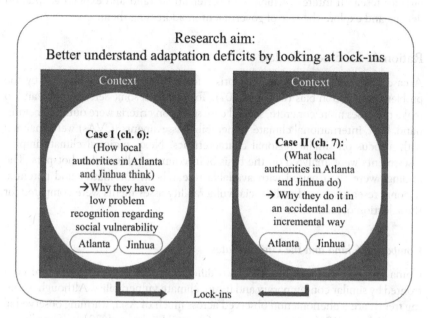

Figure 4.1 Holistic multiple case study design.
Source: The author, based on Yin (2018)

analysis of contextual conditions but may have single, or multiple case studies with unitary or multiple units of analysis.

This book applies a multiple case study design with one holistic unit of analysis. The holistic unit of analysis is decision-making processes of regional-local governmental authorities in Atlanta and Jinhua. Chapters 6 and 7 examine two different cases and present the contemporary phenomena this book researches (see Figure 4.1). The first case looks at the vulnerability-thinking and problem recognition of local authorities involved in the policymaking process regarding climate-vulnerable populations. Because problem recognition about the matter is relatively low, the framework of lock-ins is applied to explain why. The second case also examines the problem recognition of local authorities as it relates to adaptation but focuses on what type of adaptation efforts exist, and why they are occurring that way by also looking at lock-ins.

The design of multiple-case studies follows an analogous logic that is the replication of a similar reasoning. In this book, the logic evolves around the emerging analytical concept of path-dependent and deeply embedded structural lock-ins that is conceptualized in the political ecology. The analytical concept and overarching explanatory framework together aim to explain the protraction of uneven vulnerability and deficient policymaking. At the same time, the notion of "deficiency" as it relates to climate change adaptation as an evolving policy field and self-evident concept shall also be questioned. Both cases are grounded in the broader research interest, which is to better understand and explain adaptation deficits and explore the role of government in addressing them.

Rationale for case selection

A case study (small-n-studies) comprises fewer cases and is confronted by the problem of selection bias (Gerring 2007). To cope with some selection bias and to make the cases more concrete, several case selection criteria were outlined beforehand. First, international climate vulnerability assessments (CVAs) were analyzed with a focus on similar regional characteristics. Next, national climate impact assessments were analyzed on the basis of in-country vulnerability hotspots. The findings were then paired with available research on the matter and potential regional research gaps. Lastly, social vulnerability assessments were compared for the selection of local cases.

Comparing China and the United States

China and the United States are two different political systems that are confronted by similar contemporary and future climate vulnerability. Although warming trends are a phenomenon observed across much of Asia, warming observed in the period after 1979 was strongest over China (Hijioka et al. 2014). The climate trends observed for North America are similarly high and include an increase in the occurrence of severe hot weather events and heavy precipitation events (Romero-Lankau et al. 2014). In the past, both countries have been prone to multiple hazard events (Melillo et al. 2014; Zhou et al. 2014; Yang et al. 2014). The different hazards range from geophysical events (e.g., earthquakes, landslides, avalanches), hydrological happenings (e.g., floods), via meteorological events (e.g., storms) to climatological (e.g., extreme temperature, drought, wildfire) and biological hazards (e.g., epidemic, insect infestation) (Sauerborn and Ebi 2012).[2] Climate change is likely to increase the frequency, intensity, duration and spatial distribution of extreme events (IPCC 2012; Sauerborn and Ebi 2012; IPCC 2018).

Another study paired the findings of the expected annual multiple climatic hazards intensity index with a vulnerability model in order to examine the expected annual affected population, mortality and GDP loss risks (Shi et al. 2016). The findings indicate that China and the United States are among the top-15 countries with the highest total expected annual mortality risk, affected population risk and GDP loss risk for multiple climatic hazards.

Whereas China is a post-socialist, semi-authoritarian country with fragmented forms of centralized political decision-making, the United States is considered a liberal democracy with strongly federalized structures. Table 4.1 lists several important characteristics that are symbolic of the different political and economic systems as well as stances on the environment. Characteristics are a dominant one-party rule in China vis-à-vis a bi-partisan divide and two-party rule in the United States. China's shift from a developing to an emerging economy vis-à-vis the largely stagnant development status in the United States is another important

Table 4.1 Comparative characterization of environmental and climate governance

China	United States
One-party rule by an autocratic political party	Two-party rule, bi-partisan divide
Rapid transition from developing to developed economy	Economic stagnation
Projection of a developing country image when negotiating emission reductions	Projection of a developed country image with historical responsibility
Assuming a strong consistent leadership role	Inconsistent approach, conflict as defining feature of climate change and environmental politics

Source: the Author, adjusted from Edmonds and Ho (2016) and Bailey (2015)

difference that has impacted policymaking on the environment in both countries. These often-times theory-guided but practically minded selection criteria were complemented by a set of preliminary helicopter interviews in both countries. These initial interviews frenetically emphasized how the influence of higher political levels on local climate agendas was playing a determining role.[3] At the time of field research, the formal national-level-stance could not be any more different: whereas Donald Trump was a convinced skeptic, who halted all major recent efforts on climate governance and rolled back the country's already weak environmental legislation, Xi Jinping established a strict environmental agenda under the ecological civilization.

It is beyond the scope of this book to examine the influence from national to local political processes, or to what extent this is symbolic politics. The background of the very different political engagement with climate change at the federal and central levels was an important underlying issue in which the cases need to be reflected. However, in the context of a globalizing political economy, it became apparent that the function of a supposedly authoritarian state might not differ from that of a democratic state to the extent that some would assume (see Chapters 9 and 10). As the period of data collection lasted until mid-2019, more recent political changes thereafter could only be reflected upon to a limited extent.

Comparing cases within Georgia and Zhejiang[4]

Within each country, a region was chosen on the basis of national climate impact assessments. The U.S. Third National Climate Impact Assessment Report (NCA3) finds the Southeast to be among the exceptionally susceptible regions to SLR, extreme heat, hurricanes and decreased water availability (Carter et al. 2014). The Southeast is among the most susceptible regions to changes in water availability (Romero-Lankao et al. 2014) and human health risk (Gutierrez and LeProvost 2016). Georgia is located within the Southeast (see Figure 4.2).

Figure 4.2 Case study region within the United States, location as outlined by the third
National Climate Assessment

Source: Melillo et al. (2014: ix), Copyright: NOAA

China's first National Assessment Report (NAR) on Climate Change, published in 2007, later on known as National Climate Change Assessment Report
(NARCC), acknowledges a higher than the average global temperature rise in the
past century (NDRC 2007). The NAR also predicts an annual mean temperature
increase, which is above the global level. In China too, climate change is unevenly
distributed with obvious regional differences. The amount of people affected by
increased flooding and extreme heat, as well as economic assets exposed is particularly significant at the eastern coast. Coupled with rapid urbanization and economic development, intensifying climate impacts will have a particular impact in
Zhejiang province (also see Chen et al. 2018). Zhejiang is located at the center of
China's eastern region (see Figure 4.3).

The regional political environments in which both cases are embedded provide
important background knowledge. Whereas the provincial government of Zhejiang is among the environmental pioneers within China, the state government
of Georgia has been considered conservative regarding climate policies. Zhejiang province is home to cities pursuing a strict Eco-agenda, with different pilot
programs (such as eco-cities and low-carbon cities), which have been initiated
throughout the province. Aside from pilots that focus on mitigation efforts, resilience plans and efforts were also put in place in Zhejiang. Further, the Zhejiang
Government has begun the planning process for becoming an eco-province. In

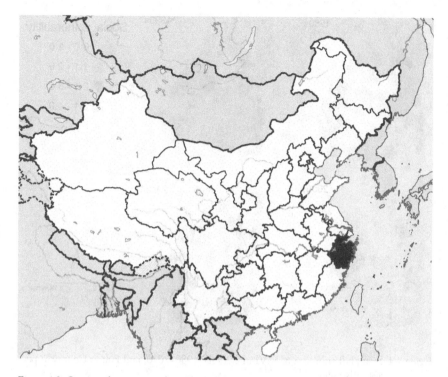

Figure 4.3 Case study region within China

Source: Wikimedia Commons, CC BY-SA 3.0, Copyright: TUBS – adapted from China edcp location map.svg (Uwe Dedering) https://commons.wikimedia.org/w/index.php?curid=16493992

contrast to Zhejiang, the political situation of Georgia's state-government is quite different, characterized by relative conservative policy- and lawmakers, who have been reluctant to move the needle on climate change.

Why Atlanta and Jinhua?

Social vulnerability to climate change and natural hazards was the primary selection criteria for a local case study. Recent vulnerability assessments indicate that the state of Georgia, the United States, and Zhejiang province, China, are particularly vulnerable to climate extremes with regard to social and climatological factors (Chen et al. 2013; Zhou et al. 2014; EPA 2014; KC et al. 2015; Guttierez and LeProvost 2016). Temporally, high climate vulnerability was detected across the state of Georgia between 1970 and 2010 (KC et al. 2015). Aside from the coastal counties and the rural black belt region, metro Atlanta counties are considered to be more socially and climatologically vulnerable (KC et al. 2015). Among the counties that make up the metro Atlanta region, Fulton County has been among the most socially vulnerable counties for over four decades (see Figure 4.4).

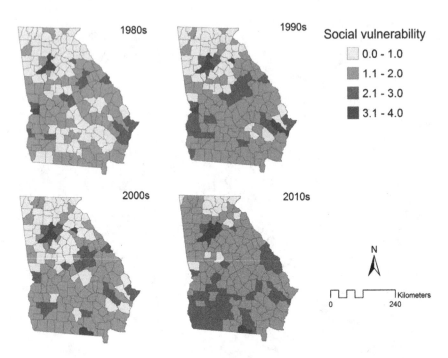

Figure 4.4 Spatial distribution of SoVI in Georgia, 1980–2010, and Fulton County as persistently vulnerable.

Source: KC et al. (2015: 69), Copyright: Elsevier Journal of Applied Geography

Chen and colleagues applied the Social Vulnerability Index (SoVI), to the YRD region in China. Trying to control for variation, they adjusted the Principal Component Analysis to the Chinese context. Within the Yangtze River Delta (YRD) region, the highest social vulnerability scores were concentrated in the southern geographic ends of Zhejiang province: Jingning, Suichang, Yunhe, Lanxi, Pan'an, and Shengsi (see Figure 4.5).

Based on a similarly high social vulnerability, Atlanta, which is located in Fulton County in southeastern Georgia and prefectural Jinhua, in eastern Zhejiang, which is formally in charge of administering Lanxi and Pan'an were selected. The chosen regions exhibit high degrees of environmental exposure, social sensitivity and lowered adaptive capacity with regard to natural and climate-sensitive hazards.

Much of the adaptation and vulnerability research focuses on coasts, due to the more direct and visible impacts of sea-level rise (SLR). Atlanta and Jinhua are two inland locations that are affected by long-term climatic changes and have been insufficiently dealt with regarding adaptation research compared to other cities in that region. Because this book is interested in the socio-political dimension of climate-vulnerable populations and because Atlanta and Jinhua exhibit the highest social vulnerability scores in the chosen regions, they were selected.

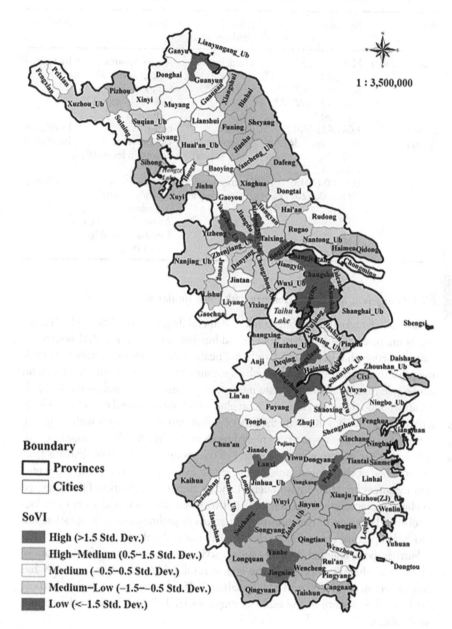

Figure 4.5 Spatial distribution of SoVI in the Yangtze River Delta region and Lanxi and Pan'an as most vulnerable within Jinhua prefecture.

Source: Chen et al. (2013: 177)

Table 4.2 Comparison of regions in which the cases are located

Area	Population	Political administration	Divisions	Population
Metro Atlanta	Metro: 5,789,700 (2010) Municipal: 420,003 (2010) / 472,522 (2016)	Municipality	Metro Atlanta: County level: 30 Cities: 140 Gov of Atlanta: 12 districts	Increasing
Fulton	1,041,423 (2017)	County	6 districts	Declining
Jinhua	Prefecture: 5,361,572 (2010) Urban: 1,077,245	Prefecture level city (*dijishi*)	County level: 9 Township level: 191	Increasing
Lanxi	666,700 (2013) 660,000 (2017)	County-level city (*xianjishi*)	7 towns (*zhen*), 6 subdistricts (*jiedao*), 2 townships (*xiang*), 1 ethnic township (*zuxiang*)	Declining

Slightly different municipal contexts with some similarities

Besides a heightened social and climatic vulnerability, municipalities in both cases are home to emerging "urban leaders" addressing climate change. Although the state of Georgia has been considered a politically unfavorable environment regarding proactive responses to climate change issues, the city of Atlanta has taken an active role regarding some climate mitigation and environmental issues. In 2015, Atlanta adopted its first climate action plan which was followed by a climate resilience plan in 2016. Although the city of Jinhua has not come up with a climate action plan yet, it has won prices regarding its innovative floodwater management concept. It became known as a pioneer of the sponge city program that was later on launched at the central government level. To varying degrees, both cases present instances of active environmental and/or climate governance. Efforts related to climate adaptation, however, have not received attention and seem to be laggardish in both cases. This non-pioneer status regarding climate adaptation was another important criterion for selecting Atlanta and Jinhua.

Further, both regional contexts show varying similar patterns of urbanization. China as a whole is undergoing the largest rural–urban transition a country has ever experienced (Tacoli and McGranahan 2008). Like other countries, parts of the United States are in the midst of major societal shifts related to the urban–rural sphere.

Data collection

The empirical foundations of this research are grounded on a combination of primary and secondary sources such as interviews, participant observation, official government documents and literature. In this book, qualitative data refers to small-n research, based upon in-field analysis.

Semi-structured expert interviews

Expert interviews aim to reconstruct social situations or processes to find a social scientific explanation for an observed phenomenon (Gläser and Laudel 2004). The expert interview may serve as a shortcut for otherwise more complex data collection processes, when experts act as "crystallization point" through inheriting practical insider knowledge (Bogner et al. 2009). Therefore, the most valuable data source used is a set of 53 interviews conducted with a total of 71 people in China and the United States. Out of the total amount, 27 interviews were conducted in China and 26 in the United States. The interview questions were divided into different blocks.

Interview phases

The interviews were divided into three phases: the pre-case selection-based interviews from July to October 2016 (n = 18, China = 13, US = 5), research phase I October 2016 to March 2017 (n = 13, China = 7, US = 6) and research phase II from April to November 2017 (n = 22, China = 7, US = 15). The majority of China-focused interviews took place in Jinhua and Hangzhou in Zhejiang province, Shanghai and Beijing. The U.S.-focused interviews were mainly conducted in Atlanta and Washington, DC. Short research trips were also taken to Chongqing, San Francisco (California) and Brunswick (southern Georgia). Interview solicitation occurred through means of cold-contacting, participation in conferences, political exchange programs and local political events, as well as snowball method. In the Chinese case, field access to political decision-makers was especially challenging. Here, a cultural exchange program by the Jinhua Municipal Government and a homestay program at a local village in Jinhua enabled the participation at local political events and field access to different levels of government. The interviews lasted 45 to 60 minutes on average in the United States and 100 minutes on average in China. The majority of interviews in research phases I and II were conducted on a one-on-one basis, whereas the pre-case selection interview phase in China mainly occurred with other interviewers and interviewees.

Experts interviewed

The target audience for the interviews either worked as government officials, academics or in professions related to climate adaptation policy- and/or local decision-making (such as emergency management, city planning or public health). One of the biggest limitations of this approach was the validation of the provided remarks. Interviewees were chosen at different administrative levels with different backgrounds (government, nongovernment, academic) in order to control for different positions in the field (for an overview of interview partners, see A.2). The interviewed academics and interstate practitioners provided essential information that critically reflected upon governmental practices. As knowledge construction and perceptions work subjectively, this method of verifying has shortcomings in its own part.

Interview coding

All interviews were conducted in confidentiality and anonymized. Due to the perceived sensitive nature of the issue in both political contexts, particular attention was paid to safeguarding the anonymity of the interviewees through the development of an interview codebook. If "G" stands after the number of the interview, this implies that there were more interviewees but only one interviewer. In case the interview code is marked with a "G" in front of the interview number, the interview was conducted by more than one person. This was the case in one research opportunity, which followed after an invite from a Chinese governmental institution to learn more about the political legitimacy approach of the Communist Party of China (CPC). This exchange program consisted of 15 European and North American researchers, who went through a ten-day program of visiting Political Leadership Academies and political research institutions across the country. Especially in China, a lot of interviews ended up being given by more than one person, either because more departmental colleagues were invited to join the interview and/or a translator could be arranged. As an interviewer, I had only limited influence on the number of people, who ended up being invited to partake in the interview.

The interview process and interview transcriptions were guided by the instructions and simple transcription rules outlined by Dresing and Pehl (2013). The data was processed with a qualitative data analysis software, which helped to organize the data schematically.

Participant observation

Participant observation and field notes from the attendance at the first Georgia Climate Conference (GCC) in November 2016 further grounded the study. The first conference brought together actors from across Georgia engaged in climate (policy) research and practice, neighboring states and other parts of the United States. The second GCC took place in November 2019 and consisted of audience of 430 experts from different levels of governance (public, private, nonprofit) and academic sectors. Participant observation is an ethnographic method that is interested in the study of (political) behavior, which is viewed as a process entangled in different social environments (Brodkin 2017). Ethnographic approaches to participant observation have only recently received attention in political science with the aim to "take a fresh look at old battles" (Brodkin 2017: 131). Despite its definitional looseness, ethnographic research contributes to political science research by not perceiving human behavior as a vacuum. The main assumption that guides ethnographic research is that people, their modes of thought and behavior develop in interaction with their lived environment. Based upon this understanding, ethnographic methods in political science take a look at the specific settings, which shape possibilities for political thought and activity (Brodkin 2017). Thereby, "informal" political channels can be examined by looking at ideas, relationships and narratives outside official formal institutions. In light of the partial absence of formal political structures and institutions engaged in climate policymaking, this method appears to offer a useful toolbox of political inquiry. Combined with interviews, observation as a social process is a useful strategy for exposing contradictions (Brodkin 2017).

Document analysis

Aside from expert interviews, participant observation and literature from different policy arenas (disaster, city planning, climate mitigation and adaptation) together with policy-relevant documents (such as Five-Year Plans (FYPs), Climate Action Plans (CAPs), White Papers or policy strategies) were cross-examined. The literature was informing the research critically throughout the process, particularly in a time when unexpected findings were generated. As this process did not follow a structured format, it is refrained from referring to it as "triangulation."

Research process

The empirical research alternated between China, the United States and Germany. Due to difficulties of field access, prolonged periods were spent in China in contrast to the relatively short research stays in the United States. This was compensated with an extensive Fulbright scholarship (2018) hosted by the Kennedy Institute of Ethics. The research stay here enabled fruitful discussions of theoretical readjustments, as well as practical and methodological implications of population-targeted policy programs. Visiting scholar positions at local educational institutions (Shanghai Jiao Tong University, Emory University Atlanta and Georgetown University in Washington, DC) helped to establish and maintain field access. In Atlanta and Jinhua, this enabled better access to not just educational resources and databases but was also a factor for establishing contact and trust. Further, they were an important precondition to travel and be invited to local events. In China, the participation in two governmental programs was of crucial importance and granted access to people in higher political positions.

Data analysis

For the analysis of content, this book uses the method prevalent in Grounded Theory. Grounded Theory both provides techniques for a structured, comparative analysis of content based on a systematic organization and comprimization of collected data. It is one method of qualitative research that is grounded on the basis of the generated data and the view of the participants interviewed (Creswell and Clark 2006). The approach takes no a priori assumptions with the aim of developing a theory (Creswell and Clark 2006). The research findings are to be understood as emerging from the work process, in which they were produced (Strübing 2014).

Open coding of data

The first step included compressing the findings from the first round of interviews by organizing them in a data analysis software (F4Analyse). The software was used to help organize the data schematically by structuring the text analysis and qualitatively coding, as well as filter and compare information. It was used to cluster the findings by looking for similar explanations related to problem recognition regarding social vulnerability and adaptation. Because categories constitute shortened and simplified points of analysis and skepticism is warranted for every type of thinking related to clear-cut and distinctive categories (Muckel 2007), the explorative act of open coding was found to be the most fitting for the research endeavor. The risk of a premature subsumption logic was addressed by constantly iterating and informing the preliminary findings with additional interviews, literature and document analysis.

Constant iteration

The literature on adaptation barriers informed the interview guideline and first set of case-unit interviews. In the process, it became evident that not only are the different categories ("barriers") interconnected, their explanations run short of exploring the root causes of deficient policymaking. Thus, the categories were iterated on the basis of findings from more interviews and by focusing more strongly on information on path dependencies and interrelated aspects.

This led to an almost complete abandonment of the predetermined indicators. Thereby, the case studies moved from an inductive to an instrumental level (in line with Mills et al. 2010). The categories were ultimately informed by one last theoretical iteration through applying the lock-in framework developed by Siebenhüner and colleagues (2017). The emerging angle of path-dependent adaptation lock-ins helped to identify a non-exhaustive list of explanations relevant for the chosen cases and was used as a main analytical framework. The literature on lock-ins is not a full-fledged theory of the policy process yet but in the making. Lock-ins seek to explain the systemic nature of policy processes and interdependence of different policy fields and political institutions. This angle turned out to be more suitable in explaining the persistence of adaptation deficits and incremental nature of decision-making in both political contexts.

Multicausal framework of analysis

This book opts for a multicausal explanatory framework, which does not aim to control for variation but embraces complexity. Due to their complex and interdependent nature, Becker and Schramm (2002) explicitly argue against viewing socioeconomic and environmental processes conventionally as independent of each other. Multicausal research aims to enhance a rich understanding of the contextual phenomena of policy problems by considering the multicausal nature of the policy process.

Limitations

Selective perception

Diekmann (2007) has laid out the major challenges of empirical social scientists, which persist in the subjective choosing of reality and what the scientists consider worth perceiving, may this happen deliberately or subconsciously. Diekmann refers to this problem as "selective perception." In the course of this book, the "selective perception" aspect clearly persisted in the choice of regions and overtly academic concepts.

Although this book is also criticizing the "scientistic" nature of social vulnerability frames in their political applicability, when conducting interviews in the field the own scientistic understanding of vulnerability and barriers became obvious. Many of the initial interviews spend significant amounts of time on exchanging over related concepts, some of which were understood or framed differently by most policymakers. In taking a longer time to explain the purpose of discussion, some of the findings were anticipated. This was an important learning process with greater caution applied throughout the interviews.

Ethical challenges

Doing research and conducting interviews in a context of limited political awareness about climate impacts and/or social vulnerability presents numerous challenges. Speaking of social vulnerability and critically confronting government officials with existing policy frameworks, which some of them may considerer (in) adequate for dealing with human susceptibility to climate change, adds to the challenge of the interview setting. Having government officials admit that vulnerable populations exist in their district is particularly sensitive in the Chinese context, because they may consider their political credibility questioned and responsibility neglected. To a lesser degree, this was also given in the United States. At the same time, the researcher carries responsibility regarding their externality to the studied setting, which is opposed by the interviewees, who in most cases were living in the region under study.

Narrow lens of "experts"

Throughout the research process, the understanding of what constitutes an "expert" and "expert knowledge" significantly shifted. Very often, the broader public is important for cross-validating the internal insights of policy processes and external evaluations of policy advisors. At the same time, professional politicians know well how to manipulate and influence the interviewer. Especially in the United States most interviews with governmental officials provided shallow insights through plain and at times exhausted policy language. The interview process reached a point at which everything sounded the same, knowledge production seemed limited and a bias was reached. Taking a break from one field-side was important to reflect upon this.[5]

Overlooking social practices

This book did not address local mechanisms and strategies of populations, nor did it address the complexity and significance of everyday social practices. Means of self-governance and cultural practices were not reflected upon but are a likewise important field of study (see, e.g., Scott 1988; Shannon 2014). Local practices of indigenous self-governance are an important but commonly overlooked and underestimated field aside from community-led forms of adaptation. Furthermore, this idea of being socioeconomically deprived and equating this with a lack of adaptive capacity neglects other sources of ideas certain groups might have. Scholars have sufficiently hinted at this problem (e.g., Ribot 2011). In line with James C. Scott's early work (1988), societies may well choose to function outside the organized state. Not reflecting upon informal (adaptation) practices outside the governmental realm and thereby overlooking important cultural factors therefore marks a severe lack of this study.

Talking about vulnerable populations

No validation studies were conducted analyzing how people view and conceptualize their own vulnerability. Follow-up surveys could investigate the perceptions and visions of local populations regarding their susceptibility, risk perception and awareness of local populations regarding extreme heat and other climate impacts and pair these with the studies at hand as well as the perceptions of local decision-makers. Being sensitive to community-based self-perception and developing a shared understanding of societal exposure is important when designing policies and developing population-targeted solutions without disempowering or colonizing.

Data limitations: quality of interviews and field access

Although it was clear right from the start that the lack of socioeconomic data coherence is a major challenge, especially in the Chinese case (on the same matter, also see, e.g., Hu et al. 2017), the practical ramifications were drastically underestimated. The same applies to local language dialects in Zhejiang province. Recording the interview was essential due to the speed of the conversation and prevalence of local dialect.

Due to matters of field access, the data collected from the interviews is severely different and marks another major limitation. On the one hand, the difficulty of field access persisted in the not as widely accepted practice of interviewing decision-makers in China, as well as the considered sensitive nature of the topic. On the other hand, alternative opportunities came up and were used to get to the perspectives of high-ranking Communist Party officials to pre-inform this research. Pre-informing the case selection was much shorter in the United States than in China. In China, the pre-research phase was more formal and also realized by the participation in an exchange program initiated by the Communist Party in China. Another program was a so-called homestay program in the Chinese countryside that ended up providing major access to local authorities involved in city development planning and emergency management of the chosen case study Jinhua. Overall, more time was spent on first examining the correct terming regarding social vulnerability to climate change in China. The understanding of this concept was not intuitively available for decision-makers here. The same was the case for the concept of climate adaptation, which was often thought of in the context of disaster risk reduction or referred to differently. Compared to the United States, more group interviews were conducted with high-ranking government officials in China. Here, it was more difficult to get local politicians to talk about local governmental decision-making practices. At the same time, the local levels of government in China (i.e., prefecture-level cities, counties) are much stronger interconnected with dynamics at the provincial level. Therefore, policymaking there received more attention. As a result of limited access to local decision-makers, academics, with expertise in local government and with insights for the city of Jinhua played a much greater role.

In the United States, individual interviews flourished but in some cases were of distinctively different quality: interviews with local politicians tended to be too short and would get lost in a great deal of superficial policy language. Gaining interviews was not as challenging, to the contrary, it reached the point where access seemed to be unlocked in its entirety as opposed by my financial and personal resources, which became severely limited. Data availability on related policy documents severely shrunk after the Trump inauguration in January 2017. Further, events that were supposed to be held on a heat-health-climate change perspective by the Georgia Department of Public Health and would have provided greater insights were cancelled in light of the political climate. Due to the otherwise open-minded research context, limitations of data accessibility and collection were underestimated.

Scope of cases

Due to the focus on multicausal explanations, the explanatory scope of the cases is rather large and aims to contribute to theory-building. The inference from the number of qualitative interviews is rather small and must be considered a starting point upon which further research can be built to explore the explanations in greater depth. Because of this, setting boundaries for collecting empirical evidence is another critical limitation. For both cases, the collection of evidence reached an analytic periphery cases when information from empirical material started to repeat itself. Yin (2018) considers this "analytic periphery" a helpful boundary to limit the scope and for concluding the collection of evidence, as information seems to be of decreasing relevance to the case. Yet, the research framework is still rather broad, because it seeks to explore the interconnections of different aspects.

Dependent variable: adaptation deficits

This research followed a strongly iterative research design. Although the regions were deliberately chosen on the basis of their adaptation deficits, and examining non-pioneers was an important argument to the study of lock-ins and policy formulation tools, the dependent variable was difficult to handle. The governmental approaches which were analyzed include accidental adaptation efforts, which were identified in the qualitative interviews to constitute unintentional adaptation measures. Based on the understanding that knowledge is culturally constructed, the importance of analyzing adaptation non-pioneers was argued for. This was confirmed in the field, where decision-makers, especially in China had a detailed understanding of the challenges related to climate change impacts but would phrase them differently or not draw conclusions for the municipality of Jinhua. In order to avoid bias in the process of collecting data, a streamlined approach would have helped to cope with this challenge.

Notes

1 This chapter follows the differentiation of "method" and "methodology" of Evans et al. (2014). Accordingly, methodology is considered a branch of knowledge that deals with the application of method in a particular field, whereas "method" refers to the research design that was applied for answering the research questions.
2 For a more detailed classification of disasters, see Sauerborn and Ebi 2012: 2.
3 Interviews I-01 to I-11 and G-01 to G-08.
4 See Appendix 1 for more information on geophysical and climatological exposure of Georgia and Zhejiang.
5 For a more thorough discussion on pitfalls of expert interview-based research, also see Lena Bendlin (2019: 75ff).

References

Bailey, Christopher J. *US Climate Change Policy. Transforming Environmental Politics and Policy*. Surrey and Burlington: Ashgate, 2015.
Becker, Egon, and Engelbert Schramm. "Gekoppelte Systeme. Zur Modellierung und Prognose sozial-ökologischer Systeme." In *Sozial-ökologische Forschung. Ergebnisse der Sondierungsprojekte*, edited by Ingrid Balzer and Monika Wächter. München: Oekom, 2002.
Bendlin, Lena. *Orchestrating Local Climate Policy in the European Union. Inter-Municipal Coordination and the Covenant of Mayors in Germany and France*. Wiesbaden: Springer VS, 2019.
Bogner, Alexander, Beate Littig, and Wolfgang Menz, eds. *Experteninterviews. Theorie, Methoden, Anwendungsfelder. 3., grundlegend überarbeitete Auflage*. Wiesbaden: Verlag für Sozialwissenschaften, 2009.

Brodkin, E. "The Ethnographic Turn in Political Science: Reflections on the State of the Art." *Political Science & Politics* 50, no. 1 (2017): 131–134. Doi: 10.1017/S1049096516002298.

Burnham, Peter, Karin Gilland Lutz, Wyn Grant, and Zig Layton-Henry. *Research Methods in Politics* (2nd ed.). New York: Palgrave Macmillan, 2008.

Carter, L. M., J. W. Jones, L. Berry, V. Burkett, J. F. Murley, J. Obeysekera, P. J. Schramm et al. "Ch. 17: Southeast and the Caribbean." In *Climate Change Impacts in the United States: The Third National Climate Assessment*, edited by J. M. Melillo, Terese (T.C.) Richmond, and G. W. Yohe, Vol. 9, 396–417. U.S. Global Change Research Program, 2014. Doi: 10.7930/J0N- P22CB.

Chen, Lin, Chunying Ren, Bai Zhang, Zongming Wang, and Mingyue Liu. "Quantifying Urban Land Sprawl and Its Driving Forces in Northeast China from 1990 to 2015." *Sustainability* 10 (2018). https://doi.org/10.1289/EHP44.

Chen, Wenfang, Susan Cutter, Christopher T. Emrich, and Peijung Shi. "Measuring Social Vulnerability to Natural Hazards in the Yangtze River Delta Region, China." *International Journal of Disaster Risk Science* 4, no. 4 (2013): 169–181. https://doi.org/10.1007/s13753-013-0018-6.

Creswell, John W., and Vicki L. Plano Clark. *Designing and Conducting Mixed Methods Research*. Thousand Oaks, CA: Sage, 2006.

Diekmann, Andreas. *Empirical Social Research. Basics, Methods, Applications.* (Original: *Empirische Sozialforschung. Grundlagen, Methoden, Anwendungen*). Hamburg: Rowohlt Taschenbuch Verlag, 2007.

Dresing, Thorsten, and Thorsten Pehl. *Practicebook Interview, Transcription and Analysis.* (Original: Praxisbuch Interview, Transkription & Analyse. Anleitungen und Regelsysteme für qualitativ Forschende) (5th ed.). Marburg, 2013. www.audiotranskription.de/praxisbuch, last accessed June 7, 2017.

Edmonds, Richard, and Peter Ho. *China's Embedded Activism: Opportunities and Constraints of a Social Movement*. London and New York: Routledge Studies on China in Transition Book, 2016.

Environmental Protection Agency (EPA). "Adaptation Implementation Plan Draft, US EPA Region 4". *EPA*. Published online, 2014. https://www3.epa.gov/climatechange/Downloads/Region4-climate-change-adaptation-plan.pdf

Evans, David, Paul Gruba, and Justin Zobel. *How to Write a Better Thesis* (3rd ed.). Switzerland: Springer International Publishing, 2014. https://www3.epa.gov/climatechange/Downloads/Region4-climate-change-adaptation-plan.pdf

Gerring, John. "What is a Case Study? The Problem of Definition." In *Case Study Research. Principles and Practices*, edited by John Gerring, 17–36. Cambridge and London: Cambridge University Press, 2007.

Gläser, Jochen, and Grit Laudel. *Experteninterviews und qualitative Inhaltsanalyse als Instrumente rekonstruierender Untersuchungen.* Wiesbaden: Verlag für Sozialwissenschaften, 2004.

Grandy, Gina. "Instrumental Case Study." In *Encyclopedia of Case Study Research*, edited by Albert J. Mills, Gabrielle Durepos, and Elden Wiebe, 474–475. Thousand Oaks: SAGE Publications, Inc. City, 2012. https://dx.doi.org/10.4135/9781412957397.

Gutierrez, Kristie, and Catherine E. LePrevost. "Climate Justice in Rural Southeastern United States: A Review of Climate Change Impacts and Effects on Human Health." *International Journal of Environmental Research and Public Health* 13, no. 189 (2016). Doi: 10.3390/ijerph13020189.

Harles, John. *Seeking Equality. The Political Economy of the Common Good in the United States and Canada*. Toronto: University of Toronto Press, 2017.

Hijioka Yasuaki, Erda Lin, Joy J. Pereira, Richard T. Corlett, Xuefeng Cui, Gregory Insarov, Rodel D. Lasco et al. *Impacts, Adaptation, and Vulnerability. Part B: Regional Aspects. Contribution of Working Group II to the Fifth Assessment Report of the Intergovernmental Panel on Climate Change.* Cambridge, UK and New York, NY, US: Cambridge University Press: IPCC, 2014.

Hu, Kejia, Xuchao Yang, Jieming Zhong, Fangrong Fei, and Jiaguo Qi. "Spatially Explicit Mapping of Heat Health Risk Utilizing Environmental and Socioeconomic Data." *Environmental Science & Technology* 51 (2017): 1498–1507. Doi: 10.1021/acs.est.6b04355.

IPCC. *Global Warming of 1.5°C, an IPCC Special Report on the Impacts of Global Warming of 1.5°C Above Pre-Industrial Levels and Related Global Greenhouse Gas Emission Pathways, in the Context of Strengthening the Global Response to the Threat of Climate Change, Sustainable Development, and Efforts to Eradicate Poverty.* Edited by V. Masson-Delmotte, P. Zhai, H. O. Pörtner, D. Roberts, J. Skea, P. R. Shukla, A. Pirani, W. Moufouma-Okia et al. Waterfield, 2018. www.ipcc.ch/report/sr15/.

IPCC. *Special Report for the Intergovernmental Panel on Climate Change Managing the Risks of Extreme Events and Disasters to Advance Climate Change Adaptation,* edited by Field, C. B., V. Barros, T. F. Stocker, D. Qin, D. J. Dokken, K. L. Ebi, M. D. Mastrandrea, K. J. Mach, G.-K. Plattner, S. K. Allen, M. Tignor, and P. M. Midgley. Cambridge, UK: Cambridge University Press, 2012.

KC, Binita, Marshall Shepherd, and Cassandra Johnson Gaither. "Climate Change Vulnerability Assessment in Georgia." *Applied Geography* 62 (2015): 62–74. http://dx.doi.org/10.1016/j.apgeog.2015.04.007 0143-6228/.

Melillo, Jerry M., Terese (T.C.) Richmond, and Gary W. Yohe (eds.). *Climate Change Impacts in the United States: The Third National Climate Assessment.* U.S. Global Change Research Program, 2014. Doi: 10.7930/J0Z31WJ2.

Mills, Albert J., Gabrielle Durepos, and Elden Wiebe. *Encyclopedia of Case Study Research.* Thousand Oaks, CA: Sage Publications, 2010. Doi: 10.4135/9781412957397.

Muckel, Petra. "Die Entwicklung von Kategorien mit der Methode der Grounded Theory." *Historical Social Research Supplement* 19 (2007): 211–231. https://nbn-resolving.org/urn:nbn:de:0168-ssoar-288620.

National Development and Reform Commission (NDRC). "China's National Climate Change Program." *NDRC,* published online, 2007. https://www.fmprc.gov.cn/ce/ceun/eng/gyzg/t626117.htm.

Newton, Kenneth, and Jan W. Van Deth. *Foundations of Comparative Politics.* Cambridge and New York: Cambridge University Press, 2005.

Ribot, Jesse. "Editorial. Vulnerability before Adaptation: Toward Transformative Climate Action." *Global Environmental Change* 21 (2011): 1160–1162. Doi: 10.1016/j.gloenvcha.2011.07.008.

Romero-Lankao, P., J. B. Smith, D. J. Davidson, N. S. Diffenbaugh, P. L. Kinney, P. Kirshen, P. Kovacs et al. "2014: North America." In *Climate Change 2014: Impacts, Adaptation, and Vulnerability. Part B: Regional Aspects. Contribution of Working Group II to the Fifth Assessment Report of the Intergovernmental Panel on Climate Change,* edited by V. R. Barros, C. B. Field, D. J. Dokken et al., 1439–1498. Cambridge and New York, NY: Cambridge University Press, 2014.

Sauerborn, Rainer, and Kristie Ebi. "Climate Change and Natural Disasters: Integrating Science and Practice to Protect Health." *Global health action* 5, no. 1–7 (2012). Doi: 10.3402/gha.v5i0.19295.

Scott, James C. *The Art of Not Being Governed: An Anarchist History of Upland Southeast Asia.* New Haven, CT: Yale University Press, 1988.

Shannon, Jerry. "Food Deserts: Governing Obesity in the Neoliberal City." *Progress in Human Geography* 38, no. 2 (2014): 248–266. Doi: 10.1177/0309132513484378.

Shi, Peijun, Xu Yang, Jiayi Fang, Wei Xu, and Guoyi Han. "Mapping and Ranking Global Mortality, Affected Population and GDP Loss Risks for Multiple Climatic Hazards." *Journal of Geographic Science* 26, no. 7 (2016): 878–888. https://doi.org/10.1007/s11442-016-1304-1.

Siebenhüner, Bernd, Torsten Grothmann, Dave Huitema, Angela Oels, Tim Rayner, and John Turnpenny. "Lock-Ins in Climate Adaptation Governance. Conceptual and Empirical Approaches." Conference Paper 2017, unpublished.

Strübing, Jörg. *Grounded Theory. Zur sozialtheoretischen und epistemologischen Fundierung eines pragmatistischen Forschungsstils*. Wiesbaden: Springer Fachmedien, VS Verlag für Sozialwissenschaften, 2014.

Tacoli, Cecilia, and Gordon McGranahan. "Rural-Urban Migration, Urbanization and Inequality in China." In *Sustainable Urban Development in China, Wishful Thinking*, edited by Marco Keiner. Münster: MV Wissenschaft, 2008.

Yang, Xuchao, Wenze Yue, Honghui Xu, Jingsheng Wu, and Yue He. "Environmental Consequences of Rapid Urbanization in Zhejiang Province, East China." *International Journal of Environmental Research and Public Health* 11, no. 7 (2014): 7045–7059. Doi: 10.3390/ijerph110707045.

Yin, Robert K. *Case Study Research Design and Methods* (6th ed.). Thousand Oaks, CA: Sage Publishing, 2018.

Zhou, Yang, Ning Li, Wenxiang Wu, and Jidong Wu. "Assessment of Provincial Social Vulnerability to Natural Disasters in China." *Natural Hazards* 71 (2014): 2165–2186. Doi: 10.1007/s11069-013-1003-5.

5 Vulnerability and adaptation governance in China and the United States

This background chapter analyzes the evolution of adaptation and vulnerability planning in China and the United States. The chapter examines how climate adaptation planning has unfolded in both countries and how this corresponds with the current state of adaptation governance more broadly. Exploring the differences of public adaptation across different political systems can help our understanding of main policy interests, gaps and challenges associated with it. Against the background of perceived high adaptation pressure (see Chapter 1), comparing adaptation efforts taken at both national and subnational levels in China and the United States has rich insights to offer how vulnerability and adaptation can be approached differently, and where similarities can be detected. Outlining the state of the adaptation field and governance efforts taken therein also helps for locating the empirical findings on vulnerability and adaptation in two municipal jurisdictions later on.

Environmental problem-solving, and disaster risk management have a long history and political relevance in both the Chinese and U.S. political systems (Mertha 2008; Kamieniecki and Kraft 2013; Young et al. 2015). Climate change adaptation (CCA) is a relatively new policy field in both countries that developed in the late 2000s. The central observation of this chapter is that while CCA planning has almost come to a standstill at the federal level in the United States under the Trump administration, it has received more attention from the Chinese central government. In the United States, this brought forward an array of non-state and local government action and also broadened the conceptual field of CCA by encompassing urban resilience. In China, despite the diversification of CCA efforts across the country, policy efforts continue to be mandated in a top-down manner with backlog demands of local expertise.

The first part of this chapter reflects upon the current status of public adaptation efforts in international comparison. Next, dominant adaptation practices are examined, by looking at recent developments in the adaptation field across both countries. To do so, main policy planning instruments, central-local dynamics and actors, as well as ideas and narratives of governmental directives related are analyzed. A brief observation about laws and regulations ends the chapter.

DOI: 10.4324/9781003183259-5

The state of climate adaptation governance

Climate adaptation has not been a focal interest of either country as part of the international climate negotiations but has, to varying degrees, received some attention at national and local government levels. Although adapting to changing environments has been part of much of humanity's history, climate adaptation as a specific domain is a new research effort and concept in China (Zhang et al. 2008; Nadin et al. 2016). Besides being a likewise nascent and experimental field, it presents a very mixed picture for the United States (Moss et al. 2014; Moser et al. 2017).

Low local capacity for adaptation planning in China

China stands out for its lack of local adaptation plans outside the designated pilot areas. Whereas researchers criticized that the Chinese government was lacking a climate adaptation strategy until recently (e.g., Zhang et al. 2008; Pan et al. 2011), central adaptation directives and policy efforts taken since 2013 have significantly changed the field of adaptation research and practice in China (Nadin et al. 2016).

Although the importance of reducing vulnerability and enhancing adaptive capacity has been recognized (see, e.g., He 2018) and the central government has invested efforts at the national level into adaptation policy, the capacity for informed adaptation planning is considered low and in its preliminary stages. Adaptation policies are quite new on the agenda of Chinese political leadership. In this context, the central nature of top-down adaptation planning and little adaptation knowledge at the local level is seen to present a gap that needs to be addressed (Li 2013; Nadin et al. 2016).[1] Climate-related policies as well as crisis management have both been guided by highly centralized principles and guidelines. Research on adaptation governance in China stresses that more localized governance principles are needed for the adaptation field, as the related climate impacts and regional circumstances have a strong local character.

More advanced local planning in the United States

The United States is more advanced regarding public adaption planning at the local level. Communities, local and state governments are experimenting with adaptation and have adopted strategies and policy frameworks.[2] In contrast to China, the lack of a nationwide approach and federal guidance has been criticized (see, e.g., Hansen et al. 2013; Moser et al. 2017). Public actors seem to have a general understanding of climate adaptation as a policy field, but widespread climate skepticism has severely hampered progress on climate adaptation and vulnerability reduction. The partisan political system is characterized by the fragmented disregard for scientific observations and future projections, which has had an influence on risk perception. At the same time, adaptation research and science have become increasingly important in the United States (Moss et al. 2014).

Many frameworks exist to evaluate existent adaptation approaches and policies. Besides the maturity of adaptation research, adaptation is a policy field that is not evolving quickly or deliberately enough (Tompkins and Eakin 2012; Moser and Boykoff 2013; Vogel et al. 2015; Moser et al. 2017).

Congruence with international level: adaptation an emerging field

When comparing the efforts of both countries with international levels, several important differences and shifts can be marked. In the late 2000s, the IPCC published AR4, which for the first time placed a stronger emphasis on adaptation. At that time, the Chinese government had only initiated the first generation of climate change planning, but still largely neglected adaptation. In contrast, the first wave of institutional adaptation policies in the early to mid-2000s flourished in the United States (Preston et al. 2011). This development coincides with the first generation of climate change plans at municipal and state levels, most of which did not address climate adaptation at that time (Wheeler 2008).

In the early 2010s and in line with the international momentum of the 2°C target, central Chinese directives launched the first generation of adaptation planning, at a time when the United States published the second generation of departmental adaptation plans. Since then, the international community has seen a rapid development of adaptation planning at different levels of government, by non-state organizations, and a multitude of autonomous efforts (see Table 5.1).

The adaptation momentum further increased in the late 2010s. The Paris Agreement (2016) became known for its fortified global commitment to reduce global warming to 1.5°C instead of 2°C. This new temperature target was further underpinned with the publication of IPCC's Special Report in 2018, which predicts that 1.5°C may be reached earlier than anticipated. Although the toothless commitment to the 1.5°C target marks an important gradual shift at the international level, the majority of CCA efforts remain in the planning stage: they are ad hoc in nature, strongly infrastructure-focused and attempt to increase the understanding of the problem (Hansen et al. 2013; Bierbaum et al. 2013; Nadin et al. 2016; Siebenhüner et al. 2017).

Dominant adaptation practices in China

China began to research the impacts of climate change during the 1990s. Up until recently, most research was in a preliminary and theoretical stage and not translated into policy responses or political strategies (Yue et al. 2007). The following sections examine how adaptation has evolved into a macro-policy framework since then.

Formalizing adaptation policy processes at the national level

Centralized efforts to address the adverse impacts of climate change have been ongoing since 2007, as part of China's first climate change program. At that time, adaptation actions existed but in most cases were not considered adaptation per

Table 5.1 Climate change policy related adaptation efforts in China, the United States and at the global level (in bold: efforts related to adaptation)

Years	Global	China	United States
Late 1980s	IPCC established by United Nations Environment Program and the World Meteorological Organization (WMO) (1988)	Ecological relocations as common disaster risk reduction practice	First generation of climate change planning (CCP)
1990s	UNFCCC (1992), Rio "Earth" Summit 1992, Adoption of the Kyoto Protocol (1997)	Passive participation in international climate negotiations	Acceleration of CCP
2000s	Kyoto Protocol comes into force (2005), AR4 WGII Report on Impacts, Adaptation, and Vulnerability published (2007), Copenhagen Summit fails (2009) → **Marks important shift to adaptation**	First generation of climate change planning (post-2007), neglected adaptation focus	First generation of institutional adaptation plans
Early 2010s	AR5 on impacts, adaptation, and vulnerability (2014), Paris Agreement (2015) → **2.0°C momentum**	First generation of adaptation planning with central directives	Second generation of departmental adaptation plans
Late 2010s	Paris Agreement enters into force (2016), IPCC 1.5°C Special Report published (2018) → **Acceleration of 1.5°C target**	Implementation of first adaptation plans through pilots at the local level	Stalled federal efforts, flourishing of urban resilience efforts by local and non-state actors
Early 2020s	Glasgow summit (COP26) postponed to November 2021 due to Covid-19, AR6 WGII Report to be published (2022)	China announces fortified reduction targets, emphasized focus on biodiversity, ecological conservation and ecological cooperation also as part of South–South cooperation and the Belt and Road Initiative	U.S. withdrawal (2017) from Paris Agreement comes into effect (2020), but is reversed by new Biden administration (2021), launch of a whole-of-government-process to establish new emission targets

Source: the author

se. Some uncoordinated actions to offset climate change were undertaken but insufficient adaptation awareness, research and lack of funding limited existing adaptation projects (Tu et al. 2012).

After a two-year preparation phase, China's first National Climate Adaptation Strategy (NAS) (*guojia shiying qihou bianhua zhanlüe*) was published in 2013. The publication presents a joint ministerial effort of 12 governmental bodies including the lead body, the National Development and Reform Commission (NDRC), the Ministry of Finance (MoF), the Ministry of Housing and Urban-Rural Development (MoHURD), the Ministry of Transport (MOT), Ministry of Water Resources (MWR), the State Forestry and Grassland Administration (SFGA), Ministry of Agriculture (MoA), Meteorological Administration (CMA), and the State Oceanic Administration (SOA).

Each of these departments fulfill different roles. The MoA, for instance, is responsible for infrastructure construction of affected farmlands. The MWR is responsible for the design of flood prevention systems and engages in more proactive forms of adaptation characterized by planning efforts that span over decades, if not centuries. The primary purpose of the NAS is to provide guidance on priority areas and act as a strategy (Heggelund and Nadin 2017). The NAS can be understood as a macro-policy framework, which marks the beginning of a formalization of the adaptation policy process (Nadin et al. 2016).

Since China published its National Adaptation Strategy in 2013, studies on climate adaptation opportunities, broader implications for China's climate policies and examinations of progress made by the central government at different levels of government are growing (Li 2013; Chao et al. 2014; Nadin et al. 2016; Dai et al. 2018; He et al. 2017; He 2018). This is in part related to several policy initiatives that were launched as pilots across the country.

Central-local dynamics

The NAS provides a framework for selected provincial governments and ministries. NDRC officials are keen to emphasize that "the strategy is not a plan, but rather an outline of work that will take time to implement" (Nadin et al. 2016: 310). NAS outlines 13 provincial pilots, which were chosen aside from Beijing. Among them are Chongqing (representing the Three Gorges Area), Guangdong, Guangxi, Hainan, Hebei, Heilongjiang, Inner Mongolia, Jiangxi, Jilin, Ningxia, Shanghai, Sichuan and Xinjiang. The pilots test different research approaches, based upon key vulnerabilities and climate risks (Nadin et al. 2016). Priority adaptation sectors are determined according to regions, such as agricultural adaptation in the Northwest (Ningxia), coastal adaptation in the Southeast (Guangzhou) or grassland and ecosystem-based approaches in the North (Inner Mongolia).

Under the most recent National Plan to Address Climate Change (2014–2020) all provinces and municipalities must develop their own adaptation plans. Anhui, Gansu, Guangxi, Jiangxi, Liaoning, Ningxia, Qinghai, Sichuan, Xinjiang and Yunnan provinces plus Chongqing and Tianjin Municipalities, among others, have already published their adaptation plans.

Planning instruments

China's main approach for adaptation planning is the use of pilots to test different adaptation strategies that correspond with local characteristics. Whereas climate adaptation efforts in the United States are mainly occurring in coastal regions and are often based on the initiative of local actors, provincial as well as urban adaptation pilots are equally dispersed across China.

The pilots are distributed between two phases of the same "research into policy project" (Nadin et al. 2016): Adapting to Climate Change in China (ACCC). The ACCC project was divided into two phases: ACCCI (2009 to 2014) and ACCCII (September 2014 to September 2017, which was later on extended to 2019), taking partial responsibility for the implementation of designated adaptation efforts in the NAS 2013. With a budget of six million U.S. dollars, the ACCCI was implemented by actors from China, the United Kingdom Department for International Development, and the Swiss Ministry of Climate Change and Energy. From the Chinese side, NDRC was initially the lead agency with technical support from the Institute of Urban Environment at the Chinese Academy of Sciences (CAS). Later on in the process, the newly founded Ministry of Ecology and Environment (MEE) took some of the functions of NDRC in 2018 and became the leading policymaking partner. Aside from the MEE, other policymaking partners include: Departments of Ecology and Environment (DEE) and Departments of Development and Reform (if DEE is not yet established) at provincial and city levels. They are part of the climate adaptation teams in addition to other working partners such as national think tanks (e.g., National Centre on Climate Change Strategy and International Cooperation) and educational institutions (e.g., Peking University, Chinese Academy of Agricultural Sciences). They together work to enhance capacity on climate change adaptation through supporting peer learning, interactive training programs and local policy dialogues. The aim is to implement a risk-based planning approach, close knowledge gaps by conducting risk assessments and understanding historical weather and climate impacts and ultimately designing suitable policy responses.

For the ACCCII pilots, adaptation planning teams were initially established in five provinces and one city: Inner Mongolia, Ningxia, Jiangxi, Guizhou, Jilin and the municipality of Qingdao. This was adjusted throughout the second phase of the ACCC project. In the end, two provincial pilots were chosen with a focus on grassland and livestock management in Inner Mongolia and agriculture Jiangxi. Here, the project supported two provinces to integrate climate risks into their Five-Year Plans (FYP) (*wunian jihua*). The nationwide FYPs are the central planning instrument that outlines major social and economic priorities and sets forth overarching development targets and more specific goals. FYPs exist on different levels of government and are important signifiers of major governmental directives.

At the city level, three municipalities were supported to carry out risk assessments: Qingdao in Shandong province, Lishui in Zhejiang province, and Baise in Guangxi province (see Figure 5.1). Based upon the assessments, adaptation options were suggested according to local sectoral planning (Huo et al. 2019). For

Figure 5.1 Adaptation progress in China.

Source: Huo et al. (2019: 13)

instance, in Qingdao, the main climate drivers in focus are floods, typhoons and storms and thus adaptation efforts were outlined for the marine and coastal zones. This includes developing a Qingdao System Risk Model in charge of identifying present and future flood risks of Qingdao city.[3] Concentrating on a distinctive climate risk and proposing adaptation measures and building capacity for climate adaptation through local workshops and surveys was a key pattern of the pilot projects.

Environmental degradation has been increasingly viewed as an important domestic issue and has become a growing concern in China's national and subnational FYPs. China's 12th FYP put an emphasis on mitigation efforts while stressing adaptation strategies theoretically. Adaptation has been gaining traction in the 13th FYP (2016–2020). Aside from climate change focused plans, China's annual White Papers aim to follow up on the progress as well as the targets outlined by the FYPs (also see Table 5.5).

Finally, targets are another important instrument for policy implementation. The Mandatory Target System (MTS) for the first time assigned environmental targets during the 11th FYP (2006–2010) and 12th FYP (2011–2015). MTS is an important political mechanism through which the environmental performance of government officials is measured. Throughout the years, environmental protection targets (*mubiao*) have increased as part of the overall targets of the FYPs. Most of these environmental targets provide imperatives for emission control strategies and broader environmental planning (Tang et al. 2016; Khan and

Chang 2018). The targets are also being used for the evaluation of government officials' performance. The targets are assigned to lower levels of government for their implementation. Although these targets are an important policy mechanism for greater governmental accountability and guiding environmental agendas, their actual effect on environmental performance has been called into question (Tang et al. 2016). To what extent these targets have included climate change adaptation needs further study.

Dominant adaptation practices in the United States

In the United States, public climate change planning accelerated in the mid- to late 1990s (Wheeler 2008). Earlier research shows how most of the 29 states that had some sort of climate change plan did not address climate adaptation (see Wheeler 2008).

Lack of a macro-policy framework

Whereas earlier periods of environmental governance were characterized by a wealth of information and science-based political decision-making, environmental and climate governance have experienced serious drawbacks. During the first year of the Trump administration (2017), web resources of many federal agencies experienced a systematic overhaul characterized by significant language shifts and the removal of documents, webpages and websites related to climate change.[4] These changes are emblematic of a greater shift in governmental priorities, which weakened the already low national emission reduction targets set by the Obama administration (2009–2017) entirely.

Inconsistency, conflict and policy stagnation are considered key features of climate policymaking in the United States (see, e.g., Bailey 2015). Dismissing the significance of anthropogenic climate change has become a major characteristic of the U.S. response to climate change and is strongly partisan, with the majority of Republican Party members rejecting climate change science.[5] Climate skepticism has become such a dominant feature that "getting skeptical about climate skepticism" has become a major platform to unveil the arguments made by notorious climate change deniers and science skeptics more broadly.[6] A recent study compared conservative parties worldwide regarding their attitude toward climate change and concluded that the Republican Party is an "anomaly in denying anthropogenic climate change" (Båtstrand 2015: 538).[7]

The limited authority and resources of the U.S. government in enabling early adaptation planning have been criticized by researchers from a variety of disciplines (e.g., see: Moser and Ekstrom 2010; Boda and Jerneck 2019). Adaptation policies have nonetheless been formed at federal, state and local levels of government. Most policy practitioners seem to agree that all "adaptation is local" and scaling up is occurring primarily from local to regional levels (Moser et al. 2017). The adaptation field has been influenced by international and federal policies and funding and advocacy from philanthropic organizations, foundations and non-governmental organizations.

Planning instruments

In light of deadlocked political institutions, federal action on climate adaptation has been mainly initiated through Executive Orders (EOs). EOs have become a primary planning instrument for climate change policies. Obama's strategy was characterized by regulation-making through the EPA and major use of Executive Orders. In order to get climate and environmental legislation passed in a hostile Republican-dominated Congress, a range of EOs were issued under Democratic Presidencies.

EOs mandate actions for Federal Agencies. During the Bill Clinton Presidency, EOs also played an important role in furthering the environmental agenda and that of vulnerable communities. One example in light of the environmental justice movement is EO 12898 *Federal Actions to Address Environmental Justice in Minority Populations and Low-income Populations*, which was issued in 1993. This EO sets out how federal agencies shall focus on human health effects of adversely affected populations in light of environmental damage. In total, Clinton issued approximately 40 EOs to achieve environment-related policy objectives.

After 2013, federal agencies began to publish their first climate change adaptation plans including an outline of their specific departmental vulnerabilities and how they are planning on addressing them. This was made possible through the EO 13514, *Federal Leadership in Environmental, Energy, and Economic Performance* from 2009.

Adaptation task force based on EO 13514

The Interagency Climate Change Adaptation Task Force (ICCATF) was established in 2009 with the intent to create a federal adaptation framework on climate preparedness and resilience. The White House Council on Environmental Quality (CEQ), the Office of Science and Technology Policy (OSTP), and the National Oceanic and Atmospheric Administration (NOAA) established ICCATF in 2009. Altogether, the Task Force is a collaborative effort by representatives from 20 different federal agencies, which work to ensure that Federal Agencies align their climate adaptation planning efforts.[8] The last progress report of the ICCATF was published in 2011.

In addition to the Adaptation Task Force, the Obama administration established "the State, Local, and Tribal Leaders Task Force on Climate Preparedness and Resilience" as part of the President's Climate Action Plan in 2013. The goal of the task force is to inform federal agencies and enable greater stakeholder engagement. Another important EO was published in November 2013. *Preparing the United States for the Impacts of Climate Change* (EO 13653) refers to vulnerability to climate change related risks in the context of natural or built systems, economic sectors, communities and natural resources.

Table 5.2 Core Executive Orders related to climate adaptation

Date	Executive Order	Administration
1993	EO 12898 "Federal Actions to Address Environmental Justice in Minority Populations and Low-income Populations"	Clinton Administration
2009	13514, "Federal Leadership in Environmental, Energy, and Economic Performance" → Creation of an Adaptation Task Force	Obama Administration
2013	EO 13653 *Preparing the United States for the Impacts of Climate Change*	
2021	EO 13990 *Protecting Public Health and the Environment and Restoring Science to Tackle the Climate Crisis*	Biden Administration
	EO 14008 "Tackling the Climate Crisis at Home and Abroad"	
	EO 14013 "Rebuilding and Enhancing Programs to Resettle Refugees and Planning for the Impact of Climate Change on Migration"	
	EO 14027 "Establishment of Climate Change Support Office"	
	EO 14030 "Climate-Related Financial Risk"	

Source: the author

In the absence of congressional support, EOs became the main instrument to promulgate climate change policy. The Trump administration has retracted several Obama-era EOs and tried to overturn policies across eight major policy domains.[9] Most of these attempts are directed against environmental policies. The adaptation harming policy shifts under the Trump administration have made visible the vulnerability of the U.S. political system (Moser et al. 2017). The already limited number of adaptation-relevant policies at the federal level came to an almost complete standstill.

The Biden administration launched two EO's in the immediate aftermath of the inauguration. These are EO 13990 "Protecting Public Health and the Environment and Restoring Science to Tackle the Climate Crisis," EO 14008 "Tackling the Climate Crisis at Home and Abroad," EO 14013 "Rebuilding and Enhancing Programs to Resettle Refugees and Planning for the Impact of Climate Change on Migration," EO 14027 "Establishment of Climate Change Support Office" and EO 14030 "Climate-Related Financial Risk" focusing on health aspects related to climate change and restoring science.

Coastal-focused adaptation efforts

Aside from vanishing departmental obligations through EOs, 18 states have issued climate adaptation plans in the United States: Alaska (2010), California (2009, 2014, 2018), Colorado (2011, 2018), Connecticut (2013), Washington D.C.

(2016), Delaware (2015), Florida (2008), Maine (2010), Maryland (2008 and 2011), Massachusetts (2011), New Hampshire (2009), New York (2010), North Carolina (2016), Oregon (2010), Pennsylvania (2011), Rhode Island (2018), Virginia (2008), Washington (2010). Most of these states are located along either the East or West Coast (see Figure 4.2). Adaptation strategies are either formulated within holistic Climate Action Plans (CAPs), as designated state-led Climate Adaptation Strategies, broader Preparedness Plans or resilience strategies.

When comparing macro policy frameworks, only punctual planning efforts exist in the United States, most of which are currently stalled at the federal level. More action has been initiated at state and local levels. In contrast, the Chinese central government has launched a series of recent efforts, which also put forward instructions for implementation. The central government has laid out the groundwork for future engagement with the issue, thereby putting climate adaptation officially on the national agenda. Under the Obama administration (2009–2017), several initiatives were launched during Obama's first term, whereas the Xi administration began to initiate adaptation efforts post-2013 (see Table 4.2).

Planning instruments related to vulnerability governance

Vulnerability assessments

When preparing strategies to adapt human systems to climate impacts, population risk assessments are another tool for assessing CCA options. In China, risk-based vulnerability research remains weak (Hunt and Watkiss 2011; Wu et al. 2012). The 12th Five-Year Plan (FYP) (2012–2015) for the first time put a formally equal emphasis on mitigation and adaptation efforts by emphasizing the risks different sectors are facing. However, besides positive trends and a heightened understanding of climate risk, China lacks experience with regard to vulnerability assessments and policy implementation at the local level (Nadin et al. 2016).

The United States, on the other hand, is well experienced in developing risk-based vulnerability assessments as well as sectoral adaptation plans. Nonetheless, there seems to be an "obvious lack of adaptation implementation" as well (Hansen et al. 2013). A further imbalance can be detected in existing adaptation planning. Globally, coastal regions are known to be more advanced than other inland parts of the country. In the United States, this can be attributed to a range of factors such as direct affectedness through sea-level rise (SLR), the propensity of majority Democratic states along the coast, or a strong network of non-state actors and non-governmental (NGO) activists. China has seen more widespread planning efforts, covering the country with its different geographic and climatological exposure more equally.

The U.S. Global Change Research Act of 1990 and following periodic national climate change assessments have contributed to a greater awareness of climate change risks in the United States. There is also growing pressure in key economic

Figure 5.2 Adaptation progress in the United States (Hawaii and Alaska not included).

Source: Georgetown Climate Center (2018) (dark grey = states with a state-wide adaptation plan, black dots = municipalities with an adaptation plan)

Table 5.3 Comparison of national CCA policy measures

United States	China
• Interagency Climate Change Adaptation Task Force (ICCATF) (2009) • President's Climate Action Plan (2013) • EO 13653 "Preparing the U.S. For the Impacts of Climate Change" (2013) • Range of EOs during Biden's first year in office	• Macro-Policy Adaptation Framework NAS (2013) • 14 Provincial Adaptation Pilots (2009–2017) • 28 Municipal Pilots (2016)

Source: the author

sectors at different local and state scales (Moss et al. 2014). Up until the inauguration of President Trump, the Environmental Protection Agency (EPA) used to measure climate change impacts by region, sector and state. Since then, most of the data was taken offline and is no longer updated by the EPA. As of January 19, 2017, some of the old data is made available through static Web Snapshots and links to external databases.

Hu Jintao's call for scientific development has influenced and informed public decision-making processes in China. The key notion, which Hu Jintao began to establish after the Third Plenary session in 2003, is the "scientific development concept" (*kexue fazhan guan*), which aims to promote environmental and social concerns as opposed to mere economic developments (Fewsmith 2004). To what extent this had an influence on science-informed policymaking and political use of climate impact assessments is an item for further study. Today, climate impact assessments are common practices in both countries. People's health has also become an increasing governmental concern for Chinese leaders. However, major tensions between short-term economic development and long-term environmental protection continue to persist.

Climate-vulnerable sectors

Both countries have taken a sectoral approach to vulnerability governance (see Table 4.3). NDRC, the main corresponding government body in China has outlined six priority sectors as per National Adaptation Strategy, which are also in line with China's Third Assessment Report (TAR3) on Climate Change. Human health has only gradually been incorporated as a concern. The fourth U.S. National Climate Assessment (NCA4), published by the U.S. Global Change Research Program and mandated by Congress outlines a refined form of what the EPA considers priority sectors. Based on the priority sectors, adaptation and mitigation options are formulated. The sectoral approach is commonly found in adaptation and vulnerability studies and is also widely applied within governmental efforts when outlining related policy challenges.

The most recent U.S. national climate assessment emphasized the interdependence of sectors and their interactions, calling for a multi-sector and complex systems approach in light of cascading impacts that create complex risks (see, e.g., Clarke et al. 2018).

Table 5.4 Climate vulnerability priority sectors in China and the United States

China	United States
Agriculture	Agriculture and rural communities
Coastal zones and maritime waters and industries	Coasts
	Ecosystems, ecosystem services, and biodiversity
	Energy supply and demand
Forestry and ecological systems	Forests
	Tribes and indigenous peoples, land cover and land-use change, built environment, urban systems and cities
	Transportation
Tourism	
Public health	Human health
Water resources	Water oceans and marine resources
	Air quality
	Climate effects on U.S. international interests
	Land cover and land-use change

Sources: NRDC, NAS 2013 and TAR 2015. NCA4 2018, largely congruent with the EPA before January 2017

National Climate Assessments in the United States

The United States published three National Climate Assessments since 2000 through their U.S. Global Change Research Program (USGCRP). The USCRP was mandated through the Global Change Research Act with the aim to develop and coordinate "a comprehensive and integrated United States research program which will assist the Nation and the world to understand, assess, predict, and respond to human-induced and natural processes of global change" (USGCRP, n.d.).

The USCRP is comprised of 13 federal agencies, including the Department of Agriculture; Department of Commerce; Department of Defense; Department of Energy; Department of Health and Human Services; Department of the Interior; Department of State; Department of Transportation; Environmental Protection Agency; National Aeronautics and Space Administration; National Science Foundation, Smithsonian Institution and the US Agency for International Development. Since the second NCA was published in 2009, the USGCRP is legally mandated to conduct National Climate Assessments (NCA) every four years. In that capacity, the third NCA was published in 2014. The first of two volumes for the fourth NCA was published in November 2017 and outlines important aspects for human safety and health. The second volume is expected to be published in late 2018, and for the first time shifted from national to regional chapters with a greater focus on risk-based framing (GCC n.d.). The NCAs, particularly the later assessments, have relied on extensive public review mechanisms based on a careful selection of authors (see Table 5.5).

Table 5.5 Overview of U.S. National Climate Assessments

Date	Name	Publishing body
2000	First National Climate Assessment: Climate Change Impacts on the United States: The Potential Consequences of Climate Variability and Change	USGCRP
2009	Second National Climate Assessment: Climate Change Impacts on the United States: The Potential Consequences of Climate Variability and Change	USGCRP
2010–2014	Third National Climate Assessment (NCA)	USGCRP and extensive public input
2017–2018	Fourth National Climate Assessment (NCA4) NCA4 Vol. I (CASR) Climate Assessment Special Report (2017) NCA4 Vol. II, Climate Change Impacts, Risks, and Adaptation in the United States (late 2018)	Subcommittee on Global Change Research (SGCR), USGCRP agencies NOAA as administrative agency for NCA4, extensive federal and nonfederal authors

Source: the author

National Climate Assessments in China

In China, the Ministry of Science and Technology (MOST), together with the China Meteorological Administration and the Chinese Academy of Science (CAS) published its first National Assessment Report on Climate Change (NARCC) in 2007. The most recent, third assessment report was released in 2015. Additionally, the State Council and the NDRC began publishing annual White Papers on "China's Policies and Actions for Addressing Climate Change." Here, climate adaptation is also addressed in a sector-specific manner. Whereas the adaptation chapter of the first white paper (2008) does not mention public health as a sector, it gradually gained importance over the years. Additionally, the Ministry of Public Health published a separate "National Action Plan for Environment and Health (2007–2015)" (see Table 5.6).

Adaptation types and priorities

In the United States and in light of the lack of federal guidance, most adaptation efforts continue to be initiated by local actors as a response to intensifying climate change, predominantly in the coastal areas. In China, governmental adaptation efforts are more dispersed and centrally orchestrated.

China's focus on urban areas, food security, water and migration

In February 2016, NDRC and MoHURD went a step further in their adaptation efforts and released the "Urban Climate Adaptation Action Plan" (*chengshi shiying qihoubianhua xingdong fan'an*). The joint statement emphasizes the increasing risks

Table 5.6 Overview of China's National Climate Assessments

Date	Name	Publishing body
2006	First National Assessment Report on Climate Change (NARCC1)	China's Ministry of Science and Technology (MOST), China Meteorological Administration (CMA), Chinese Academy of Sciences (CAS)
Oct 2008	First White Paper on China's Policies and Actions for Addressing Climate Change	State Council
Nov 2009	China's Policies and Actions for Addressing Climate Change, the Progress Report 2009	National Development and Reform Commission (NDRC)
Nov 2011	Second National Assessment Report on Climate Change (NARCC2)	MOST, CMA, CAS
Nov 2012	Fifth White Paper on China's Policies and Actions for Addressing Climate Change	NDRC
Jan 2015	China National Assessment Report on Risk Management and Adaptation of Climate Extremes and Disasters	12 government departments
Nov 2015	Third National Assessment Report on Climate Change (NARCC3)	MOST in collaboration with the CMA, CAS and Chinese Academy of Engineering (CAE)
Dec 2017	China's First Biennial Update Report on Climate Change	NDRC
Oct 2017	Tenth White Paper on China's Policies and Actions for Addressing Climate Change (2017)	NDRC

Source: the author

that cities are facing in light of climate impacts and the need to improve cities' ability to adapt to climate change. The goal of the plan is to incorporate climate adaptation planning into the overall economic and social development FYPs; into the town and industrial development planning; as well as into construction standards by 2020 (MoHURD 2016).

A follow-up notice was published in February 2017 and outlines 28 urban adaptation pilots. The chosen cities and regions are expected to carry out urban climate change impact and vulnerability assessments and introduce urban adaptation action programs. The notice further states that climate change awareness and capacity need to be strengthened and that initiatives such as the sponge city program and construction of eco-cities are intended to help cities adapt to climate change (NDRC 2017). The issued statement further emphasizes that China's overall urban adaptation efforts are still in their infancy (*zai qibu tan*). The urban adaptation plan is placed in the context of promoting and implementing the construction of an ecological civilization. The urban pilots are spread throughout provinces and autonomous regions across the country.

In China, most scientific concerns have focused on adaptation as linked to food security and environmental migration as well as the economic impacts of climate change on agriculture (Zhang et al. 2017). Another important silo is the water sector. Altogether, China follows a largely top-down initiated "developmental type" (*fazhan xing*) and "incremental type" (*zengliang xing*) adaptation approach (He et al. 2017).

Engineering and technological adaptation measures with recent shifts

The most popular types of adaptation evolve around engineering and technological measures (Pan et al. 2011; He 2018). Soft adaptation measures "such as enhancing knowledge, providing information, clarifying institutional responsibilities and developing legislation" are much more rarely used (He 2018: 96). Though human health has long been a neglected sector, the prioritization of health in China's National Adaptation Strategy is considered an "extremely positive move" (Ma et al. 2015). Yet, long-term comprehensive studies are still lacking, together with an awareness about the importance of social support and assistance systems, particularly for migrant populations (Ma et al. 2015). China's First Biennial Update Report on Climate Change which was submitted to the UNFCCC in 2016 mentions the continued gaps in China's public education and health infrastructure.

As part of floodwater management and water governance, China has shifted between hard engineering solutions and soft nature-based adaptation solutions. Meng and Dabrowski (2016), for example, show how the deeply rooted top-down planning culture has affected how different departments and local governments deal with urban adaptation to flooding in the background of a historical Fengshui-based philosophy that emphasizes the coexistence with water.

Eco-civilization perspective

Since 2003, China has actively pushed forward a socialist eco-civilization perspective. In 2007, at the 17th National Congress of the Communist Party, the former president Hu Jintao for the first time officially proclaimed the concept of an ecological civilization (*shengtai wenming*), thereby declaring ecology and sustainable development as major items on the agenda of the Chinese Communist Party (Sze 2015). In the following years, 14 eco-provinces (*shengtai sheng*), 150 eco-cities (*shengtai shi/cheng*) and 11 national eco-counties (*shengtai xian*) were rolled out. President Xi Jinping furthered the discourse when he proposed the Sponge City Scheme in 2013, which is similar to blue-green infrastructure city projects in the UK and sustainable urban drainage systems in the United States. The sponge city program aims to provide sustainable flood risk management through the enhancement of ecosystem-based flood barriers (see Chapter 6).

Ecological migration

In China, studies on climate-induced migration have developed in a variety of contexts. With the country being a largely agrarian society historically, research on the relationship of resettlement as an adaptation strategy for drought-affected communities has been gaining momentum as part of the NAS. It was exemplified in the most recent FYP for drought-prone northwestern Ningxia province (Zheng et al. 2013, 2016). Ecological resettlement programs existed throughout much of Ningxia's ancient history and as of 1983 fulfilled the dual purpose of poverty reduction and ecological restoration (Shu 2016). Few studies critically examine the relationship between resettlement and vulnerability and question relocation as an effective means of adaptation and poverty alleviation (e.g., Rogers and Xue 2015; Dubé 2015). Other studies have examined resettlement in light of China's rapid construction of hydrological power dams, particularly in the context of the Three Georges Dam (e.g., Jackson and Sleigh 2000) and urban relocations (Rogers and Xue 2015). Ecological migration has been a forefront research interest in China since the 1970s and has only recently been considered an adaptation measure with those matters being framed around the issues of food security; energy security and sustainable development. Central governmental efforts existed since the 1980s and have relocated approximately one million people so far (Wong 2016).

Lack of clear focus in the United States

Due to the lack of a macro-framework and planning from the federal level, it is hard to make a qualified statement about adaptation priorities in the United States. A recent critical assessment of the climate adaptation field detects "a lack of clear regional, sectoral, and cross-cutting priorities to drive focus" (Moser et al. 2017: 67). Most adaptation efforts are driven by climate impacts but are rather responsive and symptomatic than addressing the root causes (Moser et al. 2017). Adaptation practitioners tend to be mainly located in coastal areas. This was reflected upon by one adaptation policy advisor working for a subnational organization in California:

> When you talk about adaptation planning, [the] work usually is about impacts of climate change. There are not many cities that have a climate adaptation plan, having a plan is new itself // [It is] not widely used. The majority of work that has been done on climate adaptation, [is] mainly on sea-level rise. However, [the] coastline is usually mostly wealthy people of the country.
>
> (I-04G, respondent a: 7)

Besides practitioners' calls for sectoral investments into water resources, urban systems, infrastructure and coastal zones, the lack of a clear and all-encompassing vision is criticized by adaptation experts. More systemic changes are called for in light of the pervasive lack of urgency of the adaptation field (Moser et al. 2017). Advisors to public governmental institutions generally note the rising levels of knowledge and awareness about vulnerable populations, as mirrored in this statement:

> [Regarding the] vulnerability question: [in the] last 10 years there is a much keener appreciation of climate resilience and vulnerable populations.
>
> (I-02: 6)

The sustained theoretical emphasis on supporting equity, social justice and cohesion, systems thinking and building adaptive, transformative capacities is met by very different political practices (Moser et al. 2017). This research attempts to explain why these deficiencies persist, especially at the social level by looking at lock-ins related to vulnerability, adaptation and political institutions.

Climate-induced migration as new policy effort

In the United States, indigenous communities in northern Alaska used to be at the forefront of debates on climate-induced migration (Bronen 2013). The Isle de Jean Charles in southern Louisiana is another recent focal point of federal interest based on drastic coastal erosion and the loss of 98 percent of land since the 1950s. Community resettlement efforts have been on their way and with 48 million US dollars, it has become the most expensive relocation program initiated by public agencies. Most relocation programs are quite new. Current studies examine what lessons can be learned from African and Asian countries regarding the resettlement of rural populations in light of environmental and climate-induced change (e.g., Arnall 2019). Aside from fairly new policy efforts, domestic climate-induced migration studies are likewise recent, focusing on major inland population shifts in light of future sea-level rise projections and increased coastal vulnerability due to storm surges (e.g., Hauer 2017). The distribution of much of the U.S. population is expected to reshape tremendously in light of forced relocations, adding further pressure on already urbanized metro areas located inland such as Atlanta, Austin or Denver. These regions are known for their geographical and urban exposure to extreme heat waves and the heat-island effect (see, e.g., Stone et al. 2013), which will further exacerbate with intra-urban migration

Actors and recent developments

In addition to the actors and agencies outlined previously, this section reviews recent developments and actor constellations of interest for the adaptation field.

Growing political significance of environmental actors in China

The 13th National People's Congress (2018) experienced a major government reshuffle regarding the formation of new ministries. The Ministry of Ecology and Environment (MEE) was created, and is expected to monitor, implement and penalize polluting industrial sectors. China saw further overhauls with the creation of a new Ministry of Emergency Management (MEM). The new MEM is expected to improve emergency response mechanisms, aside from stepping up Chinese efforts in disaster preparedness. In 2017, the seven former departments of the Ministry of the Environment launched a "Green Shield mission" (*lü dun*), intended to inspect the industrial pollution of national nature reserves. As a result, 2,460 enterprises were shut down and declared illegal in 2017 (MEE 2018). MEE is said to follow up on this matter through the deployment of the "Green Shield 2018" (*lü dun* 2018). These developments are considered institutional innovations in China's environmental protection strategy. In this context, the revival of the "ecological red line" was promulgated. The growing political significance of environmental ministries and governmental bodies was also noted by several interview partners, who described how their status changed in terms of financial resources and with regard to considering them as a potentially new stepping-stone for political careers.

Importance of climate change leading groups

Based on the central government's model, which established a National Leading Committee on Climate Change (NLCCC) in 2007, local governments at different levels (provincial, prefecture, municipal, autonomous and county levels) started forming regional leading groups. The NLCCC itself is an offspring of the National Climate Change Coordinating Leading Small Group (NCCCLSG), which was originally established in 1990 to coordinate China's early climate change research and international efforts under the IPCC and UNFCCC. In the early 1990s, the NCCCLSG was considered a low-ranking government body before it was moved to the National Development and Reform Commission in 1998 (Held et al. 2011).

With the approval of the Kyoto Protocol in 2002, NCCCLSG was replaced through the establishment of the NLCCC in 2007, which is sometimes also referred to as the National Leading Group for Addressing Climate Change (NLGACC) (*guojia yingdui qihou bianhua lingdao xiaozu*). The upgrade included the integration of 20 ministries. It was spearheaded by the premier, marking a further improvement of its political power and status. Since the early 2000s, climate change governance structures changed considerably in a short period of time with comprehensive and rapid policy outputs (Held et al. 2011; Gilley 2012). Though aggressively pursuing a renewable energy policy, there has been growing criticism of the gap between targets and emission levels caused by weak enforcement and lack of actual commitment to emission reductions (e.g., Ong 2012).

Non-state and local actors

The environmental sector has been appraised as a field that enables greater involvement of non-governmental actors in China (Perry 2010). However, the western understanding is considered inadequate to describe and comprehend Chinese conceptions of "non-state actors" (Guttman et al. 2018). State-society and central-local political relations have received considerable attention from researchers from different disciplines (e.g., Gundumella 2016; Kostka and Nahm 2017). Despite recent hardening and recentralizing government patterns, the environmental governance landscape has further diversified regarding actor constellations and responsible administrations. At the same time, the 2017 launch of China's new NGO law has influenced the engagement of environmental NGOs (also see Kostka and Zhang 2018). The links to the adaptation policy field remain unclear and need further research.

Liu et al. (2017) detect for main features NGOs play in China's climate change governance: (1) they rely on governmental partnerships and have only limited political space, (2) due to practical problems such as funding they can only develop inadequate professional capacity, (3) they continue to rely on international finance but experience growing domestic support and (4) their public advocacy receives only little social recognition. They are generally more subordinate to dominant governance processes with activist being largely absent due to political sensitivity (Liu et al. 2017). Their political legitimacy remains uncertain in light of the recent NGO law (Liu et al. 2017). The concept of "environmental authoritarianism" characterized by little or no involvement of social actors and a nonparticipatory decision-making approach, defined by the presence of scientific and cadre elites remains a valid framework to describe governance processes related to non-state actors (e.g., Beeson 2010, 2016; He 2018).

One example, which needs further comparative analysis, is the role of public inputs as part of the National Climate Assessments (NCAs). No mainstreamed interest or approach exists that guarantees the input of the broader public in adaptation-related planning processes in China. In contrast to the United States, where NCAs are increasingly built on public input, non-state actors have only played fragmented roles as part of the stakeholder engagement processes in the adaptation pilots in China (Nadin et al. 2016). However, the lack of capacity to participate in environmental decision-making processes based on socioeconomic status and ethnicity has been criticized in China and the United States (Xie 2011; Freudenberg et al. 2011; Imperiale and Pian 2013).

Although local administration remains a contested environmental governance arena in light of the central government behaving like a "regulatory state" from the top (Li et al. 2011), cities nonetheless have played an increasing role in environmental problem-solving in China. In the context of rapid urbanization and environmental degradation, cities have become key actors since they emit the majority of the world's greenhouse gases (Ohshita et al. 2017).

Adaptation-specific actors: MoHURD and NDRC, CAS

It is commonly pointed out that actors are manifold in the governance of climate adaptation. They range from public actors to private individuals and households, firms and industries, non-governmental actors and community-based organizations. In both countries, actors at different levels of government are involved in the field, however in China adaptation policy is still mainly initiated by higher level government.[10]

The Ministry of Housing and Urban-Rural Development (MoHURD) and NDRC play key roles in assessing candidates for urban adaptation pilots. NDRC used to be the lead agency that had great convening power over secondary agencies, expert organizations such as the CAAS and industry (also see Hart et al. 2015: 33). In light of its continued implementation deficit in relation to environmental regulations and growing health concerns, the new Ministry of the Environment and Ecology (MEE) was created in March 2018. MEE is expected to monitor and implement the environmental laws set at the top and investigate and punish pollution at the provincial and local levels. In the case of the ACCC project, the MEE has functionally displaced the NDRC and begun to play an active role in the governance of climate change adaptation.

Condescending environmental institutions in the United States

In contrast to China's financially backed commitment of environmental ministries, the Environmental Protection Agency (EPA) had to deal with budget cuts of about 23 percent (2.5 billion USD) for the fiscal year 2019. The budget drawbacks are expected to lead to the elimination of several programs aside from shrinking the agency in general size and ambition, further outsourcing the work of environmental protection to individual states (Dennis 2018). Nonprofit organizations such as the Environmental Defense Fund (EDF) and news outlets such as CNN pointed to the health implications those budget cuts will likely result in, risking a potential public health emergency by targeting the most vulnerable communities. However, the EPA has historically been a low-funded actor, hitting an all-time low in 40 years (EDF 2020). The budget cuts are only one example of recent controversies surrounding the governmental agency. Climate skeptic Scott Pruitt, an outspoken advocate against the EPA agenda became its fourteenth Administrator and resigned after 18 months in office due to corruption scandals. He was subject to at least 13 federal investigations. Former coal lobbyist Andrew Wheeler became the acting chief thereafter and weakened environmental policy and climate change policies such as Obama's Clean Power Plan and rolled back over 100 environmental regulations. The challenge of the Biden-Harris Administration (2021–2025) is to restore the EPA and reinstate key climate policies and environmental protections, which can take up several years.

Adaptation specific actors: subnational actors, non-state actors and universities

In the United States, adaptation researchers have emphasized either leadership taken by state and regional actors (e.g., Chatrchyan and Doughman 2008) or the implementation of adaptation action at the community level (e.g., Vogel et al. 2016). Within states, which are run by climate-skeptic governors, nature conservation agencies have played a core role through initiating climate adaptation efforts based on nature conservation. However, climate-skeptic states in the Southeast and Midwest have been particularly struggling on the matter due to the politicization of the issue and an altogether absent political leadership.

Private adaptation efforts have become particularly common in light of the largely lacking federal adaptation support and soon to be outdated departmental efforts. Against the background of scientific uncertainty regarding the extent and type of climate impacts as well as suitable interventions, cities and universities have become particularly strong collaborators in the field of climate adaptation (Moser and Boykoff 2013).

Actors' understanding of adaptation and human vulnerability

The United States EPA acknowledges that climate change will affect certain population groups more than others. According to the EPA, the vulnerability of populations is determined by geographic location and their ability to cope. This perspective underlines a spatial dimension of vulnerability whereby vulnerability is understood in geographical terms. The EPA and USGCRP analyze to what extent the southern and western regions of the United States are "most sensitive to coastal storms, drought, air pollution and heat waves." On the one hand, southeastern cities such as Atlanta, Miami, and New Orleans are at high or very high risk of sea-level rise. Cities in the southwestern part of the country, such as Albuquerque, Phoenix, Las Vegas, Denver, San Diego and Los Angeles are particularly vulnerable to average temperature increases in addition to more frequent; intense and longer drought periods and therewith connected stresses regarding water availability (EPA 2016).

Apart from the geographic dimension of vulnerability, the EPA outlines how the ability to cope corresponds with different population groups. People in poverty, older adults and young children are among the population groups least able to cope with projected changes in climate (EPA 2016). The EPA particularly refers to native Americans and urban populations, whose distinct sensitivities to climate impacts may not just interfere with their cultural traditions but will also affect human comfort and health of "city dwellers" due to increased energy costs and water scarcity (EPA 2016). Overall, EPA seems to engage in a "starting-point approach" of vulnerability by understanding climate change as a potential exacerbator of pre-existent vulnerabilities (see Chapter 2).

The corresponding research body in China, the MEE recently emphasized that vulnerable groups such as children and older people should be prioritized in the context of improving indoor air quality. Vulnerable populations have become an increasing issue regarding air pollution and related calls for increasing public health services. However, in the climate adaptation context, only limited efforts exist using the socio-demographic and economic predisposition of different parts of the population to inform adaptation planning processes. Overall, uneven societal preconditions and human exposure to climate impacts have been drastically neglected.

The NAS of 2013 specifically refers to China as a developing country (*fazhan zhongguo jia*) that is ecologically vulnerable. Specifically, the document refers to the vulnerability of the ecological environment (*shengtai huanjing zhengti cuiruo*) or areas that are particularly vulnerable (*shengtai cuiruo diqu*). The plan lists different sectors that are to be addressed: agriculture, forestry, water resources, ocean, health, housing as well as city and countryside construction. The vulnerability understanding of the document refers to ecological environments and different sectors rather than people. Ecological migration (*shengtai yimin*) is specifically mentioned as a poverty alleviation measure. Though the Department of Civil Affairs is considered to be a key actor for local vulnerable populations, it has played no traceable role in environmental protection and climate adaptation governance efforts.

Besides emphasizing the country's differential vulnerability in an international geopolitical context, China domestically refers to geographical vulnerability, sector-specific vulnerability as well as ecological vulnerability in a differential sense. Published in December 2016, the "First Biennial Update Report on Climate Change of the People's Republic of China" considers the Haihe-Luanhe River basin in the north of China to be the most vulnerable region to climate change, followed by the Huaihe River basin and the Yellow River basin, both located in the urbanized parts of eastern China. Besides flood-prone river basins, China's climate impact assessment of 2007 emphasized the vulnerability of ecological and water systems in addition to stressing its strong vulnerability to the impacts of sea-level rise. The latest Biennial Update Report calls for strengthened adaptation action in the water, forest, grassland and land sectors (State Council 2016). Although the report mentions human vulnerability to climate change only as a sideline, it states that it is the responsibility of local governments to detect "sensitive and vulnerable communities to climate change" and strengthen adaptation actions in the urban sector (State Council 2016: 122).

Environmental laws and regulations

Both countries are advanced when it comes to their legal regulations on environmental matters. Compared to developed countries, China's environmental legislation was initiated late (1979) but has considerably advanced since then.[11] This is ten years after the United States published the National Environmental Policy Act (NEPA) in 1969. Compared to China, the "golden years of environmental lawmaking" happened between 1964 and 1980 in the United States, with bipartisan majorities in Congress legislating 22 major environmental acts (Klyza and Sousa 2010) (see Table 4.6 for an overview of defining environmental moments and regulations).

In 1970, the United States government founded the EPA, which until today remains the primary body addressing both environmental- and health-related concerns but only has little regulatory power. Although the United States was a driving force of modern environmental policy in the 1960s, American environmental policy has come to a still-stand lately with only little regulations passing Congress (Press 2015). The United States Congress failed to pass federal climate legislation multiple times, due to the widespread Republican opposition to acknowledging climate change is occurring. Although the United States is considered to follow a law-centered process, the last two decades stand out in stalled efforts in the environmental arena.[12]

Soft adaptation instruments

Most political adaptation instruments continue to be soft in nature. Due to the lack of legislation, administrative decision-makers have taken central roles in responding to adaptation needs in the United States. One of the major future challenges of the United States is how to preserve executive efforts in legislative authority to ensure their continuity (He 2018).

China too has focused on the policy as a primary pathway for addressing adaptation challenges (He 2018). Climate adaptation still ranks "low on the legislative

Table 5.7 Environmental consciousness, defining moments and regulations in China and the United States

Period	China	United States	Defining environmental moments & regulations
1960s	No consciousness	Developed rapidly	1962: Silent Spring (Rachel Carson) 1969: Santa Barbara Oil Spill
Early 1970s	Awareness	"Peaked"	1970: First Official Earth Day 1972: Clean Water Act
Late 1970s	Development	Declined but better than mid-sixties "Backlash"	1976: Resource Conservation and Recovery Act 1979: Environmental Protection Law
1980s	Keep pace with international situation		Environmental Justice Movements grow 1987: Air Pollution Prevention and Control Law
1990s	The environment becomes an important topic	"Boost" and higher than early seventies	1990: Clean Air Act 1992: Earth in the Balance (Al Gore book) 1997: Energy Conservation Law Environmental protests rise in China
2000s	Flourishing	Structural reflection after holding new consciousness	2000: Air Pollution and Control law 2004: *The Death of Environmentalism* (Shellenberger and Nordhaus book) 2005: The River Runs Black (Elizabeth Economy book) 2006: An Inconvenient Truth (Al Gore documentary)
Early 2010s	Seeking responsible leadership role	Widening gap between domestic political opposition and active international engagement	Environmental protests in China grow 2012: Important Revision of Env. Law 2014: Flint water crisis declared 2015: *Under the Dome* Documentary Since 2014: *Years of Living Dangerously* television series (Arnold Schwarzenegger) 2015: Air Pollution Control and Prevention Law "2.5 pm" movement develops in China
Late 2010s	Unbroken spirit	Drastic backlash, shift of political engagement to subnational and non-state levels	2016: Paris Agreement 2017: China launches six carbon trading pilots 2017: Different movements across the United States: March of Science, Earth Day, "We Are Still In"

Source: the author, adjusted based on Zhong and Shi (2010: 99ff.)

hierarchy" and lacks legally binding force (Zhou et al. 2018). Because of the nature of China's centrally mandated adaptation efforts and its largely top-down planning nature, it is difficult to examine whether planned efforts were actually implemented at the local level, as no specialized legislation exists yet (Ng and Ren 2017). At the same time, existent policies tend to lack clearly defined targets and specific guidance (He 2018).[13]

Summary

The foregone chapter found that climate adaptation policy is a relatively new research effort and concept in China and presents a very mixed picture for the United States. With few exceptions, climate adaptation policy largely remains in the planning stage in both countries. China stands out for its lack of local adaptation action outside the designated pilot areas. Here, most formal efforts are initiated in a top-down manner, whereas "all adaptation is local" appears as the dominant paradigm for the United States.

While China has made considerable improvements in the adaptation policy field since the launch of its macro-policy framework (the National Adaptation Strategy) in 2013, federal and departmental efforts in the United States were stagnating in light of the political administration under Trump. Here, governmental efforts occur mainly at the level of local government, with an ever-strong engagement of non-state actors. In China, non-state actors such as Oxfam and local NGOs do play a role in adaptation governance but are not as dominant as they are in the United States. Climate change still constitutes a relatively new topic for Chinese NGOs and political space is more restricted.

Certain policy sectors such as agriculture, air pollution and water have received more attention in China. Aside from traditional policy sectors such as water and agriculture, human health, indigenous populations and community-based adaptation are more popular policy concerns in the United States than in China. Coastal adaptation is a sector that is more progressed in both countries, but multi-sectoral adaptation efforts appear to be largely absent.

Notes

1 In this context "adaptation knowledge" refers to the formalized understanding of adaptation as a distinctive policy field and does not account for different forms of knowledge, such as indigenous and/or personal/experience-based knowledge.

2 For a comprehensive account of recent adaptation efforts across the United States, see Moser et al. (2017) "Rising to the Challenge together, a review and critical assessment of the state of US Climate Adaptation Field".

3 For more information on the last project phase report, please find Huo et al. "Adapting to Climate Change in China," available at: www.sayersandpartners.co.uk/uploads/6/2/0/9/6209349/2019_hou_li_et_al_9-30_accc_final_report_v.6.pdf. For more information on the earlier phases of ACCCI and ACCCII, please find an introductory presentation on "China: National Adaptation Programs and Strategies", available at: https://assets.publishing.service.gov.uk/media/57a09ddee5274a31e0001abe/National-Adaptation-Programs-and-Strategies.pdf, last accessed August 31, 2019, and

a brief summary "Adapting to Climate Change in China (ACCC II) Project Introduction", available at: http://documents.climsystems.com.s3.amazonaws.com/IGCI/RIDS/Adapting%20to%20Climate%20Change%20in%20China%20introduction.pdf, last accessed August 31, 2019.

4 See Toly Rinberg et al. "How Climate Change Web Content is Being Censored Under the Trump Administration," published January 2018 by Environmental Data and Government Initiative, available at: https://envirodatagov.org/wp-content/uploads/2018/01/Part-3-Changing-the-Digital-Climate.pdf, last accessed October 30, 2018.

5 For a more differentiated view on the different forms of American climate skepticism, please see "Global Warming's Six Americas" published by the Yale Climate Program on Climate Change Communication, available at: https://climatecommunication.yale.edu/about/projects/global-warmings-six-americas/, last accessed December 8, 2019.

6 See "About Skeptical Science," Skeptical Science, last modified 2018, www.skepticalscience.com/about.shtml.

7 Although this may not be in line with the most recent developments across Europe.

8 For more detailed information, see "Climate Change Resilience", Council on Environmental Quality, published at the Obama White House Archive. Available at: https://obamawhitehouse.archives.gov/administration/eop/ceq/initiatives/resilience, last accessed August 31, 2019.

9 For an earlier overview of policy domains and executive action, see "How Trump is rolling back Obama's legacy", published by *The Washington Post*, January 20, 2018, last accessed January 18, 2020, www.washingtonpost.com/graphics/politics/trump-rolling-back-obama-rules/?utm_term=.24287746ae96.

10 This observation does not mean to disregard the multifold efforts of adjusting to a changing climate that have been initiated for long periods of time in Chinese society, by a wide range of actors and outside the realm of the formalized climate change adaptation (CCA) concept. This observation only relates to formal governmental and CCA intentional strategies as they relate to the emerging policy field.

11 Zhilin Mu et al., for instance, provide a great Summary of China's Environmental Laws "Environmental Legislation in China: Achievements, Challenges and Trends," *Sustainability* 6 (2014): 8967–8979. Doi: 10.3390/su6128967.

     Another great overview is provided by Qin Tianbao and Zhang Meng, "Development of China's Environmental Legislation," In Routledge *Handbook of Environmental Policy in China*, published by Eva Sternfeld, 17–31, New York: Routledge, 2017.

12 Due to scope, see the following authors for an overview of environmental and climate change governance across the United States:
     (1) Barry G. Rabe, *Statehouse and Greenhouse: The Emerging Politics of American Climate Change Policy* (Brookings Institution Press, 2004).
     (2) Henrik Selin and Stacy D. Van Deveer (eds.), *Changing Climates in North American Politics: Institutions, Policymaking, and Multilevel Governance (American and Comparative Environmental Policy)* (MIT, 2009).

13 The legal aspects and nature of climate adaptation efforts could only be touched upon briefly in this book. See He (2018) for a comparative assessment and more detailed discussion of legal and policy climate adaptation pathways in Australia, China and the United States.

References

Arnall, Alex. "Resettlement as Climate Change Adaptation: What Can Be Learned From State Led Relocation in Rural Africa and Asia?" *Climate and Development* 11, no. 3 (2019): 253–263. https://doi.org/10.1080/17565529.2018.1442799.

Bailey, Christopher J. *US Climate Change Policy. Transforming Environmental Politics and Policy.* Surrey and Burlington: Ashgate, 2015.

Båtstrand, Sondre. "More than Markets: A Comparative Study of Nine Conservative Parties on Climate Change." *Politics and Policy* 43, no. 4 (2015): 538–561. https://doi.org/10.1111/polp.12122.

Beeson, Mark. "The Coming of Environmental Authoritarianism." *Environmental Politics* 19, no. 2 (2010): 276–294. Doi: 10.1080/09644010903576918.

Beeson, Mark. "Environmental Authoritarianism and China." In *The Oxford Handbook of Environmental Political Theory*, edited by Teena Gabrielson et al. Oxford: Oxford University Press, 2016. Doi: 10.1093/oxfordhb/9780199685271.013.14.

Bierbaum, Rosina, Joel B. Smith, Arthur Lee, Maria Blair, Lynne Carter, F. Stuart Chapin III, Paul Fleming et al. "A Comprehensive Review of Climate Adaptation in The United States: More Than Before, But Less Than Needed." *Mitigation and Adaptation Strategies for Global Change* 18, no. 361 (2013): 361–406. https://doi.org/10.1007/s11027-012-9423-1.

Boda, Chad Stephen, and Anne Jerneck. "Enabling Local Adaptation to Climate Change: Towards Collective Action in Flagler Beach, Florida, USA." *Climatic Change* 157 (2019): 631–649. Doi: 10.1007/s10584-019-02611-6.

Bronen, Robin. "Climate-Induced Displacement of Alaska Native Communities." Brookings Working Paper, 2013.

Chatrchyan Allison, and Pamela Doughman. "Climate Policy in the USA: State and Regional Leadership." In *Turning Down the Heat*, edited by H. Compston and I. Bailey, 241–260. London: Palgrave Macmillan, 2008. https://doi.org/10.1057/9780230594678_14.

Chao, Qingchen, Liu Changyi, and Yuan Jiashuang. "The Evolvement of Impact and Adaptation on Climate Change and Their Implications on Climate Policies." (Original in Chinese: 气候变化影响和适应认知的演进及对气候政策的影响) *Climate Change Research* 10, no. 3 (2014): 167–174.

Clarke, Leon, L. Nichols, R. Vallario, M. Hejazi, J. Horing, A. C. Janetos, K. Mach et al. "Sector Interactions, Multiple Stressors, and Complex Systems." In *Impacts, Risks, and Adaptation in the United States: Fourth National Climate Assessment, Volume II*, edited by D. R. Reidmiller et al., 638–668. Washington, DC: U.S. Global Change Research Program, 2018. Doi: 10.7930/NCA4.2018.CH17.

Dai, Liping, Helena F. M. W. van Rijswick, Peter P. J. Driessen, and Andrea M. Keessen. "Governance of the Sponge City Programme in China with Wuhan as a Case Study." *International Journal of Water Resources Development* 34, no. 4 (2018): 578–596. Doi: 10.1080/07900627.2017.1373637.

Dennis, Brady. "Trump Budget Seeks 23 Percent Cut at EPA, Eliminating Dozens of Programs." *Washington Post*, February 12, 2018. https://www.washingtonpost.com/news/energy-environment/wp/2018/02/12/trump-budget-seeks-23-percent-cut-at-epa-would-eliminate-dozens-of-programs/, last accessed December 4, 2021.

Dubé, François N. "Population Resettlement in China a Lose-lose Scenario." *East Asia Forum*, October 2, 2015.

EDF. "Deep EPA Cuts Pt Public Health at Risk: The Imperiled Agency Protects Our Air, Water, and Health." *Environmental Defense Fund*, January 1, 2020. https://www.edf.org/deep-epa-cuts-put-public-health-risk, accessed December 5, 2021.

Environmental Protection Agency (EPA). "Climate Change Impacts on Society." *EPA*, https://19january2017snapshot.epa.gov/climate-impacts/climate-impacts-society_.html, 2016.

Fewsmith, Roger. "Promoting the Scientific Development Concept." *China Leadership Monitor* 11 (2004): 1–11.

Freudenberg, Nicholas, Manuel Pastor, and Barbara Israel. "Strengthening Community Capacity to Participate in Making Decisions to Reduce Disproportionate Environmental Exposures." *American Journal of Public Health* 101, no. 1 (2011): 123–130. Doi: 10.2105/AJPH.2011.300265.

Gilley, Bruce. "Authoritarian Environmentalism and China's Response to Climate Change." *Environmental Politics* 21, no. 2 (2012): 287–307.

Gundumella, Venkat. "Environmental Governance in China." *Theoretical Economics Letters* 6, no. 3 (2016): 583–595. Doi: 10.4236/tel.2016.63064.

Guttman, Dan, Oran Young, Yijia Jing, Barbara Bramble, Maoliang Bu, Carmen Chen, Kathinka Furst et al. "Environmental Governance in China: Interactions Between the State and Nonstate Actors." *Journal of Environmental Management* 220, no. 15 (2018): 126–135. https://doi.org/10.1016/j.jenvman.2018.04.104.

Hansen, L., R. M. Gregg, V. Arroyo, S. Ellsworth, L. Jackson, and A. Snover. "The State of Adaptation in the United States: An Overview." Report prepared for the John D. and Catherine T. MacArthur Foundation, EcoAdapt, 2013.

Hart, Craig, Jiayan Zhu, and Jiahui Ying. "Mapping China's Climate Policy Formation Process." Development Technologies International, Working Paper, 2015.

Hauer, Matthew. "Migration Induced by Sea-Level Rise Could Reshape the US Population Landscape." *Nature Climate Change* 7 (2017): 321–325.

He, Xiangbai. "Legal and Political Pathways of Climate Change Adaptation: Comparative Analysis of Adaptation Practices in the United States, Australia and China." *Transnational Environmental Law* 7, no. 2 (2018): 1–27. Doi: 10.1017/S2047102518000092.

He, Xiaojia, Xueyan Zhang, and Xin Ma. "The Development of Climate Change Adaption on Institution and Policies in China." (Original in Chinese: 中国适应气候变化制度建设与政策发展方向研究) *Climate Change Research Letters* 6, no. 1 (2017): 40–45. Doi: 10.12677/CCRL.2017.61005.

Heggelund, Gørild, and Rebecca Nadin. "Climate Change Policy and Governance." In *Routledge Handbook of Environmental Policy in China*, edited by Eva Sternfeld, 97–112. London and New York: Routledge, 2017.

Held, David, Eva-Maria Nag, and Charles Roger. "The Governance of Climate Change in China." LSE Working Paper 01/2011.

Hunt, Alistair, and Paul Watkiss. "Climate Change Impacts and Adaptation in Cities: A Review of the Literature." *Climatic Change* 104, no. 1 (2011): 13–49. https://doi.org/10.1007/s10584-010-9975-6.

Huo, Li, Roger Street, Paul Sayers et al. "Adapting to Climate Change in China: Final Project Report Phase II." 2019. Funded by the Swiss Agency for Development and Cooperation (SDC).

Imperiale, Sara, and Wang Pian. "Waste Incineration, Community Participation and Environmental Justice: A Comparative Study of China and the United States." *Vermont Journal of Environmental Law* 14 (2013): 435–463.

Jackson, Sukhan, and Adrian Sleigh. "Resettlement for China's Three Gorges Dam: Socioeconomic Impact and Institutional Tensions." *Communist and Post-Communist Studies* 33 (2000): 223–241.

Kamieniecki, Sheldon, and Michael E. Kraft. *The Oxford Handbook of U.S. Environmental Policy.* Oxford: Oxford University Press, 2013.

Khan, Mehran, and Yen-Chiang Chang. "Environmental Challenges and Current Practices in China – A Thorough Analysis." *Sustainability* 10, no. 2547 (2018): 1–20. Doi: 10.3390/su10072547.

Klyza, Christopher McGregory, and David Sousa. "Beyond Gridlock: Green Drift in American Environmental Policymaking." *Political Science Quarterly* 125 (2010): 443–463. https://doi.org/10.1002/j.1538-165X.2010.tb00681.x.

Kostka, Genia, and Jonas Nahm. "Central – Local Relations: Recentralization and Environmental Governance in China (Introduction to Special Section)." *The China Quarterly* 231 (2017): 567–582.

Kostka, Genia, and Chunman Zhang. "Tightening the Grip: Environmental Governance Under Xi Jinping." *Environmental Politics* 27, no. 5 (2018): 769–781. Doi: 10.1080/09644016.2018.1491116.

Li, Bingqin. "Governing Urban Climate Change Adaptation in China." *Environment and Urbanization* 25, no. 2 (October 2013): 413–427. Doi: 10.1177/0956247813490907.

Li, Yuwai, Bo Miao, and Graeme Lang. "The Local Environmental State in China: A Study of County-Level Cities in Suzhou." *The China Quarterly* 205 (2011): 115–132. www.jstor.org/stable/41305197.

Liu, Lei, Pu Wang, and Tong Wu. "The Role of Nongovernmental Organizations in China's Climate Change Governance: Role of NGOs." *Wiley Interdisciplinary Reviews: Climate Change* 8, no. 6 (2017): e483. Doi: 10.1002/wcc.483.

Ma, Wenjun, Lin Hualiang, Liu Tao, Xiao Jianpeng, Luo Yuan, Huang Cunrui, Liu Qiyong et al. "Human Health, Well Being and Climate Change in China." In *Climate Risk and Resilience in China*, edited by Rebecca Nadin et al., 181–210. New York: Routledge, 2015.

Meng, Meng, and Marcin Dabrowski. "The Governance of Flood Risk Planning in Guangzhou, China: Using the Past to Study the Present." 17th IPHS Conference, Delft 2016.

Mertha, Andrew C. *China's Water Warriors: Citizen Action and Policy Change*. Ithaca and London: Cornell University Press, 2008.

Ministry of Ecology and Environment (MEE). "Seven Line Departments Make Arrangements for the 'Green Shield 2018' Inspection on Nature Reserves." 2018. https://english.mee.gov.cn/News_service/Photo/201804/t20180410_434126.shtml, last accessed December 5, 2021.

Ministry of Housing and Urban-Rural Development (MoHURD). "Cities Adapt to Climate Change Action Plan" (Original: Original:城市适应气候变化 行动方案.) MoHURD 2016. www.mohurd.gov.cn/wjfb/201602/W020160224041125.doc, last accessed July 14, 2021.

Moser, Susanne, and Maxwell Boykoff. *Successful Adaptation to Climate Change: Linking Science and Practice in a Rapidly Changing World*. New York: Routledge, 2013.

Moser, Susanne, Joyce Coffee, and Aleka Seville. "Rising to the Challenge, a Review and Critical Assessment of the State of the US Climate Adaptation Field." Kresge Foundation Report, 2017.

Moser, Susanne C., and Julia A. Ekstrom. "A Framework to Diagnose Barriers to Climate Change Adaptation." *Proceedings in the National Academies of Sciences* 107, no. 51 (2010): 22026–22031. Doi: 10.1073/pnas.1007887107.

Moss, Richard H., Thomas J. Wilbanks, and Sherry B. Wright. "The State of the Art in Adaptation Science, Policy and Practice in The United States." In *Climate Change Adaptation Manual, Lessons Learned from Europe and other Industrialized Countries*, edited by Prutsch et al., 301–308. New York: Routledge, 2014.

Nadin, Rebecca, Sarah Opitz-Stapleton, and Xu Yinlong (eds.). *Climate Risk and Resilience in China*. New York: Routledge, 2016.

NDRC. "National Development and Reform Commission and Ministry of Housing and Urban-Rural Development publish Notice on Printing and Launching Pilot Work for the Construction of Urban Adaptation Cities." (Original: "国家发展改革委 住房城乡建设部 关于印发气候适应型城市建设试点工作的通知], 发改气候"), no. 343, NDRC,

2017. www.ndrc.gov.cn/zcfb/zcfbtz/201702/t20170224_839212.html, last accessed August 31, 2019.

Ng, Edward, and Chao Ren. "China's Adaptation to Climate & Urban Climatic Changes: A Critical Review." *Urban Climate* 23 (2017): 352–372. https://doi.org/10.1016/j.uclim.2017.07.006.

Ong, Lynette H. "The Apparent 'Paradox' in China's Climate Policies Weak International Commitment on Emissions Reduction and Aggressive Renewable Energy Policy." *Asian Survey* 52, no. 6 (2012): 1138–1160.

Ohshita, Stephanie, Jingjing Zhang, Li Yang, Min Hu, Nina Khanna, David Fridley et al. "China Green Low-Carbon City Index: Report on the Performance of 100+ cities (2010–2015)." In *Lawrence Berkeley National Laboratory*. China Energy Group, published online, 2017. https://eta.lbl.gov/publications/china-green-low-carbon-city-index / https://eta-publications.lbl.gov/sites/default/files/chinacityindex052017_en.pdf.

Pan, Jiahua, Zheng Yan, and Anil Markandya. "Adaptation Approaches to Climate Change in China: An Operational Framework." *Economia Agraria y Recursos Naturales, Spanish Association of Agricultural Economists* 11, no. 1 (2011): 1–14. Doi: 10.22004/ag.econ.117619.

Perry, Elizabeth. "Popular Protest: Playing by the Rules." In *China Today, China Tomorrow: Domestic Politics, Economy and Society*, edited by Joseph Fewsmith, 11–28. Lanham: Rowman and Littlefield, 2010.

Press, Daniel. *American Environmental Policy: The Failures of Compliance, Abatement and Mitigation.* Cheltenham: Edward Elgar Pub, 2015.

Preston, Benjamin L., Richard M. Westaway, and Emma J. Yuen. "Climate Adaptation Planning in Practice: An Evaluation of Adaptation Plans from Three Developed Nations." *Mitigation and Adaptation Strategies for Global Change* 16 (2011): 407–438.

Rogers, Sarah, and Tao Xue. "Resettlement and Climate Change Vulnerability: Evidence from Rural China". *Global Environmental Change* 35 (2015): 62–69.

Shu, Xihong. "Chapter 2 The History and Present Condition of Ecological Migration in Ningxia." In *Ecological Migration, Development and Transformation*, edited by P. Li and X. Wang, 21–46. Berlin and Heidelberg: Social Sciences Academic Press and Springer, 2016. Doi: 10.1007/978-3-662-47366-5_2.

Siebenhüner, Bernd, Torsten Grothmann, Dave Huitema, Angela Oels, Tim Rayner, and John Turnpenny. "Lock-ins in Climate Adaptation Governance. Conceptual and Empirical Approaches." Conference Paper 2017, unpublished.

State Council (SC). "The People's Republic of China First Biennial Update Report on Climate Change." (Original: "中华人民共和国气候变化, 第一次两年更新报告"). SC, December 2016.

Stone, Brian, Jason Vargo, Peng Liu, Yongtao Hu, and Armistead Russell. "Climate Change Adaptation Through Urban Heat Management in Atlanta, Georgia". *Environmental Science and Technology* 47 (2013): 7780 − 7786. Doi: dx.doi.org/10.1021/es304352e.

Sze, Julie. *Fantasy Islands Chinese Dreams and Ecological Fears in an Age of Climate Crisis.* Berkeley: University of California Press, 2015.

Tang, Xiao, Zhengwen Liu, and Hongtao Yi. "Mandatory Targets and Environmental Performance: An Analysis Based on Regression Discontinuity Design." *Sustainability* 8, no. 931 (2016): 1–16. Doi: 10.3390/su8090931.

Tompkins, Emma L., and Hallie Eakin. "Managing Private and Public Adaptation to Climate Change." *Global Environmental Change* 22, no. 1 (2012): 3–11. https://doi.org/10.1016/j.gloenvcha.2011.09.010.

Tu, Kevin, Yinlong Xu, and Ye Qi. "China's National Climate Change Adaptation Strategy in an International Context." *Carnegie Endowment for International Peace*, May 18, 2012.

U.S. Global Change Research Program (USGCRP). "About About USGCRP." n.d. Published online. https://www.globalchange.gov/about, last accessed December 5, 2021.

Vogel, Brennan, and Daniel Henstra. "Studying Local Climate Adaptation: A Heuristic Research Framework for Comparative Policy Analysis." *Global Environmental Change* 31 (2015): 110–120. Doi: 10.1016/j.gloenvcha.2015.01.001.

Vogel, Jason, Karen M. Carney, Joel B. Smith, Charles Herrick, Missy Stults, Megan O'Grady et al. "Climate Adaptation: The State of Practice in U.S. Communities." Kresge Foundation Report, November 2016.

Wheeler, Stephen M. "State and Municipal Climate Change Plans: The First Generation." *Journal of the American Planning Association* 74, no. 4 (2008): 481–496. https://doi.org/10.1080/01944360802377973.

Wong, Edward. "Resettling China's Ecological Migrants." *New York Times*, October 25, 2016. www.nytimes.com/interactive/2016/10/25/world/asia/china-climate-change-resettlement.html, last accessed July 15, 2021.

Wu, Shao-Hong, Tao Pan, and Shan-Feng He. "Climate Change Risk Research: A Case Study on Flood Disaster Risk in China." *Advances in Climate Change Research* 3, no. 2 (2012): 92–98. Doi: 10.3724/SP.J.1248.2012.00092.

Xie, Lei. "Environmental Justice in China's Urban Decision-Making." *Taiwan in Comparative Perspective* 3 (2011): 160–179.

Young, Oran R., Dan Guttman, Ye Qi, Kris Bachus, David Belis, Hongguang Cheng, Alvin Lin et al. "Institutionalized Governance Processes, Comparing Environmental Problem Solving in China and the United States." *Global Environmental Change* 31 (2015): 163–173. https://doi.org/10.1016/j.gloenvcha.2015.01.010.

Yue, Li, Wei Xiong, and Yanjuan Wu. "Climate Change Impacts, Vulnerability and Adaptation in China." BASIC Project, Report 2007.

Zhang, Linxiu, Renfu Luo, Hongmei Yi, and Stephen Tyler. *Climate Adaptation in Asia: Knowledge Gaps and Research Issues in China, the Full Report of the China Team.* Chinese Academy of Sciences Institute of Geographic Sciences and Natural Resources Research (IGSNRR) Centre for Chinese Agricultural Policy (CCAP). Kathmandu: Format Printing Press, 2008.

Zhang, Peng, Junjie Zhang, and Minpeng Chen. "Economic Impacts of Climate Change on Agriculture: The Importance of Additional Climatic Variables Other than Temperature and Precipitation." *Journal of Environmental Economics and Management* 83 (2017): 8–31. https://doi.org/10.1016/j.jeem.2016.12.001.

Zheng, Yan, Meng Huixin, Zhang Xiaoyu, Zhu Furong, Wang Zhanjun, Fang Shuxing, Sarah Opitz-Stapleton et al. "Ningxia." In *Climate Risk and Resilience in China*, edited by Rebecca Nadin et al., 213–262. New York: Routledge, 2016.

Zheng, Yan, Jiahua Pan, and Xiaoyu Zhang. 30. *Relocation as a Policy Response to Climate Change Vulnerability in Northern China.* Paris: ISSC and UNESCO, World Social Science Report 2013: Changing Global Environments, 2013.

Zhong, Qiu and Guoqing Shi. *Environmental Consciousness in China, Change with Social Transformation.* Sawston: Chandos Publishing, November 2020, published in an earlier manuscript, 2010.

Zhou, Hongjian, Xi Wang, Changgui Wang, Yi Yuan, Dandan Wang, Yinlong Xu, Jie Pan et al. "Adapting Against Disasters in a Changing Climate." In *A Critical Approach to Climate Change Adaptation Discourses, Policies and Practices*, edited by Silja Klepp and Libertad Chavey-Rodriguez, 63–95. London: Routledge, 2018.

6 Regional backgrounds and contextual lock-ins

This chapter provides some background information on the broader developments of the political economy in both countries and local sociopolitical factors. Due to scope, climatological and regional geographic factors could not be reflected upon in greater detail (see A.1). These contextual factors shape some of the path dependencies in which local decision-making processes are embedded. They also restrict climate adaptation options based on the way they shape vulnerability and the fact that they have developed over long periods and are very difficult, if not impossible to shape (e.g., topography). These factors came up throughout the interviews but could not be systematically assessed as part of the main study and need to be therefore considered as background information.

Reflecting upon the broader developments of the political economy and prevailing inequality trends is in so far important as social vulnerability, one of the main items under study appears to be a result of these broader trends. Instead of treating social vulnerability as a socioeconomic condition of certain populations, it is important to keep in mind the larger political economic contexts that impact local vulnerability patterns and thereby affect the governance of adaptation.

Political decision-making processes within the metropolitan areas of Atlanta and Jinhua were chosen as case studies due to their climatological, geographic and social exposure. This chapter (section "Inequality in China and the United States") starts with a brief introduction of recent inequality trends in China and the United States. The second section (section "Sociopolitical introduction and critical infrastructure lock-ins") examines the sociopolitical backgrounds and matters related to critical infrastructure of the two chosen local regions, Atlanta in Georgia and Jinhua in Zhejiang. Non-climatic factors, such as local political circumstances and cultural factors significantly shape the governance of adaptation and need to be considered when explaining related policy processes. The third part (6.3) compares the regional geographical and biophysical contexts in which climate vulnerability of Atlanta and Jinhua must be located. Characteristics of local climate, topography and impacts also shape the larger context in which decision-making occurs.

DOI: 10.4324/9781003183259-6

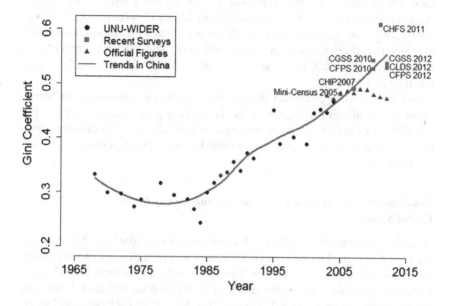

Figure 6.1 Increasing inequality in China and the United States.

Source: Xie and Zhou (2014: 6930), Copyright: PNAS, National Academy of Sciences, the graph depicts trends in Gini coefficient of family income.

Inequality in China and the United States

China and the United States head global comparisons of unequal distributions of income and wealth. Broader inequality patterns matter to understand the full picture of local social vulnerability to climate change.

Successful poverty reduction but sharp increase of inequality in China

In China, the late 1980s saw the beginning of a paralleled development characterized by drastic poverty reduction on the one hand and a sharp increase in income inequality on the other (Ravallion and Chen 2007; Li 2016; Jain-Chandra et al. 2018). Although the Chinese government managed to reduce its extreme poverty rate of 88 percent from 1981 down to 2 percent by 2013 (World Bank 2016), new patterns of hierarchy and social stratification have emerged since the introduction of economic reforms in 1978 (Li 2016).[1]

China's income inequality has not just surpassed that of the United States but also ranks among the highest in the world (Xie and Zhou 2014) (see Figure 6.1).[2] This particularly worrisome trend is reflected in the Gini coefficient, which nearly doubled between 1980 and 2012 from 0.3 to 0.55.[3] In 2010, China's income inequality as measured by a Gini coefficient of 0.5 exceeded that of the United States with 0.45 (Xie and Zhou 2014). All independently conducted

Gini coefficients uniformly surpassed official figures published by China's National Bureau of Statistics (Xie and Zhou 2014). Despite the challenges of data authenticity, accessibility, reliability and comparability most China scholars agree that there has been an indisputably large-scale rise in China's income inequality (Harvey 2005; Xie and Zhou 2014; Knight 2014; Xie 2016; Jain-Chandra et al. 2018).

In China, poverty reduction through major economic transitions has been a major source of political legitimacy for the Chinese government (see, e.g., Holbig 2006). However, this governmental perception falls short of examining worsening inequality patterns that have resulted in widespread political and social marginalization.

Steady inequality and little success in reducing poverty in the United States

In the United States, inequality has been on the rise since the late 1970s (see Figure 6.1). Today, the United States is the richest and most unequal country among industrialized economies. This development is reflected in most OECD countries, with particularly worrisome trends in North America and English-speaking nations in this group (Saez and Zucman 2014; Harles 2017; Hacker and Pierson 2018). Among affluent democratic countries, the United States has few rivals when it comes to the degree of inequality (Harles 2017). In addition to income inequality, income segregation also increased severely and is characterized by an uneven geographic distribution of poor and affluent income groups within certain areas (Reardon and Bischoff 2011). Governmental transfers and social insurance programs in the United States have the least effect on relative poverty rates (only 28 percent) compared to other industrialized economies (which reduce relative poverty by 64 percent) (Smeeding 2005). This observation suggests that the policy instruments used by the United States are largely inefficient in reducing poverty and income inequality.[4]

The United States is expected to push inequality to unseen levels and will exacerbate extreme poverty across the country. In early May 2018, the United Nations Human Rights Council published a report on extreme poverty and human rights in the United States, concluding that the country "is now moving full steam ahead to make itself even more unequal" besides already leading "the developed world in income and wealth inequality" (UN 2018: 19). The report was met by serious criticism of then U.S. ambassador to the UN, calling it a waste of resources, diverting attention away "from the world's worst human rights abusers and focusing instead on the wealthiest and freest country in the world."[5]

Sociopolitical introduction and critical infrastructure lock-ins

The following paragraphs introduce some of the sociopolitical path dependencies that stand out as challenges in both regions. Critical infrastructure is one of them. Critical infrastructure refers to the basic physical structures required for the functioning of a society and economy. Its planning and construction reflect the ideas, cultural values, habits and knowledge of a certain period (Siebenhüner et al. 2017). Critical infrastructures are planned and usually built to last for long periods and are therefore prone to lock-ins (Siebenhüner et al. 2017). Built urban infrastructure is particularly prone to lock-ins.

Atlanta

Atlanta has become one of the largest metropolitan areas of the country. Atlanta's urban population is about 4.5 million, with 5.6 million living in the metropolitan area.[6] Urbanized and densely populated areas are generally at high exposure to climate change (see Stone et al. 2013; Stone et al. 2014).

Political administration: strong degree of decentralization

Within Metro Atlanta, Atlanta is the biggest city; it is geographically located in Fulton County, which also saw the largest share in population growth in the state. Fulton County and bordering DeKalb to the Southeast are said to have experienced the worst warming across the Atlanta Metropolitan region. Fulton County is divided into South and North with Atlanta being located in the middle (see Figure 6.2). Whereas the county is considered to be a political subdivision of the state, many municipalities are formally considered subdivisions of the county. Fulton County is composed of 15 cities with Atlanta located in the North. Historically, the county has been divided along racial lines with Caucasian communities living in the northern districts 1, 2 and 3 and African American communities living in the southern districts, 4, 5 and 6. Most of Fulton County belongs to District 4, which is located at the center of the six districts, further dividing the North and the South. Aside from the socio-demographic North–South divide, the local political administration is divided regarding the representation of different constitutional interests by income level: low, middle and high.

Side by side technological development and innovation and being one of the fastest growing economies, Atlanta is declared to be the most unequal city of the United States (Foster and Lu 2018). Dealing with racial segregation is one of Atlanta's longest standing challenges. According to the latest Census in 2010, Atlanta's population was composed of predominantly Black or African American (54 percent), White (38.4 percent), Hispanic of any race (5.2 percent), Asian (3.1 percent) and Native American (0.2 percent). Though the African American population has severely declined in recent decades and been pushed to the outer parts of the city, Metro Atlanta has the second largest African American population after New York. This is so far relevant as the rest of Georgia state remains predominantly white and inequality an ethnicity-based phenomenon.

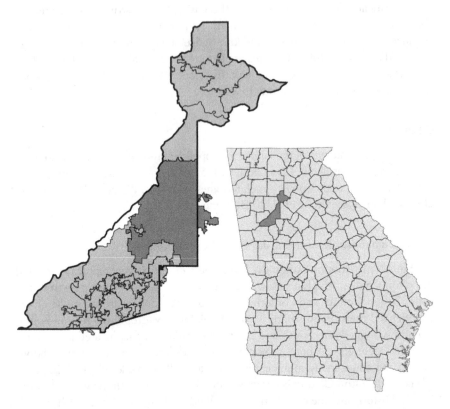

Figure 6.2 Location of Atlanta in Fulton County (left) and location of Fulton County within Georgia (right).

Source: Droidman 1231, Copyright: CC BY-SA 4.0, Wikimedia Commons

Weak critical infrastructure: transportation and housing

In 2014, Atlanta experienced a major snowstorm also known as "Snowmaggedon" or "Snowpocalypse." Here, the outfall of and dependence on critical infrastructure became particularly visible. Surprisingly, the heart of the debate did not focus on the disruption of daily lives based upon the cancellation of flights and schools, but on how relatively little snow (two inches) brought down an entire metro region because of a breakdown of the transportation system. One state actor, working for the federal environmental protection division in Georgia, Atlanta, expressed his thoughts:

> I am sure you have heard about our snow event. Looking at it from the outside, or as someone who spent thirteen hours trying to get home, and I only live nine miles from here, um, the issue was not one of poor emergency management or poor preparation by anybody, it was the fact when you have all the people in the city of Atlanta trying to leave at once – it is not going to happen.

And I do not think anyone has come up with a plan for how to deal with that. Cause we are so automobile dependent, and unfortunately, it just became an incredible traffic emergency, which is what it really was.

(I-27: 45)

The pairing of urban development, rapid population growth, infrastructural dependence and public transport deficiencies were provided as major reasons for heightened exposure. The cultural dependence on cars was more rarely reflected upon as part of the interviews. Another actor, who works at the regional planning authority and intergovernmental coordination agency, the Atlanta Regional Commission (ARC), drew direct parallels between the snowstorm in 2014 and the collapse of a major bridge and traffic junction (I-85) in 2017. The bridge collapsed right before the second phase of U.S.-based interviews had started. Naturally, the acute nature of this interstate transport hub emergency was brought forward by the interviewees and helped to uncover some of the underlying issues the city of Atlanta has been struggling to deal with. This actor, who also advises the Mayor's Office of Sustainability and Resilience on climate change and transport states:

Regional equity is a big issue. Especially in the North-South divide, north of I-20, south of I-20, I am sure you have come across that // like a real difference in equity incomes, like quality of housing and pretty much everything across that barrier.

(I-31: 120)

The racial divide of the city came up as a phenomenon deeply interwoven into the fabric of the city that shapes peoples' opportunity to escape poverty. Racial segregation did not just appear as a sociopolitically protracted phenomenon but likewise as a factor that is maintained through the way roads and transportation hubs were built. The Metropolitan Atlanta Regional Transport Authority (MARTA) is the principal public transport operator and known to be chronically underfinanced despite its core role in maintaining racial segregation. Those living in certain parts of Atlanta have no means to reach their jobs, which are usually located outside of where they live. Inaccess to public transport and not being able to afford a car are two core components, which impact the opportunities of groups with lower socioeconomic background. Two researchers on public health and environmental issues, who were also advisors to Atlanta's Climate Action Plan express:

Respondent a: But the whole city still is divided along racial lines. Biarcliff and Moreland Ave for example. It's one street. But one side is rich, the other poor. That's why there is a cut off and a new street name. People didn't want to belong to the other side of the street. MARTA is solely based on segregation.
Respondent b: Those people who do not have a car can rarely move out [of] the poverty level.

(I-20G: 36ff.)

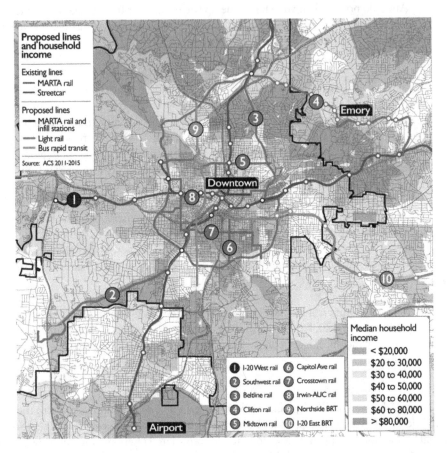

Figure 6.3 Income segregation along public transport routes and interstates in Atlanta.

Source: Freemark (2017), Copyright: MARTA Urbanization as core component of and catalyst for vulnerability

These interviewees pointed out that MARTA and public transportation lines as well as big interstates often act as a wall, which separates the affluent from the less affluent districts with lower standards of living. Affluence is likewise segregation-based. The racial composition of the city is reflected in the latest 2010 Census and the Racial Dot map designed by Dustin Cable in 2013. The majority of White-Caucasians live in the North and northeastern part of the Metro Atlanta region, whereas the Northwest and South are predominantly African American. This is also mirrored in the distribution of median household income and access to public transit, as well as interstate highways (see Figure 6.3). People in the northeastern parts of Atlanta have a median annual income of between 40,000 and 80,000 US dollars, whereas the income of people in the South is largely below 30,000 US dollars (Freemark 2017).

Skewed income distribution and access to adequate transportation are also reflected in Fulton County, which divides the city of Atlanta into two halves. These factors were also considered to be proxies for increased criminality coupled with segregation-based transportation infrastructure, as the ARC employee states:

> Problematic neighborhoods are south of the I-20 and west of the I-75. So, the center of Fulton County [District 4] and the eastern part [District 5].
>
> (I-31: 20)

In consideration of the prevalence of widespread gentrification and suburbanization of poverty in Atlanta, public transport was commonly reiterated as a necessary condition for lifting people out of poverty. Conforming with this, "Opportunity Deferred" is a report by the Partnership for Southern Equity (PSE), an Atlanta-based nonprofit working on issues related to equity. The publication investigates the impact of MARTA's transportation legacy on regional prosperity and health. Accordingly, decades of transportation and housing policies at different political levels supported residential segregation (PSE 2017). The report explains how race shaped most of Atlanta's transportation infrastructure and continues to underlie key decisions on urban renewal programs. The transportation planning failed on the basis of the continued inaccess to regional public transit networks and limited (economic) opportunities for the region's southern populations, which are mainly communities of color (PSE 2017).

The interaction of urbanization and the city's growing vulnerability to weather–climate extremes has become an increasing matter of concern and research interest. On the one hand, urbanization is a component that will lead to more drastic climate impacts based on the modification of natural landscapes to suit the human need. On the other hand, human exposure to impacts such as gradual warming or extreme heat events will increase. One academic, working on urban environmental planning and climate change in Atlanta reflects upon the irony of increased urban vulnerability:

> The city physically is very vulnerable to heat, and very vulnerable to flooding. I sometimes . . . scapetry[7] cities as machines for flooding. I mean they are really // we put a lot of resources into designing cities to move water away, but most of what cities do is collect water, by waterproofing the environment. So, cities are machines for flooding, they are machines for heatwaves, and that is probably due to their design. It is a design that has been put into place over many decades, and it is gonna take many decades to reverse it.
>
> (I-29: 61)

The modification of land surfaces through increased density and environmentally unfavorable built physical environments, as well as increased energy usage leads to the urban heat island effect. Urban heat islands are signified by significantly warmer areas than their rural counterparts and are expected to intensify heat waves in the future.[8] This in turn has a cascading effect on crop production and heat-related mortality as well as morbidity (KC et al. 2015). Stone and his colleagues (2014) attribute the projected heat exposure in cities to two major factors: 1) climate forcing and 2) the urban heat island effect. At the same time, the physically built urban environment is said to have a significant impact on floodwater runoff caused by precipitation, due to its altered physical environment, making urban areas particularly vulnerable to urban flooding (Debbage and Shepherd 2019). A state actor working for the Georgia Environmental Protection Division in Atlanta touches upon the delicate issue of people moving into floodplain areas:

> It floods all the time. And the real answer should be, people do not really need to be living there. Uhm, that is not an answer accepted by the people who live there. And for many years, there have been attempts to try and convince folks that this is not a good place to continue to have homes, every time they get flooded, people come back and rebuild. Or they have already built their homes, so they are ready to be flooded and then it is not a big deal to clean up afterwards.
>
> (I-27: 56)

At the same time, the urban center and county jurisdictions therein are supposedly more vulnerable to having to administer a larger portion of the criminal justice system. A couple of interviewees outlined the prison sector as a particular Georgia challenge and sensitive population groups therein due to the large number of prisons throughout the state and their dependence on being let out in case of an emergency (I-23, I-27, I-30, I-33). Georgia is heading national statistics regarding the number of incarcerated people. This population group, however, is a politically sensitive group when it comes to climate vulnerability (see next chapter). One elected government official critically reflected on the underlying issue of racism:

> Well, in the United States, especially in the urban core cities, we are faced with the need to address criminal justice reform and to truly shift ourselves out of, what I think still many people would agree, grew from the vestiges of slavery and how policing came about in this country.
>
> (I-33: 12)

These interviewees also point out racism as a long-standing path-dependency that until today shapes policymaking processes. The political attention the criminal justice system is receiving based on segregated forms of policymaking was oftentimes considered inadequate. Another temporally sensitive population group is those residing in nursing homes and hospitals, as those people depend on special care.

Heat waves are characterized by higher temperatures and humidity, which are associated with increased mortality due to heat-related illnesses such as heat exhaustion, heatstroke or aggravated cardiovascular and pulmonary conditions (Chen et al. 2017). Areas in rural Georgia are more exposed to hazardous heat events with less access to health care and a higher percentage of single-person households

(Manangan et al. 2014). Aside from lack of rural access to adequate healthcare, Atlanta is characterized by its "food deserts," areas where there is an absence of nutritious and fresh food within a one-mile radius.[9] Food deserts coincide with a higher amount of people living below the poverty line. The geographic concentration of limited access to healthy food has also been outlined as a major concern:

> Access to food is this big public health issue. In the West it's only food marts // [it is] all pre-packaged food. A lot of people don't feel empowered to change their situations.
>
> (I-20G, respondent b:39)

Food deserts critically coincide with the state's high rates of obesity and chronic diseases. Concurrently, Metro Atlanta and North Georgia are considered food insecure, with 1 in 7.5 people and 1 in 4 children across the state lacking consistent access to adequate food.[10] Within the Metro Atlanta region food insecurity rates are particularly high in Fulton, Dekalb and Clayton counties (Shannon et al. 2014). Projections to 2020 estimate an increase in suburban food insecurity due to poverty rates and limited or no public transport. At the same time, urban cores are considered to be more exposed, as an Atlanta-based academic and policy advisor outlines the geographic specificity of heat exposure:

> When we talk about vulnerability, there are two dimensions to that. . . . One is going to be where is your exposure to extreme heat likely to be greatest, and that would certainly be in the most urbanized parts of the regions, that would be Fulton and Dekalb County, mainly Fulton County, and then there is sensitivity of the population. So, and that is also gonna fall into Fulton and Dekalb counties, certainly the urban core is most vulnerable due to greatest exposure and the greatest sensitivity.
>
> (I-29: 3)

The greatest sensitivity was defined by the lack of access to adequate healthcare, as well as healthy food, making the vulnerability of urban institutions more apparent. Water scarcity was another major concern. Interviewees connected rapid population growth with challenges related to limited water resources, as expressed by a Georgia state actor:

> So, when we have periods of droughts, it becomes a serious concern about the availability of water, as Atlanta continues to grow, that is the other problem. Atlanta unlike of other cities, there are no geographic boundaries to stop Atlanta from continuing to expand. . . . When I first came to the city, we had fewer than two million people, now we have over six million. The city of Atlanta effectively extends to the Tennessee border, because it is just constant growth outward, there is not anything like a mountain range to stop the city from growing.
>
> (I-27: 15f.)

Aside from declining drinking water, the effect of water shortage is apparent in the agricultural sector. Georgia is among the states that produce annual specialty crops[11] such as fruits and vegetables (Shannon et al. 2014). At the same time, Georgia has been a state, which has been affected by long-term droughts (Stooksbury 2003). Over the years, drought has been considered a somewhat "normal condition" of the Southeast. This was also reflected by some of the interviewees. Until 2004, Georgia's population had nearly doubled compared to the 1960s, however the water resources remained constant (Stooksbury 2003). One policy practitioner and academic working on environmental issues stresses the economic impacts of water scarcity:

> I have been let to believe that Georgia cares about its peanuts and its cotton and its corn. And just to see how they are bright red, forty-million dollars of losses. I do not know how that compares to the total productivity, but I feel like forty-million dollar of losses in a county probably burns a little bit.
>
> (I-21: 133)

In the Southeast, the human and financial damage of weather and climate-related events have been tremendous and the highest in the country (Carter et al. 2014). The damage caused by Hurricane Katrina in Louisiana is certainly a major factor of greater awareness. It is seen as an icon of the vulnerability of modern societies toward climate-related events (e.g., Sommer 2015). The 2017 Atlantic hurricane season broke previous high-cost records with an estimated $306.2 billion in economic damages (Carter et al. 2018). Hurricane Matthew in 2018 is currently the costliest hurricane in history causing an economic damage of $30 billion.[12]

In the United States, the projected sea-level rise in the Northeast and Southeast is expected to reshape the U.S. population distribution significantly (Hauer 2017). Atlanta will be the number two gainer in terms of population changes induced by SLR (Hauer et al. 2016). The direct vulnerability of coastal areas due to sea-level rise-induced displacement will add pressure on inland regions (Hauer et al. 2016). In 2005, Atlanta experienced an influx of approximately 100,000 people after hurricanes Katrina and Rita.[13] The changing demographics and extra pressure on the city also became apparent throughout the interviews, as reflected by a city planner in Atlanta:

> For Atlanta, I mean I think, understanding especially the impact of climate refugees and seeing that as sort of just to amplify that // our exaggerated problem of day-to-day rush hour traffic, you know? I mean if our city cannot absorb, you know, I do not know what you call that, the migration of people within a short period of time, for a short or for an extended period of time. . . . When Katrina hit in New Orleans, a lot of people came here uhm some people stayed, some people came for a few months or a few years.
>
> (I-32: 56)

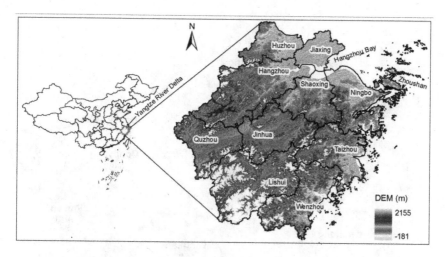

Figure 6.4 Location of case study unit, major cities and elevation river rich Zhejiang province.

Source: Yang et al. (2014: 7047), Copyright: MDPI, Creative Common CC BY license

To conclude, urbanization pressure is already evident and will further exacerbate, impacting already vulnerable populations more than others.

Jinhua

Jinhua is located in Zhejiang province in the eastern part of China and is part of the Yangtze River Delta Region (YRD). YRD is known as the world's largest agglomeration of metropolitan areas with a population of approximately 115 million inhabitants spanning over 210,700 square kilometers (USDC 2016). The region is also known as the golden triangle, made up of Jiangsu province to the North, Shanghai to the East and Zhejiang in the South (see Figure 6.4). The entire YRD region has experienced rapid rates of economic growth and urbanization with Zhejiang at the center. Today, it is one of China's most densely populated provinces with 56 million inhabitants in 2016[14] and with 67 percent has one of the highest urbanization rates across the country.

Political administration: strong influence of the provincial government

Within Zhejiang, the administrative division of Jinhua is not as densely populated with 5.5 million inhabitants in 2016 and an area of 11,000 square kilometers, as compared to Zhejiang's urbanization hotspots: northwestern Hangzhou (9.1 million), northeastern Ningbo (7.8 million) or southern Wenzhou (9.1 million). Jinhua is located in the southwestern part of the province, which is considered the

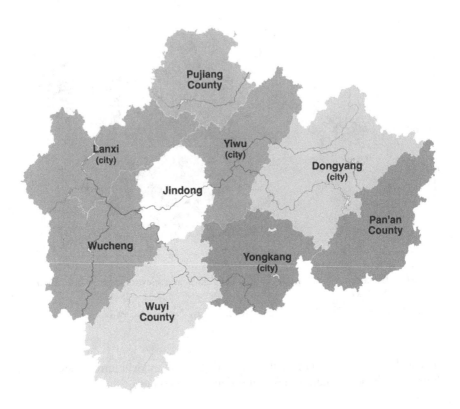

Figure 6.5 Administrative division of Jinhua.

Source: Chk2011, Copyright: CC BY-SA 4.0, Wikimedia Commons

counterpart of the traditionally wealthier Northeast. The Northeast is also known for a comparatively more developed and industry-oriented flat landscape, while the Southwest is, with few local exceptions, less developed and characterized by its mountainous landscape (Ye and Wei 2005).

Regarding its administrative status, Jinhua is a prefecture-level city.[15] Prefectural cities are directly administered by the provincial government and have drastically increased since the late 1970s (Chung and Lam 2010). They are considered the second level of subnational authority (Chien 2010). The prefecture-city oversees administering the counties, which in Jinhua's case includes nine county-level divisions: the four county-level cities Lanxi, Yiwu, Dongyang and Yongkang; the two districts Jindong and Wucheng; and the three counties Pan'an, Pujiang and Wuyi (see Figure 6.5).

Whereas Jinhua's residential population size is comparable to that of Metro Atlanta, the urban population is twice the size of urban Atlanta. In line with national as well as provincial population growth and rapid urbanization, Jinhua has seen a steady increase in its population throughout the years with a population increase of 6.8 percent from 2016 to 2017 alone. Over this period, only the county-level city Lanxi and Chun'an County have faced a demographic decline by approximately 15 percent.[16] These parts of the city are considered rural and not as popular as the central parts of Jinhua.

Not as rapidly developing as the provincial capital Hangzhou to the North, people living in Jinhua emphasize its slower pace and close proximity to ancient villages, smaller townships and the mountains. Economically, Jinhua is a fast-growing region with a GDP growth of 6.5 from the previous year (HKTDC n.d.).[17] Compared to Atlanta, where housing prices skyrocketed over the past years, Jinhua remains an affordable city. During the past two decades, rapid urbanization occurred in three major areas of Zhejiang province: Hangzhou Bay, the Jinhua-Quzhou Basin and the Wenzhou-Taizhou coastal zone (Yang et al. 2014). In Jinhua, this can partially be explained through the city's proximity to major city hubs (Hangzhou, Ningbo and Shanghai to the North, Yiwu to the East) and the Jin-Wen railway route, which connects the East with the South.

Inequality manifesting as rural–urban divide

Zhejiang is traditionally an agricultural province. The southern part, in which Jinhua is located, is among the least developed. Zhejiang is also among the states with the highest inequality (Bhattacharya et al. 2018). Of its 11 major cities, the most severe income gaps are found in Jinhua and Lishui (Wang and Zhao 2016). The impression prevailed that certain regions of Jinhua were prioritized in their foresight development and environmental planning. This observation regards particularly rural regions with high inequality and county-level cities that were identified in the social vulnerability assessments (see section "Inequality in China and the United States").

Though formally under the administration of Jinhua, the county-level city of Lanxi is considered a suburban offshoot, with a steady proportion of rural populations still engaged in agriculture (85 percent), low access to education and a population density of 491 people per square kilometer in 2010. Within Jinhua, Lanxi is one of the most flood-affected regions and the second most important grain area in the province. Lanxi produces mainly rice and bamboo in an industry area of 10,926 square kilometers. Although Lanxi is more exposed climatologically and socially, no known efforts exist to tailor Jinhua's resilience efforts particularly to this region. More data is needed to confirm this impression and is of particular importance, as county-level cities may become the next key battleground for further urbanization and economic output.

Though the per capita net income of farmers in Zhejiang is almost double the national average, the income of farmers in Lanxi was half to a third lower (6277 RMB) than the provincial average (9258 RMB) in 2008 (FAO 2012). The prevalent industrial sectors in Zhejiang include mechanics, hardware, agricultural products, textiles and clothing (Shi and Ganne 2009). In Lanxi, approximately 85 percent of the population is working in cash crop production, including vegetables, fruits, tea, silk farming (sericulture) and aquaculture (FAO 2012). The trends within Zhejiang province are in line with the country's general shift from an agricultural to a market-based economy. The relative rurality of Jinhua and Lanxi was commonly pointed out as infrastructural backlog. One policy planner in the environmental sector in Jinhua states:

> This, this is a global problem. In some areas, we may not be ideal, but [there is other areas in which] we may do better. For example, in Jinhua there is the problem of rural home pension [for the elderly]. That is to say, those who are over the age of sixty and over seventy years old are not very well-off, and then it is not very convenient for them to support their children. In the rural areas, it is centralized, and they are provided with food. In this practice, I think, we do better.
>
> (I-G39, respondent a: 208)

The interviewee is referring to centralized government support, which in matters of crisis, the rural area of Lanxi seemed to have received by the provincial and not the municipal government, although Lanxi is formally administered by the Jinhua government. In political practice however, Lanxi and Pan'an are considered cultural offshoots.

This is also reflected in their exposure to environmental hazards and inequality. The population most exposed to potentially hazardous substances, such as e-waste, lives in Lanxi (also see Wang et al. 2012). In Lanxi, not only is environmental exposure geographically sensitive but so too are regional inequality patterns. In their analysis of the dynamics of multi-scale intra-provincial inequality patterns, Yue et al. (2014) found wealth inequality throughout Zhejiang from 1990 to 2010. According to their findings, there is a significant North–South divide in provincial development. At the county scale, Lanxi belonged to the main inland counties that experienced a decline of GDP per capita. Yue and colleagues attribute this to a "club effect," implying that counties adjacent to neighboring counties with higher economic development are likely to move up in development because they profit from the gains of their neighbors. Further, their findings indicate that almost all poor and less developed counties are, like Lanxi, located in the southwestern region of the province, implying inland areas are falling into a "poverty trap" (Yue et al. 2014).

Little political attention on migrant workers

Zhejiang is one of the coastal epicenters where rural to urban migration has occurred at a large scale and ranks second after Guangdong province regarding the largest number of immigrants (Zhang 2019). Transient populations are commonly referred to as floating populations (*liudong renkou*). In line with the skyrocketing national developments of urbanization and rise of floating populations throughout China, Zhejiang is home to approximately 20 million transient people (Luo et al. 2017). With approximately 54.4 million provincial inhabitants in 2010, the percentage of transient populations ranges from 30 to 48 percent (Liang et al. 2014).

Table 6.1 Floating population in Zhejiang, China, 2000 and 2010

Intra-provincial floating population		Interprovincial floating population	
2000	2010	2000	2010
32%	23%	68%	77%

Source: Liang et al. (2014).

Table 6.2 Number of rural-urban populations, population share (2000 and 2010 compared), area, and population density (2010) in Lanxi and Jinhua

Variables	Lanxi	Jinhua
Rural population (2010)	352,242	2,199,771
Urban population (2010)	208,272	3,161,801
Total population (2000)	607,196	4,571,881
Total population (2010)	560,514 (−0.80%/y)	5,361,572 (+1,61%/y)
Area	1,310 km²	10,926 km²
Population Density (2010)	427.7/km²	490.7/km²

Source: NBS 2012: 2010 population census

Whereas the intra-provincial floating population decreased slightly between 2000 and 2010, the interprovincial floating population increased by almost ten percent (see Table 6.1).

The rural–urban gradient in Jinhua and Lanxi is precisely the opposite, with a significant amount of people residing in rural areas in Lanxi and more people in urban areas in Jinhua altogether. The same holds true for population growth, which was over the past ten years negative in Lanxi and marginally positive in Jinhua (see Table 6.2).

Interestingly, migrant workers seemed to not receive as much political attention, whereas the issue of "water reservoir migrants"[18] was more prominent. This is partially due to high degrees of societal discontent and voiced criticism regarding large-scale governmental infrastructure programs, which resulted in forced displacements. This problem was commonly pointed to, as mirrored by one Zhejiang-focused academic:

> In the area of natural disaster … and water reservoir construction, a lot of people are affected, who have to move. This is a particularly big issue in our province, as many villages are affected; thousands of people; populations in new villages. There are many policies for this. [However], an office for these other people, the migrant workers, does not exist, but for those water reservoir migrants.
>
> (I-14: 46)

Aside from little attention on migrant workers in light of weather-induced events, ecological migrants were also considered a problematic issue, as they occupy only a small part of the deliberation besides being a "major problem" (I-13). However, this topic seemed to have received much greater attention from governmental actors across China and within Zhejiang province in particular. State-led displacement programs have been initiated throughout China for a long time. The different notions of displacement range from displacement in light of disaster events, infrastructure projects (e.g., hydropower dams) as well as drought-focused agricultural forms of relocating local farmers. Considering declining food productivity, migration has become an important issue under study in northwestern China with a predominant focus on Ningxia province (e.g., Shu 2016). Here, resettlement programs have also been serving the function of poverty reduction. Resettlement programs in light of natural hazards and/or climate change have been a wide governmental practice, and only as early as the 1970s referred to as "ecological resettlement" but more recently as "adaptation policy." In 2017, China was the country with the highest levels of new and all disaster-related displacements worldwide.[19] The floods in June 2017 in southern China led to widespread displacement within different provinces, out of which Zhejiang was one.

Summary

The chapter started by taking a brief look at inequality trends in China and the United States. Atlanta and Jinhua are municipalities, where vulnerability to climate change is highly unevenly distributed within the population and intersects with a wide range of other factors, such as rising inequality, racial segregation and an advancing urban-rural divide. The chapter argues, this social vulnerability must be understood also as an outcome of the political economies of both countries. The subsequent chapter section examined some of the sociopolitical trends and challenges both municipalities are facing.

Atlanta is shaped by two major coinciding trends: (1) rapid economic growth coupled with urbanization and (2) segregation-based inequality. Inequality manifests in unequal access to transportation, healthcare, affordable housing as well as fresh and healthy food. Additional pressure on the city because of water scarcity and migration were further concerns, which are expected to exacerbate. Jinhua is likewise experiencing rapid economic growth and urbanization. At the same time, some less favorable parts of the city have experienced population decline. In Jinhua, inequality is reflected as a phenomenon of a widespread rural–urban divide of a former rural-agrarian society. Affected counties of Jinhua such as Lanxi are also those parts, where the most vulnerable populations are outlined (see section "Inequality in China and the United States") and where lower housing quality is commonly detected aside from only limited access to healthcare. Transient populations are an issue across the province, but political attention has been rather focusing on water reservoir migrants than migrant workers.

Notes

1 Extreme poverty is commonly defined by the international poverty line of the share of people living below $1.90 (PPP) per day.

2 Xie and Zhou (2014) cross-validated micro-level data with six independently conducted surveys because of the long-standing practical difficulties with government statistics and concealment practices of China's National Bureau of Statistics (NBS).

3 The Gini coefficient or Gini index is a common statistical measure for the distribution of wealth and inequality. A Gini coefficient of 0 expresses perfect equality whereas a Gini coefficient of 1 means maximal inequality.

4 For more information about the limitations of current policy approaches, which are intended to address poverty, see the 2014 report by the Hamilton Project, edited by Kearny and Harris, "Policies to Address Poverty in America," available at www.hamilton project.org/assets/files/policies_address_poverty_in_america_full_book.pdf, last accessed January 8, 2020.

5 Comment made by Nikki Haley; article published by the Hill: Nikki Haley: 'Ridiculous' for UN to analyze poverty in America. Available at: http://thehill.com/policy/international/un-treaties/393659-nikki-haley-ridiculous-for-un-to-analyze-poverty-in-america, last accessed August 23, 2018.

6 For a more detailed analysis of Atlanta's urban development, please find the Brookings Report "Moving Beyond Sprawl", published 2016 online: www.brookings.edu/wp-content/uploads/2016/06/atlanta.pdf/.

7 The respondent refers to treating cities and urban structure as scapegoats for the dilemma we find ourselves in.

8 The EPA defines an urban heat island as a built-up area that can be 1–3°C (1.8–5.4°F) warmer than its surroundings. For more information see: "Heat Island Effect", accessed February 22, 2019: www.epa.gov/heat-islands.

9 Shannon (2014) from the University of Georgia has articulated a critique on why food deserts constitute a spatialized form of 'neoliberal paternalism', prevalent in aiming to discipline poor bodies, who are viewed to be mismanaging their own lives. See: Shannon, Jerry. "Food Deserts: Governing Obesity in the Neoliberal City." *Progress in Human Geography* 38, no. 2 (2014): 248–266, http://phg.sagepub.com/content/38/2/248. The critique is reflected in a recent publication by Parke Wilde et al. (2017), who study whether people associated the 1-mile radius to consider themselves "food secure". Their findings indicate that this was not the case but rather the mode of transport used or available played a dominant role for people to consider themselves "food insecure", see Parke Wilde, Abigail Steiner, Michele Ver Ploeg. "For Low-Income Americans, living ≤1 mile (≤1.6 km) from the Nearest Supermarket is not Associated with Self-Reported Household Food Security." *Current Developments in Nutrition* 1, no. 11 (2017). https://doi.org/10.3945/cdn.117.001446.

10 Information according to the Atlanta Community Bank, Facts and Stats, last accessed February 8, 2019: www.acfb.org/facts-stats.

11 According to U.S. law, annual specialty crops are defined as "fruits and vegetables, tree nuts, dried fruits, and horticulture and nursery crops, including floriculture" (Walthall et al. 2013: 75).

12 See Michael Bastasch, "Hurricane Michael Did Billions Worth of Damage, Including Reducing A Quiet Beach Town to Rubble." *The Daily Caller*, October 12, 2018, https://dailycaller.com/2018/10/12/hurricane-michael-damage/, last accessed February 9, 2019.

13 According to Molly Samuel, "Atlanta's Population Could Rise with Sea Levels." *Wabe*, Apr 17, 2017, www.wabe.org/atlantas-population-could-rise-with-sea-levels/, accessed February 9, 2019.

14 According to data by the General Office of Zhejiang Government: www.zhejiang.gov.cn/col/col931/index.html, last accessed August 26, 2017.

15 Prefecture-level cities are administrative division that ranks below the provincial and above the county level.

16 For more information on population development in the region, see a report released by Yiwu Think Tank: Not bad! 2017 Jinhua resident population rankings released,

Yiwu first (*zhong bang! 2017 Jinhua shi changzhu renkou paihang bang fabu, Yiwu di yi*), published May 8, 2018. Available online: www.jrpp.com.cn/news/76180/, accessed March 2, 2019.

17 For more information on GDP growth, see the Jinhua Municipal Government: www.jinhua.gov.cn/, last accessed March 2, 2019.

18 The term usually refers to those who are replaced in light of big infrastructure development such as hydropower dams. The three Georges Damn is certainly the most popular issue of contention in this regard.

19 See the Internal Displacement Monitoring Centre (IDMC) is aware of the data restrictions and research gaps that apply in China in the context of development projects, and displacements regarding politically as well as ethnically sensitive issues.

References

Bhattacharya, Prabir, Javier Palacio-Torralba, and Xinrong Li. "On Income Inequality within China's Provinces." *Chinese Studies* 7 (2018): 174–182. https://doi.org/10.4236/chnstd.2018.72015.

Carter, L. M., J. W. Jones, L. Berry, V. Burkett, J. F. Murley, J. Obeysekera, P. J. Schramm et al. "Ch. 17: Southeast and the Caribbean." In *Climate Change Impacts in the United States: The Third National Climate Assessment*, edited by J. M. Melillo, Terese (T.C.) Richmond, and G. W. Yohe, 396–417. U.S. Global Change Research Program, 2014. Doi: 10.7930/J0N- P22CB.

Carter, Lynne M., Adam J. Terando, Kirstin Dow, Kevin Hiers, Kenneth E. Kunkel, Aranzazu R. Lascurain, Doug Marcy, Michael J. Osland, and Paul Schramm. "Southeast." In *Impacts, Risks, and Adaptation in the United States: Fourth National Climate Assessment, Volume II*, edited by David Reidmiller, C. W. Avery, D. R. Easterling, K. E. Kunkel, K. L. M. Lewis, T. K. Maycock, and B. C. Stewart, 743–808. U.S. Global Change Research Program, 2018. https://doi.org/10.7930/NCA4.2018.CH19.

Chen, Tianqi, Stefanie E. Sarnat, Andrew J. Grundstein, Andrea Winquist, and Howard H. Chang. "Time-series Analysis of Heat Waves and Emergency Department Visits in Atlanta, 1993 to 2012." *Environmental Health Perspectives* 125, no. 5 (2017): 057009. https://doi.org/10.1289/EHP44.

Chien, Shiuh-Shen. "Prefectures and Prefecture-Level Cities: The Political Economy of Administrative Restructuring." In *China's Local Administration, Traditions and Changes in the Sub-National Hierarchy*, edited by Jae Ho Chung and Tao-chiu Lam, 127–148. Oxon, New York: Routledge, 2010.

Chung, Jae Ho, and Tao-chiu Lam. *China's Local Administration*. London: Routledge, 2010. https://doi.org/10.4324/9780203871065.

Debbage, Neil, and J. Marshall Shepherd. "Urban Influences on the Spatiotemporal Characteristics of Runoff and Precipitation During the 2009 Atlanta Flood." *Journal of Hydrometeorology* 20, no. 1 (2019): 3–21. https://doi.org/10.1175/JHM-D-18-0010.1.

Food and Agriculture Organization of the United Nations. *Information Services in Rural China, an Updated Case Study*. Bangkok: Regional Office for Asia and the Pacific, 2012.

Freemark, Yonah. "Atlanta's Raising $2.5 Billion to Invest in Transit. Will It Be Money Well-Spent?" *Streetsblog USA*, June 2, 2017. https://usa.streetsblog.org/2017/06/02/atlantas-raising-2-5-billion-to-invest-in-transit-will-it-be-money-well-spent/, last accessed February 22, 2019.

Hacker, Jacob S., and Paul Pierson. "Winner-Take-All Politics: Public Policy, Political Organization, and the Precipitous Rise of Top Incomes in the United States." In

Inequality in the 21st Century, A Reader, edited by D. B. Grusky and J. Hill, 58–68. Boulder: Westview Press, 2018.

Harles, John. *Seeking Equality. The Political Economy of the Common Good in the United States and Canada*. Toronto: University of Toronto Press, 2017.

Harvey, David. *A Brief History of Neoliberalism*. Oxford: Oxford University Press, 2005.

Hauer, Matthew. "Migration Induced by Sea-Level Rise Could Reshape the US Population Landscape." *Nature Climate Change* 7 (2017): 321–325.

Hauer, Matthew, Jason M. Evans, and Deepak R. Mishra. "Millions Projected to Be at Risk from Sea-Level Rise in the Continental United States." *Nature Climate Change* 6 (2016): 691–695. Doi: 10.1038/nclimate2961.

Hong Kong Trade Development Council (HKTDC). "Jinhua, the Industrial Cosmopolitan Heart of Zhejiang Province." HKTDC, published online, n.d. https://sourcing.hktdc.com/en/info/featured-suppliers/ZhejiangJinhua/index.html, last accessed December 4, 2021.

Holbig, Heike. "Ideological Reform and Political Legitimacy in China: Challenges in the Post-Jiang Era." *GIGA Research Program: Legitimacy and Efficiency of Political Systems*, no. 18, 2006.

Jain-Chandra, Sonali, Niny Khor, Rui Mano, Johanna Schauer, Philippe Wingender, and Juzhong Zhuang. "Inequality in China – Trends, Drivers and Policy Remedies." *IMF Working Paper 18/127*, June 2018.

KC, Binita, Marshall Shepherd, and Cassandra Johnson Gaither. "Climate Change Vulnerability Assessment in Georgia." *Applied Geography* 62 (2015): 62–74. https://doi.org/10.1016/j.apgeog.2015.04.007.

Knight, John. "Inequality in China: An Overview." *The World Bank Research Observer* 29, no. 1 (2014): 1–19. https://doi.org/10.1093/wbro/lkt006.

Li, Chunling. "Class and Inequality in the Post-Mao era." In *Handbook on Class and Social Stratification in China*, edited by Yingjie Guo, 59–82. Cheltenham and Northampton: Edward Elgar Publishing, 2016.

Liang, Zai, Zhen Li, and Zhongdong Ma. "Changing Patterns of the Floating Population in China during 2000–2010." *Population and Development Review* 40, no. 4 (2014): 695–716.

Luo, Jiaojiao, Xiaoling Zhang, Yuzhe Wu, Jiahui Shen, Liyin Shen, and Xiaoshi Xing. "Urban Land Expansion and the Floating Population in China: For Production or for Living?" *Cities* 74 (2017): 219–228. https://doi.org/10.1016/j.cities.2017.12.007.

Manangan, Arie Ponce, Christopher K. Uejio, Shubhayu Saha, Paul J. Schramm, Gino D. Marinucci, Jeremy J. Hess, and George Luber. "Assessing Health Vulnerability to Climate Change: A Guide for Health Departments." *Climate and Health Technical Report Series, Climate and Health Program*. Atlanta: Centers for Disease Control and Prevention, 2014.

National Bureau of Statistics (NBS) of the People's Republic of China (2012). 中国2010人口普查分乡,镇,街道资料 (1 ed.). Beijing: China Statistics Print, Census Office of the State Council of the People's Republic of China; Population and Employment Statistics Division of the National Bureau of Statistics.

Partnership for Southern Equity (PSE). "Opportunity Deferred: Race, Transportation, and the Future of Metropolitan Atlanta." *Partnership for Southern Equity*, 2017. https://psequity.org/wp-content/uploads/2019/10/2017-PSE-Opportunity-Deferred.pdf, last accessed July 15, 2021.

Ravallion, Martin, and Shaohua Chen. "China's (Uneven) Progress Against Poverty." *Journal of Development Economics* 82 (2007): 1–42. https://doi.org/10.1016/j.jdeveco.2005.07.003.

Reardon, Sean F., and Kendra Bischoff. "Income Inequality and Income Segregation." *American Journal of Sociology* 116, no. 6 (2011): 1934–1981.

Saez, Emmanuel and Gabriel Zucman. *Wealth Inequality in the United States Since 1913: Evidence from Capitalized Income Tax Data.* NBER Working Paper Series 20625, 2014. Cambridge: National Bureau of Economic Research.

Sarah Foster, and Wei Lu. "Atlanta Ranks Worst in Income Inequality in the U.S. 2018." *Bloomberg*, October 10, 2018. www.bloomberg.com/news/articles/2018-10-10/atlanta-takes-top-income-inequality-spot-among-american-cities, last accessed July 15, 2021.

Shannon, Jerry. "Food Deserts: Governing Obesity in the Neoliberal City." *Progress in Human Geography* 38, no. 2 (2014): 248–266. Doi: 10.1177/0309132513484378.

Shi, Lu, and Bernhard Ganne. "Understanding the Zhejiang Industrial Clusters: Questions and Re-evaluations." In *Asian Industrial Clusters, Global Competitiveness and New Policy Initiatives*, edited by Bernhard Ganne and Yveline Lecter, 239–267. Singapur: World Scientific Publishing, 2009.

Shu, Xihong. "Chapter 2 The History and Present Condition of Ecological Migration in Ningxia." In *Ecological Migration, Development and Transformation*, edited by P. Li and X. Wang, 21–46. Berlin and Heidelberg: Social Sciences Academic Press and Springer-Verlag, 2016. Doi: 10.1007/978-3-662-47366-5_2.

Siebenhüner, Bernd, Torsten Grothmann, Dave Huitema, Angela Oels, Tim Rayner, and John Turnpenny. "Lock-Ins in Climate Adaptation Governance. Conceptual and Empirical Approaches." Conference Paper 2017, unpublished.

Smeeding, Timothy M. "Public Policy, Economic Inequality, and Poverty: The United States in Comparative Perspective." *Social Science Quarterly* 86, no. 1 (2005): 955–983. https://doi.org/10.1111/j.0038-4941.2005.00331.x.

Sommer, Bernd (ed.). *Cultural Dynamics of Climate Change and the Environment in Northern America.* Leiden: Brill, 2015.

Stone, Brian Jr., Jason Vargo, Peng Liu, D. Habeeb, A. DeLucia et al. "Avoided Heat-Related Mortality through Climate Adaptation Strategies in Three US Cities." *PLoS One* 9, no. 6 (2014): e100852. Doi: 10.1371/journal.pone.0100852.

Stone, Brian Jr., Jason Vargo, Peng Liu, Yongtao Hu, and Armistead Russell. "Climate Change Adaptation Through Urban Heat Management in Atlanta, Georgia". *Environmental Science & Technology* 47 (2013): 7780–7786. Doi: dx.doi.org/10.1021/es304352e.

Stooksbury, D. E. "Historical Drought in Georgia and Drought Assessment and Management." Proceedings of the 2003 Water Resources Conference, April 23–24, 2003.

United Nations. *Report of the Special Rapporteur on Extreme Poverty and Human Rights on His Mission to the United States of America, Note by the Secretariat.* A/HRC/38/33/Add.1. United Nations, Geneva, May 4, 2018.

U.S. Department of Commerce (USDC). "Fact Sheets, Yangtze River Delta Region 2014." *USDC, International Trade Administration*, published online, 2016. https://2016.export.gov/china/build/groups/public/@bg_cn/documents/webcontent/bg_cn_075985.pdf, last accessed December 4, 2021.

Walthall, Charles L., Christoper J. Anderson, Lance H. Baumgard, Eugene Takle, Lois Wright-Morton, et al. "Climate Change and Agriculture in the United States: Effects and Adaptation." Geological and Atmospheric Sciences Reports, USDA Technical Bulletin 1935. Washington, DC, (2013). https://lib.dr.iastate.edu/ge_at_reports/1.

Wang, Xiaofeng, Greg Miller, Gangqiang Ding, Xiaoming Lou, Delei Cai, Zhijian Chen, Jia Meng et al. "Health Risk Assessment of Lead for Children in Tinfoil Manufacturing and E-Waste Recycling Areas of Zhejiang Province, China." *Science of The Total Environment* 426 (2012): 106–112. https://doi.org/10.1016/j.scitotenv.2012.04.002.

Wang, Xinxin, and Shengshan Zhao. "The Current Status of Income Distribution in China – the Case of Zhejiang Province." *International Journal of Mathematics and Statistics Invention* 4, no. 9 (2016): 9–14.

World Bank. *Poverty and Shared Prosperity 2016: Taking on Inequality.* Washington, DC: World Bank, 2016. Doi: 10.1596/978-1-4648-0958-3.

Xie, Yu. "Understanding Inequality in China." *Chinese Journal of Sociology* 2, no. 3 (2016): 327–347. https://doi.org/10.1177/2057150X16654059.

Xie, Yu, and Xiang Zhou. "Isncome inequality in China." *Proceedings of the National Academy of Sciences* 111, no. 19 (2014): 6928–6933. Doi: 10.1073/pnas.1403158111.

Yang, Xuchao, Wenze Yue, Honghui Xu, Jingsheng Wu, and Yue He. "Environmental Consequences of Rapid Urbanization in Zhejiang Province, East China." *International Journal of Environmental Research and Public Health* 11 (2014): 7045–7059. Doi: 10.3390/ijerph110707045.

Ye, Xinyue, and Wei Yehua. "Geospatial Analysis of Regional Development in China: The Case of Zhejiang Province and the Wenzhou Model." *Eurasian Geography and Economics* 46 (2005): 342–361. Doi: 10.2747/1538-7216.46.6.445.

Yue, Wenze, Yuntang Zhang, Xinyue Ye, Yeqing Cheng, and Mark R. Leipnik. "Dynamics of Multi-Scale Intra-Provincial Regional Inequality in Zhejiang, China." *Sustainability* 6, no. 9 (2014): 5763–5784. https://doi.org/10.3390/su6095763.

Zhang, Yinghua. "Improving Social Protection for Internal Migrant Workers in China." International Labour Organization and International Organization for Migration Report, 2019.

Cited interviews

I-13, March 7, 2017
I-14, March 7, 2017
I-20G, April 28, 2017
I-21, May 2, 2017
I-23, May 3, 2017
I-27, May 12, 2017
I-29, May 16, 2017
I-30, May 16, 2017
I-31, May 17, 2017
I-32, May 18, 2017
I-33, May 19, 2017
I-G39, November 21, 2017

7 Protracted vulnerability

This chapter examines how local decision-makers in Atlanta, and Jinhua perceive of vulnerable populations to climate change, and whether they are generally aware that climate impacts are unevenly distributed. Social vulnerability assessments (SVAs) are taken as a baseline to compare some of the local specifics of population exposure with the view of decision-makers. SVAs are a type of climate vulnerability assessment that aims to look at the social dimension of vulnerability and present one approach, which attempts to quantify elements that affect the predisposition of populations to deal with climate impacts. When looking at population susceptibility to growing climate impacts, the concept of social vulnerability has been established as one of the dominant discourses and presents a form of policy knowledge. Policy knowledge is intended to be used politically and to affect the policy process (Daviter 2015). Despite the severe shortcomings of quantified vulnerability approaches, especially when it comes to social vulnerability, the social dimension of climate change impacts has been regaining traction in political practice as of late. With it, the urgent need to desigining (adaptation) policy responses that address uneven vulnerability to climate change has been formulated. In the policy field of climate change adaptation, making use of policy knowledge is especially tricky, because adaptation is such a cross-cutting issue that cuts across domains. As a result, there is a strong disciplinary orientation, with different sectors contributing an own understanding as per sector specialization (e.g., health, infrastructure or water) (Ford et al. 2018). Only few research endeavors have looked at social vulnerability from a non-sectoral lens and compared place-based social vulnerability across local cases. Even fewer studies pair the findings of SVAs with the prevailing local political practices related to climate adaptation and resilience. This chapter addresses these gaps by responding to the following research questions:

(1) What is the prevailing connotation of local human vulnerability to climate change in Jinhua and Atlanta as per local vulnerability assessments?
(2) How does this materialize with local decision-makers?

    a   Are understandings of SVAs applicable in political practice?
    b   How has it mattered as part of policies related to resilience and adaptation?

DOI: 10.4324/9781003183259-7

The first part of this chapter looks at the "human dimension" in the two chosen case studies by examining local connotations of socially vulnerable populations to climate change (section "Dominant constructions of human vulnerability to climate change"). This section lays out some of the findings of dominant vulnerability conceptualizations of local and regional vulnerability assessments. This includes an examination of a standardized regional climate vulnerability assessment and social vulnerability assessments. Against their background, the next section looks at the prevailing cultural connotations of human vulnerability reflected in decision-makers' perceptions. The section "Political practices of vulnerability perception and acknowledgment" proceeds through an analysis of dominant political practices as reflected in decision-makers' perceptions, policy discourses and dominant ideas. Analyzing problem perceptions and political attention helps to understand the internal logic of (political) appropriateness. They also provide insights on the scope of adaptation types, which are considered desirable and viable. This helps to understand the policy process.

Despite the varying interpretations of what constitutes the "human dimension," or social aspects of vulnerability, the findings demonstrate a clear awareness gap of considering uneven human vulnerability to climate change as part of local public policy planning related to climate adaptation and resilience. More importantly, the findings suggest that the political utilization of SVAs is problematic at different levels, including methodological, cultural and political levels. Existing social vulnerability frameworks insufficiently explain the origin, maintenance and endurance of vulnerability and can lead to (further) social stigmatization. The chapter finds that the way vulnerability categories and acknowledgment function in political practice further protracts the vulnerability of groups, which are outlined by the assessments. Processes of cultural devaluation and only selective acknowledgment of certain vulnerability indicators suggest this. Therefore, chapter section "Lock-ins related to knowledge and politics" complements the initial research questions by the following:

(3) Which lock-ins can explain this protracted form of vulnerability?

The analysis reveals protracted vulnerability as an interconnected lock-in of discourses, epistemology and political power in both Atlanta and Jinhua. The chapter draws upon the analysis of government documents, secondary literature and interviews. Some of the interviewees drew broader conclusions of social vulnerability patterns. Because of that, the clear differentiation of observed phenomena could not be clearly attributed to the county, municipal or provincial/state level.

Dominant constructions of human vulnerability to climate change

Scientific vulnerability perceptions of the case units

There is a broad-based consensus that the way climate impacts are being felt and how they unfold are highly localized phenomena. Social vulnerability assessments

(SVAs) or Social Vulnerability Indices (SoVIs) present one attempt and academic practice to quantify and capture the precondition of populations to cope with, prepare for and recover from natural hazards and climate impacts. In the case units examined, SVAs were used to enable a comparison of two localities that stand out in their unevenly distributed vulnerability to climate change, that is having population groups that are more drastically affected by climate impacts than others due to a range of different components. In Atlanta and Georgia, SoVIs are partially known and play a role in decision-making. This was not the case in Jinhua, China. The following section briefly summarizes the main findings of the social vulnerability assessments, as they were applied to examine the social dimension of vulnerability in Atlanta and Jinhua. The chapter takes a closer look at the indicators that aim to explain uneven community exposure. Despite informing the case selection and grounding the empirical study, this chapter critically questions the "social" connotation. Rather, vulnerability must also be seen as a contextual and political condition of limited adaptive capacity environments.

Regional-local connotations of vulnerability in Atlanta and Georgia

At the regional level, a collaborative effort of the Georgia Department of Public Health, the Georgia Emergency Management Agency (GEMA) and the University of North Carolina at Chapel Hill published the *Georgia 2010 Social Vulnerability Index Atlas*. At-risk populations across Georgia jurisdictions and county districts were identified on the basis of the following characteristics: (1) lower socioeconomic status (defined through income, poverty, employment, education), (2) household composition (age, dependency, single-parenting), (3) minority status and language (minority status, non-English speaking), as well as (4) housing and transportation (type of housing, crowding, transportation). The composite vulnerability was particularly high in Fulton County and Atlanta, based on the combined effects of different factors such as age, medical coverage and hospital insufficiency (see Figure 7.1).

KC et al. (2015) outline similar factors to correlate with heightened social exposure to climate vulnerability. In their study, social vulnerability variables are split into climatic exposure, population sensitivity and adaptive capacity (see Figure 7.2).[1] Sensitivity is defined through factors such as age, poverty, occupation and language. The adaptive capacity is measured through physician to population ratio, education, per capita income and irrigated land. Altogether, historically high amounts of African American populations; an increasing concentration of Hispanics; low-skilled workers; elderly without adequate health insurance and less educated populations with a lowered socioeconomic status are the significant factors for high socio-climatic vulnerability in Metro Atlanta, Fulton County and coastal sections of Georgia (KC et al. 2015). High amounts of inmate populations (prisoners) and female-headed households were also mentioned at the side to account for greater vulnerability to climate impacts.

The findings offer distinct parallels with the adaptation plans prepared by the Environmental Protection Agency (EPA) in 2014 and 2015. Georgia is a

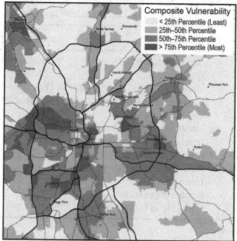

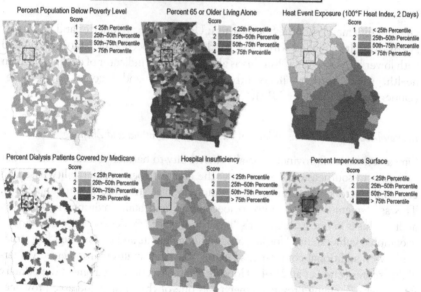

Figure 7.1 Composite vulnerability in Fulton County and Metro Atlanta.

Source: CDC, Manangan et al. (2014: 23)

designated EPA region 4, which sees low-income households at special risk to climate change. Generally speaking, the EPA considers "children, the elderly, the poor, the infirm, and tribal and indigenous populations" to be among the most vulnerable populations (EPA 2014: 34).

Climate justice scholars emphasize higher degrees of rurality, lower socio-economic status and place-based forms of geographic vulnerability as primary

Table 7.1 Overview of social vulnerability variables

Variables to measure exposure, sensitivity and adaptive capacity to climate change.

Exposure	Sensitivity	Adaptive capacity
Temperature change	Age group > 65	Physician to population ratio
Precipitation change	Age group < 5	Education
Drought	Poverty	Per capita income
Flood	Racial/ethnic minorities	Irrigated land
Heat wave	Occupation	
	Urban/rural population	
	Female-headed household	
	Inmate population	
	Non-English speaking	
	Unemployment	
	Renter population	
	Dwelling in mobile homes	

Source: KC et al. (2015: 66)

vulnerability markers of the Southeastern United States (Gutierrez and LeProvost 2016). Rurality, coupled with race, gender and socioeconomic status is equated with lower life expectancy. Rurality is also taken as an indicator of declined human health, which further affects urban–rural, racial and income discrepancies (Gutierrez and LeProvost 2016).

Regional-local connotations of human vulnerability in Jinhua and Zhejiang

One assessment of provincial social vulnerability to natural hazards observes that eastern and southeastern China show the greatest social vulnerability based on economic factors, rural status, urbanization and age structure (Zhou et al. 2014). This assessment is divided into socioeconomic and built environmental vulnerability. This is in line with Cutter's "hazards of place" model and understanding of geographic vulnerability. Housing age, building heights and types of building structures were additional components leading to higher built environmental vulnerability scores (Zhou et al. 2014). Housing age is also considered one of the primary factors why Zhejiang province emerges as one of the most vulnerable provinces throughout China (Zhou et al. 2014). The authors go a little further by concluding that the lack of access to resources and political power, as well as risk awareness levels, likewise contribute to higher population vulnerability.

Within Zhejiang province, the most common item under study when looking at population susceptibility used to be flood risk probability based on sea-level rise (see, e.g., Yin et al. 2012). Due to a significantly aging population throughout Zhejiang province, heat and health risk have become a growing concern. Only few studies assess the heat vulnerability and health risk in "developing countries" and China in particular (Hu et al. 2017). One of the view studies detects a greatly heterogenous heat health risk throughout Zhejiang during the summer period (June to August) of 2008 and 2013. In total, 141,401 heat-related deaths were recorded

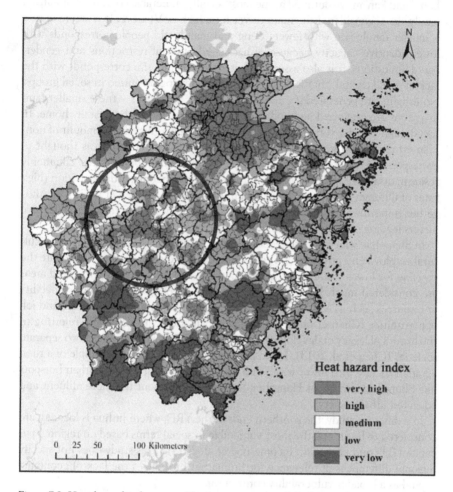

Figure 7.2 Heat hazard index across Zhejiang province with Jinhua at the center (summers 2008–2013).

Source: Reprinted (adapted) with permission from Hu et al. (2017: 1501). Copyright (2017) American Chemical Society.

during the study period. Several pockets within the province have a higher heat hazard index, with Jinhua at the very center (see Figure 8.3). The highest risk was found to concentrate in the inner-city areas, mainly due to the Urban Heat Island Effect (UHI) (Hu et al. 2017). However, in line with other studies, the authors find that the rural population may be more vulnerable to heat due to their lower capacity to adapt based on their conceived lower socioeconomic status and limited availability of medical resources (Hu et al. 2017).

Chen et al. (2013) measure social vulnerability to natural hazards in the Yangtze River Delta Region (YRD). High social vulnerability scores are detected at the southern ends of the YRD and in Jinhua prefecture in particular. The counties of

Lanxi and Pan'an are outlined as the most socially vulnerable counties. For Pan'an, family size is considered the main driver for a high SoVI score. It is assumed that a smaller family size with fewer young and mainly old people corresponds with lower adaptive capacity because of age-based physical restrictions and gender-based limitations. It is also assumed that smaller family size corresponds with the lack of social capital, that is, living together and being organized in social groups. No insights are generated if – aside from smaller family size – these smaller family groups are organized and collaborate socially outside of their family home. In Lanxi, higher social vulnerability is detected based upon the high amount of non-Han ethnic minorities living in the area. The minority condition is thought to correspond with higher SoVI scores based on the assumption that they commonly reside in disadvantageous living conditions with low-income, little education (high rates of illiteracy) and fewer employment opportunities (Chen et al. 2013). Altogether, minorities, small housing size and small family size are considered decisive factors leading to higher vulnerability across the study unit (Chen et al. 2013).

In Shanghai, which is outlined as the least vulnerable study region, low agricultural employment, poverty, high education levels and good housing quality are the baseline for lower social vulnerability scores (Chen et al. 2013). Urbanized areas are considered much less vulnerable than rural areas, based on the availability of resources such as education, medical services, governmental subsidies and job opportunities (Chen et al. 2013). Some explanations are delivered, referring to the historical legacy of development patterns in China which created two separate societies (Chen et al. 2013). Lanxi is oftentimes considered an example of a rural agrarian society with lower wages and educational levels, whereas urban hotspots like Shanghai, Ningbo or Hangzhou are seen to represent the more affluent and educated urban society.

In a different study, the southern ends of the YRD, where Jinhua is located, are considered to be among the most vulnerable in social terms based on regional per capita GDP and per capita income (Ge et al. 2013). This study does not specify regional differences of relevance for Jinhua but introduces the lack of economic capital as a broader vulnerability contributor.

Commonalities and differences of vulnerability constructions

With slight differences, the previous regional and local studies have in common that they measure social vulnerability mainly based on interdependent socio-demographic characteristics and economic factors. The social vulnerability assessments, which detect high social vulnerability at the county level (Fulton in Atlanta, Lanxi and Pan'an in Jinhua), and high rates of social vulnerability to climate change and natural hazards, are explained through high amounts of minorities, occupation/employment, lower economic prosperity and educational attainment (Chen et al. 2013; KC et al. 2015). In the United States, the number of ethnic and historic minorities was provided as primary factor, whereas employment coupled with poverty is presented to show the greatest significance in the Chinese case. Either way, lower educational attainment is supposed to weaken the adaptive capacity and lessen a population's ability to recover from the effects of climate disasters despite lowering peoples' employment opportunities (Chen et al. 2013; KC et al. 2015). Employment in climate-dependent industries, such as agriculture and timber, is considered to exacerbate the existing "precariousness" of already vulnerable Hispanics in Atlanta, who have gradually come to replace African Americans in various rural, low-skilled industries across Georgia (KC et al. 2015). In Jinhua, agricultural employment coupled with poverty is also considered a dominant force driving social vulnerability.

Different definitions of housing quality but similar problems of access

The quality of housing plays a role in both Atlanta and Jinhua but was defined differently. Chen et al. (2013) define poor housing quality through (1) the average number of occupied rooms per household, (2) per capita building area, (3) piped water access, (4) houses without kitchen, (5) without toilet and (6) without a bath. Here, access to sanitation and clean water were commonly articulated development concerns. In contrast, poor housing quality was equated with mobile homes in Atlanta. However, it was not made clear how this correlates with higher social vulnerability.

In Jinhua, the problem of low housing quality was commonly perceived to contribute to greater flood-risk vulnerability of local communities but was also considered to matter across the country (e.g., I-38, I-39G). There was outstanding awareness on this particular matter, however, the causal link between low-housing quality, deteriorating climate impacts as well as population susceptibility is far from well established. The underlying problem is similar in both cases: access to affordable and adequate housing is only limited to a certain part of the population. Reasons were not provided as to why.

Another similarity was the inaccess to adequate healthcare, which was considered to correspond with higher population vulnerability. In Jinhua's city pockets Lanxi and Pan'an, this was explained through rurality and distance to medical institutions. In Atlanta, inaccess to healthcare was a broader vulnerability driver. The issue of continued inaccess and the underlying drivers was not touched upon.

Political practices of vulnerability perception and acknowledgment

With a population of 5.8 million people in 2016, the Atlanta metro region has roughly the same size that prefecture-level Jinhua with 5.5 million inhabitants. Geographically, Jinhua is more densely populated with roughly half the size of metro Atlanta and 490 people per square kilometer opposed to only 243 people per square kilometer living in Atlanta (see Table 7.2). In local political practice of both case units, awareness about and the perception of human vulnerability to climate change significantly differed from the previous assessments. The chapter proceeds by looking at the awareness political decision-makers had about uneven human vulnerability to climate change (section "Dominant constructions of

Table 7.2 Summary of similarities and differences between Atlanta and Jinhua

Variable	Metro Atlanta	Prefecture-level Jinhua
Population	5.8 million (2016)	5.5 million (2016)
Geographic	21,694 km^2 (8,376 sq mi)	10,926.16 km^2 (4,218.61 sq mi)
Population density	243/km^2 (624/sq mi)	490/km^2 (1,300/sq mi)
Acknowledged populations	Elderly, disabled, children, low-income, (homeless)	Elderly, disabled, children, the poor, villagers, people in risk houses
Contested populations	Communities of color, homeless, incarcerated people	Migrant workers

human vulnerability to climate change"). In Atlanta and Fulton County, varying degrees of awareness existed regarding the exposure of particular parts of society. Social vulnerability seemed to be an accepted perspective more broadly but problematic when discussed in the context of climate change. Differential societal exposure to climate impacts seemed to be a relatively new lens in Jinhua with very limited to no awareness among the interviewed politicians, public officials, policy practitioners and advisors. Local academics seemed to have a better understanding and recommended to interlink matters of social stratification with human exposure to climate change.

The second chapter section ("Political practices of vulnerability perception and acknowledgment") examines the local patterns of vulnerability acknowledgment and perception. In both cases, certain parts of the populations were a matter of contention. Whereas the elderly, disabled people, children and low-income people were widely agreed upon to qualify for being considered vulnerable, communities of color, homeless populations and incarcerated population were controversially debated in Atlanta. Migrant workers were the highly disputed group in China. The main similarities and differences between Atlanta and Jinhua are summarized in the following Table 7.2.

Awareness

Fragmented awareness in Atlanta

Although the function of existent SoVIs is to help planning officials at different political levels identify at-risk populations to "better prepare communities to respond to emergency events such as severe weather, floods, disease outbreaks, or chemical exposure" (Manangan et al. 2014: 4), and although, complex analyses were conducted for the state of Georgia in terms of assessing social vulnerability to climate change, there was no indication that the findings had been processed or used by political decision-makers. This points to a science–policy disconnect. At the same time, how matters of human vulnerability to climate change were perceived on the ground significantly differed from the assessments. The interviewed local officials in the United States had some understanding of the matter in general, but showed a clear bias toward certain populations and their constituents.

One distinctive characteristic of the political system in the United States was the strong degree of decentralization and responsibility that was dispersed across different levels of government (county, municipal) (also see Chapter 5). Within the adversarial political environment in Georgia, vulnerability as it relates to climate change was considered a nontraditional form of vulnerability that is not as commonly accepted or where problem recognition is biased. An Atlanta-based policy advisor expressed that how the city institutions perceive their responsibility toward human vulnerability depends on the type of vulnerability. Accordingly, the city governments were considered well aware of vulnerability to poverty, vulnerability to drug addiction, vulnerability to crime and vulnerability to disease and in a much better position to respond to infectious disease:

> I do not think municipal governments are very attune to emerging threats, such as vulnerability to extreme heat and vulnerability to flooding. And you know, probably one of the more interesting overlaps there is, vulnerability to emerging climate induced disease. So, Zika, West Nile virus will probably get a response of the city of Atlanta, well before heat does. I mean, and that may make sense, but that is because that is a traditional // that is the exacerbation of a traditional problem. Heat is not a traditional problem. It is a nuisance, seen as a nuisance but not a problem. Zika is a problem, and . . . the counties will play a major role in dealing with that, but the resources will come from the federal government.
>
> (I-29: 92f.)

This assessment was confirmed by an official within the Fulton County Government, who indicated that human vulnerability plays a role in the policymaking of the elected officials but not as it relates to climate change. This was not seen as a priority of the Fulton County Board of Commissioners (I-24).

A majority of interviewed policy advisors and academics supported the impression of low awareness of local officials about human vulnerability to climate change, which was considered altogether absent. Elected "officials do not consider climate change to be something that people are vulnerable to" (I-31: 120). Government officials and policy advisors alike pointed out that this was also because it has not yet been made clear to local decision-makers at the county and city levels how communities will be impacted by climate change (I-24, I-31).

Different factors were mentioned to hamper increased awareness and action that is, for instance, how knowledge travels to local decision-makers, fragmentation of responsibility in light of decentralized political institutions, and perceiving only limited responsibility on this matter. The complexity of climate change vulnerability was also pointed to, as this is considered a much messier "pathway forward to implement change" (I-21: 115). The lack of attention to human vulnerability to climate change was also explained with advanced forms of climate skepticism as "a lot of people do not believe in climate change still" (I-31: 56).

Simultaneously, it was indicated that this was becoming a minority opinion "even among more conservative people" against the background of rapidly advancing climate change but that decision-makers would hide behind "the, it is not man-made, it is natural vulnerability, there is nothing we can do about it" (I-31: 56).

Aside from limited awareness, the problem of selective acknowledgment and disconnect from action was pointed to. Whereas some referred to this issue as a deliberate form of unwillingness, others attributed the lack of action to financial restraints in light of political dependence on the federal level. A former employee at the Atlanta-Fulton emergency management department indicated that some communities are recognized, "but being recognized and then being able and willing to do something about this, is two different things" (I-10: 22f.). High problem recognition and acknowledgment were considered to not necessarily lead to actual commitment.

There was certain awareness about groups being vulnerable to weather-induced events, but little reflection on growing climate impacts. The necessity to anchor vulnerability to other policy concerns was made explicit by one environmental policy practitioner at the state level:

> We have had some events, like the snowstorm a few years ago, where every-thing was just paralyzed for days. And then the I-85 collapse [note: important interstate highway route]. So, when ARC [the Atlanta Regional Commission] comes to them and say: "Oh we need performance measures and we need to address resiliency in our transportation plan." They understand that that has value. But if you say, "we need, you know, address the human health impacts that the homeless population of Atlanta would deal with climate change, they [are] probably not going to be interested in talking about that.
> (I-31: 133)

This impression was also confirmed in a group interview with two academics working on public health in the climate context, who advised the Climate Action Plan in the city of Atlanta and pointed out that "there really is no discussion on human vulnerability to climate change" and that governmental decision-makers do not talk about vulnerability in a social sense. "They talk about food desserts, yes. You could map vulnerable neighborhoods and food desserts one on one. But social vulnerability is not a focus" (I-20G: 30).

Yet, aside from the different degrees of awareness, social vulnerability assessments were partially known by some decision-makers and sporadically used in Atlanta and Georgia. In contrast to Jinhua, SVAs and SoVIs are political tools that have been used by emergency managers in the metro Atlanta region and across Georgia. Governmental agencies include GEMA, the Georgia Department of Public Health (GDPH) and the federal Centers for Disease Control and Prevention (CDC), which is headquartered in Atlanta. The CDC, for example, states that SoVIs "provide a population level approach to planning for at-risk populations before and during a disaster" (CDC 2015: 18). Nevertheless, the use of SoVIs was exclusive to the emergency management sector.

Lack of understanding in Jinhua and China

In Jinhua and China more broadly, there was a clear lack of understanding of the human dimension of climate impacts regarding two aspects: human vulnerability (*renlei cuiruo xing*) more generally and differential vulnerability (*butong de cuiruo xing*) as understood in most SVAs. It seemed that human exposure to climate impacts was generally a new lens for local officials and most policy advisors, aside from lacking an understanding of the differential means of resilience. Data on how and if at-risk populations are identified and used by governmental agencies could not be retrieved in Jinhua. The findings suggest that this type of population-level approach is not a dominant framework.

In China, experts at different levels of government mentioned the little attention, the human dimension received as part of formalized policy-planning related to adaptation and/or management of natural hazards (e.g., I-35). This is reflected in one statement of an academic and policy advisor to the provincial government, who pointed out the lack of research, as well as the lack of implementation at the political level:

> When I received your email, I was giving a class to students. I told them there is a German PhD student doing this research. I said to them, how can our students not think of such a problem, climate change and vulnerable groups. I say this issue is very important, but few Chinese scholars have done this research. From the policy point of view, which means the Chinese government, it is concerned about this issue, but it does not have any specific measures or corresponding policies to focus on the impact of climate change on disadvantaged groups.
>
> (I-12: 23)

Besides a clear lack of understanding at the local level, the central government was reported to have gradually grown an interest in the matter as part of the human health sector. The study of health impacts of climate change constitutes a very new agenda item that was only mentioned in the 13th FYP (2016–2022). The government was criticized to talk "about it very roughly, but not very specific how to develop some policy to help solve this problem" (I-12: 35). This marks a stark contrast to the U.S. case: although the awareness about human vulnerability and human health was fragmented in the climate change context and in Atlanta in particular, there was a general understanding across the United States regarding population affectedness and differential capacity to cope with climate impacts. In China, this lens was reflected in discourses on air pollution, but was more rarely touched upon as it relates to climate change.

In China and at the local level, the challenge was more profound that is, finding a shared understanding of the social dimension of human vulnerability to climate change. Finding the right expression for describing human exposure and vulnerability was a significantly complex undertaking. China's National Climate Adaptation (NAS) strategy of 2013 mentions "cuiruo qunti" (**脆弱群体**) to indicate vulnerable crowds, which literally translates into "fragile crowds." Despite infre-quent mentions, the topic of vulnerable populations was largely absent in the overall strategy. Although "cuiruo qunti" seems to be the most common term in the policy documents related to climate vulnerability and disaster risk reduction, most of the interviewees indicated that the term does not fit for describing vulner-able populations. The somewhat established term was "ruoshi qunti" (弱势群体) but was not mentioned in any of the adaptation-related documents at central or local levels.

Local conceptions of human vulnerability

*Permissible conceptions in Atlanta: lower socioeconomic status,
sick and old people*

Aside from the common perception of lacking a sense of social vulnerabil-ity to climate change, a select acknowledgment of certain populations being vulnerable to climate change was found among decision-makers at the local level. Admissible perceptions often related to people with lower socioeconomic resources to be perceived as more vulnerable because of restricted choices and limited financial leeway. Yet, interviewees more rarely established the lack of opportunity these people have in influencing political choices and how this relates to issues of racial segregation. However, the sentiment reflected in most conversations was the causal relationship of purchasing and bargaining power. It seemed that the less obvious political and economic processes, which drive change in the city, were more rarely touched upon. One city planner referred to the suburbanization of poverty as a broader phenomenon and hints at underly-ing factors such as cultural displacement in light of gentrification and rising property prices:

> There is two kind of categories, one is sort of economic, obviously people in the economic spectrum that have less choice. You know middle-income peo-ple are also being affected here, but they have more choice in where they go or what they do, how they manage that. The other affected group would be, sort of the cultural side of that, cultural displacement which is much harder to kind of identify (incomprehensible) but the composition of the city is really changing, [the] racial and cultural composition. And that is sad, more than anything. And uhm, the suburbanization of poverty is going to have really significant and negative impacts on the region and the country. It is not just in Atlanta.
>
> (I-32: 23f.)

Despite low problem recognition about the social dimension of vulnerability, certain notions appeared to be more accepted and politically less sensitive. This includes old and sick people.

Contentious populations in Atlanta: communities of color

In Metro Atlanta, it was mainly communities of color that were an issue of much contention. Population groups mentioned on the side include Atlanta's vast prison population, homeless people and those residing in Atlanta's suburban and rural regions. Aside from the dominant and much contested issue of African American populations, within Georgia, several interviewees that work in emergency management or for the federal environmental agency pointed to prisoners as a distinctively vulnerable population in Georgia, and Atlanta in particular that has received only very little attention as part of urban resilience and environmental planning (I-10, I-23). Despite the attention on some select minorities, rural populations were also outlined to not receive enough attention in the context of climate adaptation. The impact of climate change on food production across the United States and Georgia in particular would be pointed to (I-06: 22). Additionally, the aspect of suburban groups, who live in "pockets of poverty" around city areas, was likewise emphasized by policy practitioners working on urban resilience in Atlanta (I-04G, respondent a: 15). Political neglect was explained with the nature of the political system with only certain constituency interests being represented due to taxation. One person in the county government expressed:

> But you have to be sensitive to the fact that all of the citizens of Fulton County pay taxes. And so, citizens in the North, are not going to probably be willing or comfortable with paying a rate of tax that basically subsidized only the vulnerable population, depending on what their particular political philosophy is, they might be very against that and do see it as necessary to be more consistent with the allocation of resources across the county.
>
> (I-24: 67)

At the county level, the impression prevailed that politics was tailored particularly toward high tax districts, which also appeared to be more outspoken on political matters and are predominantly white. The power of certain constituencies to have their interests represented was also brought forward by an academic and long-standing policy advisor, who worked at different levels of government in the United States and Atlanta. This interviewee expressed that local decision-makers perceived their responsibility toward a more outspoken local constituency. As a result, it would be less likely that they would act upon the vulnerability of other constituents (I-21).

Permissible conceptions in Jinhua: at-risk houses, poverty and age

In China, the issue of vulnerable groups was not accessible right away. Whereas adaptation experts had some background knowledge on the matter and academics working on related issues could come to terms with related concepts more directly, it seemed to be a relatively novel discussion for government officials and those working in state agencies. In all cases, the issue was not discussed proactively. After explaining the concept of human vulnerability to climate change, poverty would be perceived as primary issue determining people's vulnerability to natural hazards and climate change. Additionally, there seemed to be consensus around villagers being particularly vulnerable, as one policy practitioner in Jinhua notes:

> In rural areas, the family conditions are really not good, they have no ability to renovate the house.
>
> (I-39G, respondent a: 212f.)

Very often, interview partners correlated the higher vulnerability of villagers with their lower socioeconomic statuses. In light of fewer financial means, at-risk houses were commonly mentioned as a significant factor of peoples' higher vulnerability. Low-quality housing was coupled with age and state of health, as well as the location. Thus, housing vulnerability to flooding or landslide accidents in light of heavy precipitation and mountain sludges was commonly outlined (I-12, I-13, I-37, I-39G). Ironically, interview partners in Jinhua and Hangzhou also explained how people with these characteristics could end up being less vulnerable due to governmental subsidies that would help them repair and renew their houses every year. One policy practitioner working at the Five Water Treatment Department in Jinhua explains:

> There is this practice of government renovation of dangerous houses. The renovation of dangerous buildings takes place for vulnerable groups, and it is indeed annual.
>
> (I-39G, respondent b: 213)

Apparently, in other jurisdictions of Zhejiang province, the local populations started perceiving typhoons as an annual opportunity for renewing their houses and getting the renovations paid through governmental programs. The post-disaster cash inflow into certain regions was noticed with shaking heads and some sort of discontent. Another Hangzhou-based resident and academic referred to two Zhejiang districts, which profited from government-initiated city redevelopment in light of the annual typhoon season:

> Wenzhou und Cong'nan became rich because of typhoons. Every year, they would experience urban renewal.
>
> (I-14: 13)

Contentious groups in China: "external populations"

Similar to vulnerable populations in Atlanta, there was contention on the question who qualifies for being considered vulnerable. The most contested groups in this context included migrant workers. They were sometimes not considered an official part of the local population, as one policy practitioner and academic in the field of emergency management and disaster politics states:

> Generally speaking, the migrant workers should belong to the vulnerable populations. However, there is some city in China, there are more migrant workers than local people, or even their population is equal. For this condition the government will not pay particular attention to that group of people. However, if the amount [sic] of migrant workers is less than the local citizens, the government will support the migrant workers, who are living in bad life conditions. So, I think it depends on different conditions.
>
> (I-13: 44)

Interestingly, interviewed government officials as well as state agency employees did not touch upon the issue of migrant workers, whereas most academics and adaptation experts also outlined migrant workers as vulnerable group. However, they would also view the conceptualization of migrant workers as vulnerable population problematic and addressing it politically highly unfeasible. Very often, migrant workers were considered an "external group" rather than belonging to the domestic population. One academic reflects upon the complexity of defining external groups:

> The external population, that is, maybe, when we do research, we need to be more prudent, we need to distinguish between external populations from Hangzhou, from big cities, and external populations from farming countrysides. The reason why I separate external populations cautiously so that is why this is why I put peasant workers (*nongmin gong*) after external population (*wailai renkou*). That is to say, the external populations like those in Hangzhou City, there are a lot of external people, but some external populations they get very good social welfare, for example, if the population is from overseas, from abroad, they are very high-level talents, so they can get a very high salary, with the preferential policy in plan. So, they are not the nongmin gong we are talking about. So, the external population I mean, is not vulnerable from the medical perspective. It should be more detailed, like the migrant worker people (*wailai de nongmingong*).
>
> (I-12: 61f.)

According to this understanding, peasant workers are considered more vulnerable than urban external populations from abroad, or those from bigger cities, due to their commonly higher socioeconomic backgrounds; greater access to resources such as education, healthcare as well as social security and preferential working rights (see Figure 7.3). In China, this dichotomy of urban and rural residents is reflected in the Household Registration System (hukou), which was implemented in 1958. Introduced under Mao Zedong, this system was thought to prevent large-scale urban–rural migration.

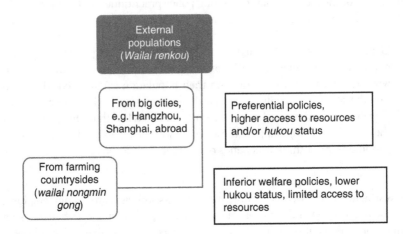

Figure 7.3 Differentiation of external populations and causal links to policy factors that influence population vulnerability.

Source: The author

According to the interviewee, the external urban population should be further differentiated into "high-level talents" from abroad and migrant workers from within China. These different groups of external populations clearly have different backgrounds and resource means. In this context, it was complained that very often all these different groups are lumped into one category rather than distinguished more carefully. Further, the problem of vulnerable groups was seen as a social and a political issue, as one academic and policy advisor notes:

> The definition of vulnerable populations should depend on different dimensions. For example, China will define the vulnerable populations according to [different] aspects. The first one is politics, the second one is the social structure.
>
> (I-12: 43)

Such a process usually serves two important functions: on the one hand, it acknowledges this is a major but overtly complex issue of great significance for the broader parts of society. On the other hand, this procedure shirks responsibility away to other, oftentimes higher jurisdictions. Aside from the complex and increasingly problematic task of classifying vulnerable populations into different groups, the aspect of political enfranchisement in light of (perceived) scarcity was reflected upon. One policy practitioner working for an environmental and economic Think Tank in China and the United States explained:

> So, there are the migrants, who obviously are in some ways a bit similar to, you know immigrants in the United States. That is another parallel that is

worth exploring. You know, there is the elites, who are, who really are not interested in franchising the migrants into // to get entitled to the public goods and social services. I kind of lumped all that into the concept of scarcity, because there is a scarcity supply for a lot of public goods in China. There is just not enough. And so, the elites know that, and they want // and there is only so much, and the elites want all of it. And they do not want to // and so as soon as you enfranchise more people, that enfranchisement is essentially redistribution. That is what it is. Political enfranchisement is economic redistribution, those are the same things.

(I-36: 78)

The political entitlement to resources was a reiterated phenomenon and an important aspect for the analysis of causal factors that may explain differential societal exposure. One Chinese academic working on vulnerability within child welfare legislation notes more generally:

Human vulnerability has not mattered much, not been a policy focus. How to look at vulnerable institutions and how they provide access to overexposed groups, I think that is more important for the Chinese analysis.

(I-26: 3)

Main finding: protracted vulnerability based on different patterns

On the ground, the social vulnerability perspective faced several political difficulties. The findings suggest that some of the vulnerability categories were not feasible in political practice. It was found that the thinking about and perception of vulnerable populations, may even lead to a protraction of vulnerability that is characterized by select vulnerability acknowledgment and social stigmatization of "vulnerable populations."

Patterns of biased vulnerability acknowledgment

Different patterns of vulnerability acknowledgment could be detected (see Table 7.3). These were in line with the issue of contention, that is, which group is "deserving" of being considered vulnerable. There were certain uncontentious groups defined by age, status of health, economic, and geographic location of people's occupation. Selected awareness existed about people with disabilities and tribal backgrounds. In contrast to acknowledgment, tribal backgrounds and historical minorities in the United States were in part considered a politically delicate issue when looking at policy implementation efforts (I-11). Contested populations, such as homeless people or migrant workers faced underestimation and de-emphasis by local decision-makers. These groups only experienced very selected acknowledgment and in some cases were not considered worthy of "vulnerability entitlement."

Table 7.3 Vulnerable groups through the eyes of local officials and policy advisors (grey = viewpoint of local officials, white = academics and policy advisors)

Pattern	Jinhua/Zhejiang	Atlanta/Georgia
Bias: Politically motivated acknowledgment of uncontentious groups	"It is those over 60 years old, 70 years old. Their health is not very good, and then, their children may also not be able to support them." (I-39G: 204f.)	"Well, the elderly, the poor and the children, I do not know. What else? Those who have to work outside?" (I-30: 99f.)
Bias: Underestimation and deemphasis of contentious groups	"Disadvantaged groups are all the same. Jinhua has the same disadvantaged groups [than other places]. It is where the income is relatively low, low-income people, the amount of people with disabilities. Ah, most of these groups are relatively small." (I-38: 59)	"Well, I mean, we got, poor people in the district, homeless people, and certainly they are subject to the same stresses than anyone else." (I-28: 55)
Stigmatization	"Those who cannot take care of themselves." (s.a.)	"Bottom of society" (s.a.)
Selected Awareness	"Who are the vulnerable? In our province that is pretty clear: there is a strong awareness about the elderly, disabled people and children." (I-14:11)	"That is a decision-cutter classic: folks have a sense for minority groups and that there needs to be outreach engagement with that group. . . . I do not hear enough talk about the disabled. There is a pretty decent discussion about the elderly." (I-09: 8)
Emphasis on local characteristics	"The definition of vulnerable population should depend on their wealth. So according to the wealth, the people get divided into three levels, the poor, the rich and the middle-class. And as for the poor group, they also can be divided into three levels, the first one is urban, the second one is rural. [The third one is the subdivision of poor into suburban and rural poor]." (I-12: 53)	"Definitely historical Gullah Geechee. Historic slave families, socially and historically vulnerable population in most of the southern states, Georgia, North and South Carolina. [. . .] uneducated, unemployed that live in the low-income housing areas, somewhere in the middle county." (I-11: 28)

Cultural stigmatization furthers population marginality

In other instances, and particularly prevalent in China, certain groups and people who qualify for being considered "vulnerable" faced stigmatization. Here, all interviewees agreed that the term for vulnerable people (*ruoshi qunti*) covers children, elderly, and disabled people. Low-income households, one-parent families, pregnant women and blue-collar workers were the extended definition provided by some interviewees (e.g., I-17, I-18). Here, the biggest issue of contestation was migrant workers. Some considered them to be part of the vulnerable, as "their situation and payment is always at the bottom of the society" (I-17: 3).

But in Atlanta, too the issue of vulnerable populations unveiled forms of stigmatization. Many interviewees exemplified an understanding of vulnerability as the incapacity of people to take care of themselves. When being asked how to define vulnerable populations, one state actor and policy practitioner expressed:

> It is the classic: the people who cannot take care of themselves, or have issues taking care of themselves.
>
> (I-27: 72)

Externally disputing that a person can provide for themselves culturally devaluates certain parts of the population. This aspect will be examined in greater detail in the following two sections.

The cultural connotations in both English and Chinese language and views of local decision-makers implied a form of weakness and codependence through being considered in need of protection. Lacking adaptive capacity was considered a form of incompetence in dealing with injury. This complies with the research of adaptation scholars, who have equated vulnerable nations, regions or people commonly with "those that are least well equipped to cope with the impacts of climate change" (see, e.g., Brooks et al. 2005: 151). Environmental justice scholar and political economist Jessie Ribot points to the problematic connotation of a social-Darwinistic ethic in the context of human adaptation to climate change. Accordingly, those who do not adapt; those who are said to have low or no adaptive capacity; those who did not survive, were not fit. This kind of victim framing and blame assertion is shifting attention away from the sources, may they be social, political or economic in nature, which (re)produce and maintain marginality (Ribot 2011).

In China, the terming of "vulnerable populations" and wording itself was problematic. The main wording, *ruoshi* signifies vulnerable, and of weak or inferior status, whereas *qunti* refers to crowds or populations. Paired together, the term can refer to disadvantaged social groups, as well as economically and politically marginalized groups. Nonetheless, the way it was used usually implied "weak people." Most interviewees stated that the term is related to the weak part of society "the sick people who cannot pay their bills, not the rich who are sick" (I-15: 23). Overall, most terms carried a stigmatizing connotation through framing socially vulnerable people as an inferior and dependent group. It seemed that it was particularly villages that were considered "backward" and under general suspicion rather than viewing them as independent. Besides stigmatizing and disempowering the considered "vulnerable," the aspect of differential recognition was highly problematic and did not quite translate into the local Chinese context.

In both cases, the underestimation and de-emphasis of certain vulnerable groups appeared to be linked to concerns of political legitimacy. Whereas population vulnerability in the context of climate change and long-standing historical segregation of African American communities was a politically sensitive issue in Atlanta, the long-standing divide of rural–urban populations, and migrant workers (that were commonly considered "external populations") were a politically sensitive issue in China.

Selected awareness

In addition to biased forms of select acknowledgment based on the politically motivated contention of distinctive groups, selected awareness about other vulnerable populations that appeared not to be driven by a political agenda could also be detected. In Zhejiang, and not exclusively related to Jinhua, an academic and policy advisor reflected upon the matter that the broad understanding of vulnerable groups relates to the elderly, children and people with disabilities. In Georgia, an academic implied that the discussion about the elderly and certain minority groups was more advanced than that of disabled people (see Table 7.3). To what extent this was the result of dominant knowledge patterns and discourses from within society was not examined. However, the findings may relate, to what has been said before, that these groups are the commonly recurring ones in the context of other debates on vulnerability.

Emphasis on local characteristics

Another pattern that was observed in both cases is the reference and emphasis of distinctive local characteristics that interrelate with other socio-spatial factors. In Zhejiang, the status of wealth was strongly related to the house registration system (hukou) and urban–rural status. For instance, a person with an urban hukou is less likely to be poor. Still, the growing disparity in the context of different degrees of urbanization was detected. In Georgia, historically marginalized groups were located in different parts of the state, and thus, in the southern part of Georgia, there was a greater awareness of the Gullah Geechee, a historical slave family that settled in that part of the country.

Lock-ins related to knowledge and politics

The third question, which guided this chapter, is: which lock-ins can be detected to explain protracted vulnerability? Protracted vulnerability was characterized by select awareness and bias in acknowledgement, as well as stigmatization of vulnerable populations. In the examined cases, lock-ins became particularly evident at the level of past-dependent knowledge patterns and processes of sense-making. These appeared to be strongly influenced by the way political institutions operate,

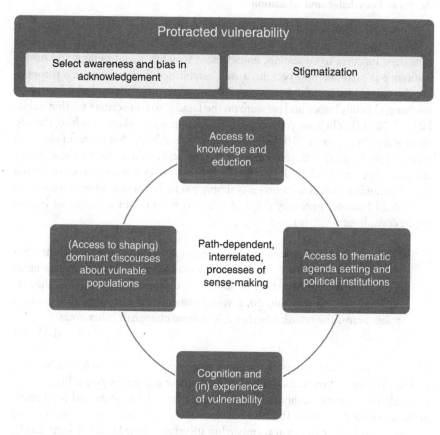

Figure 7.4 Lock-ins related to knowledge and politics that explain protracted vulnerability

also in terms of (not) providing access at different levels, i.e., access to shaping dominant discourses, access to thematic agenda setting, as well as knowledge and education. At the same time, the inexperience of decision-makers working for political institutions with vulnerability was often pointed out. These factors intersect with processes of cognition: who experiences vulnerability to climate impacts, and why? Being able to feel what it means to have little access to information on, e.g., extreme heat waves and proper cooling opportunities or limited healthcare is not readily intuitive to most actors in the political administration. Strikingly, it appeared, the populations with limited access to public goods are with few exceptions, commonly not involved in processes of sense-making and often belong to (historically) contested populations in terms of biased vulnerability acknowledgment. It became also evident that past-dependent patterns of segregation strongly interrelate with questions of access to knowledge. These factors combined can explain, why vulnerability is protracted, awareness selected and acknowledgment of population groups biased (see Figure 7.4)

Access to knowledge and education

The previous analysis showed that migrant workers were a contested group in terms of vulnerability recognition in Jinhua. Policy advisors and academics noted that their inaccess to education, employment benefits, social welfare and health insurance is systemic and became a well-established political practice throughout the years. Migrant children were emphasized to suffer particularly, due to the unchanged family legacy and pressure on the family "to compensate for that public lack" (I-26: 12). Zhejiang province has many migrant workers, reaching the top three status in China in 2010 (Zhang et al. 2016). Although it is one of the most affluent provinces, the related economic burden and medical cost for these groups are considered particularly high. In this context, several interviewees indicated that the number of migrant workers is simply too high in a time of scarcity, so that they could become a primary political issue, as reflected in a statement by this policy consultant:

> Once you politically enfranchise the migrants to equal urban citizens, that gives them the right to demand services and public goods that the elites in urban areas feel like there is already, you know, scarcity of and they do not want to hand that out. So, I would imagine that that idea, that concept applies across the board, whether it is climate change or other areas.
>
> (I-36: 79)

Ma and Adams (2014) have referred to scarce public goods such as education and healthcare as symptoms of intensifying Chinese inequality. According to their research, educational advancement presents a particular migrant and rural grievance. In a country of more than 1.4 billion people, education has become a scarce resource and barrier to equal membership in urban life (Ma and Adams 2014). The educational system has become increasingly competitive and biased toward urban-registered households and presents a "systematic discrimination against the migrant class" (Ma and Adams 2014: 8).

Whereas Chinese interviewees indicated that education and knowledge are some of many other public goods, to which the migrant worker communities only have limited access, they were considered key elements for moving out of poverty and being able to exert political power in Atlanta. But they too, interviewees argued, need to be combined with early access to other public goods, such as good jobs, due to poverty and the need to support family. One elected government official working for the Atlanta Municipal Government commented on education and workforce development as effective means to move out of marginality:

> I say education and workforce development, coupled together. Because not everyone is going to go to college but having valuable skillsets that offer you the ability to provide them to the marking [is important too], with or without a college degree, or even without even a Highschool diploma, or even before you are out of high school, is the name of the game. Because if you are living

in a margin, [and] your parents are not making a lot of money, you need income in your household. And if you have a way to feed yourself and help your parents, or help your home, your brothers and your sisters, at sixteen, [that] is a game changer for you.

(I-33: 17)

Aside from indicating the importance of education for moving out of poverty, this statement also makes visible other challenges related to communities in poverty and the need for immediate financial support through generating an income. Another interviewee, working for the Fulton County Government noticed Atlanta's divide into a richer North, with significant privileges in terms of access to public goods in contrast to the poorer South:

So, if you look at [Fulton] county as a whole, the city of Atlanta sits right in the middle. And the county is somewhat divided by communities north of Atlanta and then communities south of Atlanta. And uhm, once you start looking through our performance information, you will see that the cities that are north of the city of Atlanta and the northern part of the city of Atlanta are thriving. And the cities that are south of the city of Atlanta are the ones with the lower, medium incomes, lower access to healthy foods and stuff like that, lower education rates, et cetera.

(I-24: 62)

The recurring sentiment was that knowledge and education were key brokers for a better life, but that this was distributed unevenly across society. This is in line with political economist John Harles (2017), who finds that university education has become one of the more significant contemporary factors leading to income inequality in advanced economies. The examined case of Jinhua reveals similar trends for China and will be touched upon in greater detail in Chapters 8 and 9. An elected government official, who is also an environmental activist concludes about vulnerable populations in Atlanta and the United States:

There is a certain segment of the population that is not able to experience the American dream.

(I-08: 14)

(Access to shaping) dominant discourses about vulnerable populations

Throughout the research process, it became evident that existing social vulnerability assessments are only a narrow representation of a constructed reality that is problematic on different levels. There was a gap between what the assessments consider to constitute legitimate knowledge regarding population vulnerability and what can actually be processed in political practice. It also became apparent that the concept of social vulnerability knowledge is embedded within political institutions. In line with Daviter (2015), the role of knowledge in the policy process is bound by structural constraints. In the cases examined, some of the structural constrains relate to a rather exclusive discourse of vulnerability categories and severe methodological shortcomings of vulnerability knowledge. Further, the issue of differentiality appears to be politically problematic in addition to cultural connotations with vulnerability.

Social vulnerability assessments as laboratory findings

The examined social vulnerability assessments constitute laboratory findings that did not match the prevailing, local connotations and decision-makers' awareness of human vulnerability on the ground. It seemed that the understanding of human vulnerability to climate change, as laid out in the SVAs, did not trickle down to the local political level, nor did it seem advisable that they should. In both cases, divergent opinions on what constitutes human vulnerability and different means to politically acknowledge them turned out to be problematic. Whereas factors such as poverty and age were more likely to be acknowledged politically and constituted unproblematic indicators of vulnerability, other components such as ethnic background or hukou status were politically contentious. Further, the politicization of vulnerability or "who deserves to be considered vulnerable?" needs to be taken into consideration when operating related concepts on the ground. When confronted with population vulnerability in their jurisdiction, officials naturally took a more defensive stance.

The type of knowledge produced through SVAs and calls for recognition present a shortcut of thinking, which is interconnected with cognitive and psychosocial levels. The findings indicate that the vulnerability categories outlined in the vulnerability assessments are not very helpful to understand matters of protracted inaccess of vulnerable populations to public goods. The assessments carry intransparent assumptions and do not touch upon the underlying political patterns. Some Chinese interview partners also emphasize that poverty is not necessarily the major determinant of climate affectedness and exposure. Rather, vulnerability appears as a function of inequality based on hukou-status, which is a political path-dependency, as this Hangzhou-based academic and policy advisor to the Zhejiang provincial government explains:

> Low-income people are not necessarily the vulnerable, but it is more of a problem related to their hukou status.
>
> (I-12: 59)

Methodological shortcomings of social vulnerability knowledge

Social vulnerability assessments were considered to have introduced the "human dimension" to a previously infrastructure-tinted and largely technical field. Yet, the foregone empirical findings expose that this growing body of literature and related policy practice is vulnerable to critique on different levels: 1) Methodologically, SVAs constitute a questionable scientific practice, as the underlying assumptions are not made visible and certain status quo is unquestioned and reproduced. 2) SVAs are ethically problematic, as they constitute a political act and externally ascribe vulnerability onto a person without validation of the considered factors. Aside from the stigmatizing notion of social vulnerability as a pre-existent condition of certain populations, this external construction of vulnerability not just marginalizes certain communities further, but also imposes a status quo related to lifestyle norms and development. 3) This lens can crucially oversee local practices of self-governance. Recent validation studies suggest that cultural practices and daily habits may significantly impact how the considered "vulnerable parts of society" end up being less vulnerable. More research is needed on the science–policy interface, how and why those studies (do not) travel to decision-makers and whether this is a procedural problem, or more deeply rooted in the culture of stigmatization.

Vulnerability assessments also serve several important political functions: they intend to inform and advise decision-makers on their problem perceptions and policy planning. Simultaneously, they are an instrument of political communication, which is related to political legitimacy: assessments are considered to send an important signal that the problem of population vulnerability is acknowledged and being addressed. But very often, vulnerability recognition remains at that, despite being selective in political practice.

The politically problematic function of differentiality

Calling someone vulnerable constitutes an act of power. Through judging certain socio-demographic characteristics and making predictions about how people will react under (uncertain) conditions engages in a very subjective perception of vulnerability that stigmatizes. Aside from stigmatization the practice of naming someone vulnerable also engages in a bodily possession through externally ascribing vulnerability onto a person. Overall, the process of naming socially vulnerable groups has not changed why those parts of the community do not have access in the first place. Pointing the finger at vulnerable groups does not change the baseline conditions of inequality and exploitation. This conception too narrowly takes a rest in naming, boxing and judging the different characteristics, which deem a person vulnerable. Blame is assigned to potentially marginalized people, whose circumstances may not have been of their own making in the first place. The stigmatization becomes perfect through creating a differential sense of human vulnerability (also see Fraser 1995), instead of approaching vulnerability as an omnipresent phenomenon (Fineman 2017). Aside from stigmatizing and engaging in a culture of blame, the determination of certain vulnerable groups performs an act of disempowerment and further engages in political marginalization. An Atlanta City Council official expressed the difficulty of a cultural sense of belonging:

> It is just harder for everyone to feel like they are on the same team, when there is that level of disparity.
>
> (I-33: 101)

Emerging from the results given previously is one key factor that shapes uneven community exposure to climate impacts: the political disenfranchisement and marginalization of certain populations. Calling out vulnerable people based on exclusive and unverified scientific insights is yet another form of political disenfranchisement.

Etymological and cultural legacies of vulnerability

When the term "vulnerable" was first used in the seventeenth century, it carried the connotation of a warrior and precisely meant "having the power to wound."[2] This is tied to the Latin roots of the word "vulnerare," which used to mean "to wound" and was understood as "the capacity to be wounded." The meaning gradually shifted to a less figurative and empowering sense by focusing on the aspect of exposure. Environmental geographer Kirstin Dow significantly coined the vulnerability concept in the context of climate impacts at the beginning of the 1990s. Nestled within the geography and natural hazards literature of the 1970s and 1980s, the term "vulnerability" originally referred to the susceptibility of a system to risk. The concept of vulnerability has always been closely tied to the concepts of "resilience" and "risk" and gradually expanded to include human susceptibility to intensifying climate impacts. Apart from terminological widening, the empirical findings disclose how the initial meaning of vulnerability collapsed from "powerful wounds" to implying weak social status. In Atlanta, a structural bias against vulnerable groups was emphasized by experts from different backgrounds, who noted this as a path-dependent feature of the political system. One Atlanta-based academic and policy practitioner elaborates:

Here is a systemic bias against the vulnerable, and that our system is not cut out right now to systematically benefit those who are most in need (I-21: 118).

Cultural stigmatization through language in China

In China, three main terms are operated to describe vulnerability in the context of climate change and vulnerable populations: the first term is "cuiruo" (脆弱), which is usually paired with physical surroundings (for example, coastal vulnerability or ecological environments).[3] The term carries a materialistic connotation of things, which are seen to be fragile. The precise translation is "fragile, brittle, crisp." The use of the term comes closest to what most researchers consider climate vulnerability (*qihou bianhua cuiruoxing*) (also see Fang et al. 2009). The second term is "mingan" (敏感) and describes the sensitivity toward something. Like "cuiruo," "mingan" is an interdependent term, which is often placed next to sensitive areas, such as sensitive industries, sensitive environments or sectors. It is considered a relatively neutral term. A combination of the two terms, which signify "climate-sensitive" (*qihou mingan*) and "vulnerable areas" (*cuiruo qu*) is frequently mentioned in related policy documents. The expression "cuiruo qunti," which translates as "human vulnerability," but precisely refers to vulnerable crowds, is often mentioned next to vulnerable places or indicates regional vulnerability (*qu huhe renqun de cuiruoxing*). The term that corresponds most closely with social vulnerability is the term "ruoshi qunti." It is commonly translated as "disadvantaged," "underprivileged group" or "vulnerable community." However, the cultural meaning of the concept carries considerable judgment. In the United States, a path-dependent structural bias in terms of the way language operates culturally against designated people who are designated as "vulnerable" was noted as well.

Access to thematic agenda setting and political institutions

Historically path-dependent thematic priorities in the United States

When asked about the perception of vulnerable populations in the United States and whether they are recognized as part of official decision-making and policy planning structures, a majority of interview partners emphasized the significance of tribal and ethnic minorities, which appeared to be driven by growing historical knowledge on matters related to segregation in the South, but also through a federal directive. A state actor working for the Department of Natural Resources in southern Georgia outlined the Gullah Geechee, a historic slave family in the southern states as particularly vulnerable group that has received some acknowledgment in the South (I-11). Yet, when emphasizing vulnerable groups in a social sense in political practice, bias toward certain groups appeared to be driven by questions of social intuition. The elderly and children appeared to be the most intuitive categories, whereas categories such as ethnic minorities or migrant workers were not as obvious for some decision-makers. To what extent this is the result of segregated historiography, and decision-makers relying on different sources of knowledge based on their own background need further research.

Apart from certain thematic priorities, a discrepancy among different government officials in terms of awareness was detected. Some were highly aware in terms of vulnerable populations and were also cherished for their particularly conscious stance and engagement in that area: "[W]e have commissioners who are hyper-sensitive to vulnerable populations" (I-24). Other officials deemphasized the issue of vulnerable groups, claiming they are "subject to the same stresses than anyone else" or would guess blindly on vulnerability matters (see Table 7.3). Since data on party affiliation was not obtained, and whether this coincides with political bias toward certain groups, no inference on this matter can be made. However, among the officials interviewed, it seemed that those working in southern districts and/or in socially marginal neighborhoods had greater empathy toward vulnerable populations, as outlined in the SVAs.

Inaccess to decision-making as an extension of inaccess to education

It became obvious that access to thematic agenda-setting in the context of political decision-making strongly correlates with access to education. The historical underrepresentation of certain communities and exclusive decision-making present a core historical path-dependency that was noted by one policy advisor and adaptation planner:

> There are groups that have traditionally not been part of the regional planning. There is a push to get other regional planning communities for regional engagement // communities [that are] more excluded [are] communities of color, immigrants, lower-income communities [who are] historically not well-represented in this.
>
> (I-03: 9)

The outlined communities are those, which have had very limited access to education. It became obvious again that the analyzed SVAs deal too superficially with the underlying causes of long-term and uneven societal exposure. Local historical, cultural and political dimensions constitute important factors that further influence how sectors matter for local populations, how populations have the capacity to deal with changes (and want to deal with them) and how vulnerability is conceptualized in a way that is culturally acceptable.

Prevailing vulnerability assessments have limited reach to influence how vulnerability conceptions operate at local political levels and which groups are deemed "deserving" of being considered vulnerable. Policy advisors also linked the lack of awareness of human vulnerability to a disconnected science–policy interface. In some cases, however, it became clear that knowledge had been deliberately used or ignored. Several reasons were provided but evolved mainly around issues of political legitimacy.

In Atlanta, ethnic minorities and Afro-American populations were considered marginalized populations. Here, the structural bias against vulnerable populations was explained by decision-makers' lack of insight into these parts of the population. One policy advisor and adaptation expert notes:

> There is a structural bias across the board: who works for city government? People who were probably never poor. Is it people who have no college degrees? No. Those people without high-school degrees do not exist in city government. [There is a] total lack of knowledge about a black uneducated teenager, which makes governments somewhat resistant. They lack experience. The political machine represents one group of the society. Then there is the financial aspect of political campaigning, who are your funders, who you have to please.
>
> (I-06: 34)

In addition to the underrepresentation of some groups in decision-making due to path-dependent patterns of (in)access to knowledge, this statement points to a more sophisticated form of inaccess: privileged decision-makers, who cannot readily access the experience-based knowledge of vulnerable populations. The inexperience of vulnerability furthered knowledge restrictions and led to select interest representation of the more acknowledged and noticed population groups.

Summary

This chapter examined local characteristics that are considered to correspond with higher social vulnerability to climate change based on the comparison of vulnerability assessments. After reviewing the dominant vulnerability constructions, the chapter analyzed how local decision-makers in Atlanta, and Jinhua perceive of vulnerable populations to climate change, and whether they are generally aware that climate impacts are unevenly distributed. A significant difference between what the vulnerability assessments portrayed and actual degrees of awareness about climate change and adaptation needs, especially regarding the exposure of particular groups in society, was found. In both cases, political bias toward certain populations persists, and though the problem recognition about the social dimension of vulnerability was quite different when comparing Jinhua and Atlanta, local decision-makers' problem recognition about this was rather low in the context of adaptation in both cities. It was argued that select awareness and bias toward certain groups, coupled with select acknowledgment and stigmatization further protract the vulnerability of local populations. The last part of the chapter analyzed the lock-ins, which help to explain these findings. The dominant lock-ins detected are related to knowledge and politics. Past-dependent-knowledge patterns and processes of sense-making, coupled with the way political institutions operate in terms of access to different types of knowledge (e.g., knowledge on adaptation, vulnerability and historical segregation) can explain low-problem recognition about the social aspects of adaptation and vulnerability maintenance. These types of lock-ins become especially evident with regard to the limited access of shaping the dominant discourses and thematic agendas on vulnerable populations. This appeared to correlate with limited access to experience-based forms of knowledge but also education.

Notes

1 For regional geographic and biophysical lock-ins, also see appendix A1.
2 See "The History of Vulnerable" Merriam Webster Dictionary, available at: www.merriam-webster.com/dictionary/vulnerable, last accessed July 6, 2019.
3 The policy documents analyzed include China's 2013 National Adaptation Strategy, the latest Climate Change Response Plan (2014–2020), the Urban Climate Adaptation Action Plan (2016) and the 1st Biennial Review on Climate Change (2016).

References

Brooks, Nick, W. Neil Adger, and P. Mick Kelly. "The Determinants of Vulnerability and Adaptive Capacity at the National Level and the Implications for Adaptation." *Global Environmental Change* 15, no. 2 (2005): 151–163. https://doi.org/10.1016/j.gloenvcha.2004.12.006.

Centers for Disease Control and Prevention (CDC). Planning for an Emergency: Strategies for Identifying and Engaging At-Risk Groups. A Guidance Document for Emergency Managers: First Edition. Atlanta (GA): CDC, 2015.

Chen, Wenfang, Susan L. Cutter, Christopher T. Emrich, and Peijun Shi. "Measuring Social Vulnerability to Natural Hazards in the Yangtze River Delta Region." *China International Journal of Disaster Risk Science* 4, no. 4 (2013): 169–181. Doi: 10.1007/s13753-013-0018-6.

Daviter, Falk. "The Political Use of Knowledge in the Policy Process." *Policy Sciences* 48, no. 4 (2015): 491–505. Doi: 10.1007/s11077-015-9232-y.

Environmental Protection Agency (EPA). *Adaptation Implementation Plan Draft, US EPA Region 4*. EPA, 2014. https://www3.epa.gov/climatechange/Downloads/Region4-climate-change-adaptation-plan.pdf, last accessed January 16, 2020.

Fang, Yiping, Dahe Qin, and Yongjian Ding. "A Review of Adaptation Research on Climate Change." *Research on Arid Areas Journal* 26, no. 3 (May 2009). (Original: "气候变化适应性研究综述".)

Fineman, Martha Albertson. "Vulnerability and Inevitable Inequality." *Oslo Law Review* 4 (2017): 133–149.

Ford, James D., Tristan Pearce, Graham McDowell, Lea Berrang-Ford, Jesse S. Sayles, and Ella Belfer. "Vulnerability and its Discontents: The Past, Present, and Future of Climate Change Vulnerability Research." *Climatic Change* 151 (2018): 189–203. Doi: 10.1007/s10584-018-2304-1.

Fraser, Nancy. "From Redistribution to Recognition? Dilemmas of Justice in a 'Post-Socialist' Age." *New Left Review* 1, no. 212 (1995): 68–93.

Ge, Yi, Wen Dou, Zhihui Gu, Xin Qian, Jinfei Wang, Wei Xu, Peijun Shi et al. "Assessment of Social Vulnerability to Natural Hazards in the Yangtze River Delta, China." *Stochastic Environmental Research & Risk Assessment* 27, no. 8 (2013): 1899–1908. https://doi.org/10.1007/s00477-013-0725-y.

Gutierrez, Kristie, and Catherine E. LePrevost. "Climate Justice in Rural Southeastern United States: A Review of Climate Change Impacts and Effects on Human Health." *International Journal of Environmental Research and Public Health* 13, no. 189 (2016): 1–21. Doi: 10.3390/ijerph13020189.

Harles, John. *Seeking Equality: The Political Economy of the Common Good in the United States and Canada*. Toronto: University of Toronto Press, 2017.

Hu, Kejia, Xuchao Yang, Jieming Zhong, Fangrong Fei, and Jiaguo Qi. "Spatially Explicit Mapping of Heat Health Risk Utilizing Environmental and Socioeconomic Data." *American Chemical Society, Environmental Science & Technology* 51, no. 3 (2017): 1498–1507. Doi: 10.1021/acs.est.6b04355.

KC, Binita, Marshall Shepherd, and Cassandra Johnson Gaither. "Climate Change Vulnerability Assessment in Georgia." *Applied Geography* 62 (2015): 62–74. https://doi.org/10.1016/j.apgeog.2015.04.007.

Ma, Damien, and William Adams. *Introduction to in Line Behind a Billion People: How Scarcity Will Define China's Ascent in the Next Decade*. Upper Saddle River, New Jersey: FT Press, 2014.

Manangan, Arie Ponce, Christopher K. Uejio, Shubhayu Saha, Paul J. Schramm, Gino D. Marinucci, Jeremy J. Hess, and George Luber. "Assessing Health Vulnerability to Climate Change: A Guide for Health Departments." *Climate and Health Technical Report Series*. Climate and Health Program, Centers for Disease Control and Prevention, Atlanta, 2014.

Ribot, Jesse. "Editorial. Vulnerability Before Adaptation: Toward Transformative Climate Action." *Global Environmental Change* 21 (2011): 1160–1162. Doi: 10.1016/j.gloenvcha.2011.07.008.

Yin, Jie, Zhane Yin, Jun Wang, and Shiyuan Xu. "National Assessment of Coastal Vulnerability to Sea-Level Rise for the Chinese Coast." *Journal of Coastal Conservation* 16 (2012): 123–133. Doi: 10.1007/s11852-012-0180-9

Zhang, Weifang, Dingwan Chen, Huan Zhou, Yanhua Xu, Zhuopu Xu, Ying Ying, and Zhengyan Zhao. "Regional Health-Care Inequity in Children's Survival in Zhejiang Province, China." *International Journal for Equity in Health* 15, no. 188 (2016): 1–9. Doi: 10.1186/s12939-016-0470-1

Zhou, Yang, Ning Li, Wenxiang Wu, and Jidong Wu. "Assessment of Provincial Social Vulnerability to Natural Disasters in China." *Natural Hazards* 71 (2014): 2165–2186. Doi: 10.1007/s11069-013-1003-5.

Cited interviews

I-03, October 25, 2016
I-04G, October 25, 2016
I-06, October 28, 2016
I-08, October 31, 2016
I-09, October 31, 2016
I-10, November 2, 2016
I-11, November 4, 2016
I-12, March 7, 2017
I-13, March 7, 2017
I-14, March 7, 2017
I-15, March 9, 2017
I-17, March 14, 2017
I-18, March 22, 2017
I-20G, April 28, 2017
I-21, May 02, 2017
I-23, May 10, 2017
I-24, May 10, 2017
I-26, May 11, 2017
I-27, May 12, 2017
I-28, May 12, 2017
I-29, May 16, 2017
I-30, May 16, 2017
I-31, May 17, 2017
I-32, May 18, 2017
I-33, May 19, 2017
I-35, August 10, 2017
I-36, August 29, 2017
I-37, October 20, 2017
I-38, November 20, 2017
I-39G, November 21, 2017

8 Accidental adaptation policy

This chapter looks at local policy choices and examines the nature of existing adaptation efforts. It answers the following set of interrelated questions:

(1) What type of adaptation efforts exist?

    a   What is the problem recognition regarding adaptation (and the uneven distribution of climate impacts)?

    b   How is adaptation occurring and how has human vulnerability mattered in the context of existing government efforts?

(2) Why is it occurring that way? (Which lock-ins explain the accidental, reactive and incremental nature?)

Since the adaptation field is a relatively new policy endeavor in both local political contexts, the analysis of adaptation efforts was guided by the questions set up by Kingdon (1984, 2011) for examining the maturity and nature of public problem recognition. The questions seek answers for what is considered important and deserving government attention and identifying factors, which ultimately lead to higher problem recognition. The first part of this chapter looks at the process of adaptation policymaking as reflected in the local problem understanding and related planning. The chapter finds that the problem recognition is relatively immature and characterized by unintentional adaptation projects. The second part of the chapter explores the underlying political processes to better understand the content and state of policymaking related to adaptation, resilience (and human vulnerability). The immaturity of the problem recognition is explained by looking at path-dependent lock-ins, which suggests a low adaptive capacity of governance institutions.

Local climate adaptation by accident

The state of adaptation efforts in Atlanta

For the past 20 years, Atlanta was based in a climate-skeptic state, signified by unfavorable political conditions and little support for designing governmental

DOI: 10.4324/9781003183259-8

responses that address climate change in a deliberate manner. Thus, most climate-related efforts in Georgia have relied on municipal, local initiatives, which have gradually become attuned to its external political environment. Local actors, who are engaged in climate policy are used to maneuvering around the issue of climate change, which has often been considered a politically sensitive issue. Further, they have developed distinctive coping strategies in dealing with their adversarial political environment (Teebken fc.). Although local municipalities, like Atlanta, appear to have grown largely independent of climate-related support from the state level throughout the years, widespread climate skepticism, the lack of scientific consensus and lack of agreement on (ambitious) climate policy have had their detriments for climate governance in Atlanta as the following chapter section will show.

Against this background, the most recent political changes at the federal and state levels are worthy of some attention, as they are likely to impact Atlanta's external opportunity structures for the years to come. Georgia became a powerful swing state during the 2020 presidential elections, leading to an effective majority of U.S. Congress. With the most recent elections, some have focused on the question whether Georgia is a tipping point for a new U.S. policy (e.g., Cronin 2021). Overall, hopes are high that after 20 years of blockage on the issue, a new era of climate action will start. Simultaneously, skepticism persists in light of the large number of conservative voices within Georgia's populace, the back-and-forth tradition with political majorities and lack of substantive policy debate (e.g., see Aton 2021).

Lack of a substantive climate change adaptation strategy

The lack of effective and coherent policy development related to climate change adaptation follows this pattern. Although Atlanta has been experiencing the effects of climate-sensitive natural hazards, and although this has been increasingly pointed out by local political leaders, as reflected in a recent Senate Committee hearing on climate change,[1] it appears that the majority of political decision-makers at county and municipal levels have generally not regarded climate adaptation as an own policy field deserving government attention. Instead, climate change impacts have been addressed as part of other policy efforts, policy fields and strategies. The absence of a dedicated climate change adaptation strategy and policy efforts is reflected in a statement by a local policy advisor and academic in Atlanta, when being asked how they would characterize the current state of adaptation planning in Atlanta:

> Virtually non-existent. Virtually non-existent. You can quote me directly.
> (I-29: 37)

Since the interview took place in 2017, major political developments and intensifying climate impacts have occurred, leading to greater problem awareness about rapidly occurring climate change. Yet, intentional climate change adaptation

Table 8.1 Policy efforts addressing climate vulnerability – Atlanta

Year, Policy	Main objective	Actors
2018: City of Atlanta, Green Infrastructure Strategic Action Plan	Advance green infrastructure, address localized flooding and water quality, increase quality of life and community resiliency	City of Atlanta, Department of Watershed Management, Green Infrastructure Task Force
2017: City of Atlanta Resilience Strategy	Outline Atlanta's major resilience challenges and how to address them	City of Atlanta, Mayor's Office of Resilience
2015: City of Atlanta Climate Action Plan	Proposes a strategy for GHG emissions reductions per sector, mentions adaptation as co-benefit	Mayor's Office of Sustainability
2014: Assessing Health Vulnerability to Climate Change, A Guide for Health Departments	Consists of the Building Resilience Against Climate Effects (BRACE) framework that seeks to help health departments prepare for and respond to climate change	Center for Disease and Control Prevention (CDC), Atlanta
2013: Transit Climate Change, Adaptation Assessment/Asset Management Pilot	Outline principles of how MARTA can adapt to extreme weather events and/or a changing climate based upon the "Assed Management Guide"	Federal Transit Administration (FTA)

Source: The author

action and strategy still appear to be absent for the most part. Efforts related to adaptation were undertaken nevertheless, but were not necessarily framed that way. Among them are the city's most recent Green Infrastructure Strategic Action Plan (2018), Atlanta's Resilience Strategy (2017), neighborhood revitalization programs such as floodwater management parks, and the Climate Action Plan (see Table 8.1). One policy practitioner at the municipal level reflects upon this:

> We have one of the largest green infrastructure projects in the entire country, that is adaptation process, we are working on that, to avoid floods. We are trying to move forward with renewables, that is adaptation process. So, we already are doing some things that have to be analyzed in the [climate action] plan that you need to call assets.
>
> (I-22: 126)

Existing policy efforts are either characterized by an overtly broad focus (e.g., the Resilience Strategy), or they treat adaptation merely as a co-benefit without further specifications (e.g., in the Climate Action Plan). The Green Infrastructure Strategic Action Plan is an example, which is commonly found in adaptation policymaking, characterized by a sectoral focus, in this case water, and creating blue-green infrastructure for dealing with extreme precipitation patterns, localized flooding and stormwater runoff. Within the field of adaptation policy, climate impacts are commonly addressed as part of city planning and infrastructure.

In Atlanta, the main actors involved in adaptation planning (broadly understood) are the former Mayor's Office of Sustainability (MOoS), which transitioned into the Mayor's Office of Resilience (MOoR) in 2016/2017 and recently underwent further changes (see later). The Office is an overarching division that supports the Mayor and city government (also see Tsai et al. 2018). Other City of Atlanta agencies, which work on climate-related matters include the Department of Watershed Management, the Department of Planning and Community Development, the leading national health institute and the Centers for Disease Control and Prevention (CDC), which is headquartered in Atlanta. In addition, nongovernmental actors such as the Nature Conservancy or Chattahoochee Riverkeeper have supported related strategic processes as partner organizations.

One key actor constellation that stands out in Georgia is the Georgia Climate Project (GCP), a state-wide consortium that was founded in 2015 to build a network of experts across the state to advance the science, exchange and solutions in the context of intensifying climate impacts (GCP n.d.). GCP consists of nine colleges, universities and people working on climate change in Georgia. The Fourth National Climate Assessment (2018) referred to the "cross-disciplinary group" of the GCP as an example of best practice by developing research roadmaps that can help policy practitioners to prioritize research and action (p. 756).

The project is exemplary for the outstanding role educational actors fulfill throughout Georgia. A majority of these actors are based in Atlanta. University actors have not just taken on the classical role of providing science and actively informing policy efforts at different levels of government and the non-state sector, but also play a key role in reconvening and engaging local communities. This includes a new generation of changemakers and students, who, for instance, provided advise for local municipal actors on their climate action planning in Atlanta or running their own community projects.

Another key actor is the Federal Transit Authority (FTA) which published a Climate Adaptation Assessment for the Metropolitan Atlanta Rapid Transit Authority (MARTA). The transport sector together with the Atlanta Regional Commission (ARC), an intergovernmental coordination agency and metropolitan planning organization (MPO) for the Metro Atlanta Region has been quite active in addressing resilience in the transport sector. Aside from fragmented action by a few select actors, there is no central stakeholder that has focused and articulated dedicated adaptation efforts for the city of Atlanta. An Atlanta/based adaptation expert and policy planner puts it in a nutshell:

> If I wanted, the most effective person to work with, the most effective office to work with, in Atlanta, it would be the resilience offices. That is where this should be done. But it is just not part of their focus right now. [. . .] So, there is not a clear policy player in most U.S. cities for what we do. The clearest policy

player is the chief resilience officer and that is what the position is designed to do in my mind, is adaptation.

(I-29: 28ff.)

Due to the multi-sectoral nature of the adaptation policy field and strongly local-ized demands, adaptation requires action by different types of actors and all levels of government. It follows that engagement of different actors is not uncommon but even important for mainstreaming climate adaptation as a policy strategy at all levels of government (Russel 2019). However, in Atlanta one major finding stands out: adaptation is not a stand-alone field and is far from being a central tenet of county, municipal and state-led policymaking. In line with adaptation policymaking elsewhere, most efforts undertaken occur as part of urban planning and infrastructure development, e.g., greening the environment. No information could be retrieved to what extent Atlanta's municipal government is working on addressing social and household vulnerability, more disadvantaged city pockets and whether the city provides resources for addressing self-provisionary efforts for climate adaptation.

Broadening the focus of Atlanta's Resilience Strategy

In addition to the most recent Green Infrastructure Strategic Action Plan, which also addresses neighborhoods disproportionately affected by stormwater issues (CoA 2018), the Resilience Strategy is the closest effort the city currently has to address human vulnerability to climatic changes. Initially, the "100 Resilient Cities" (100 RC) challenge, an initiative run by the Rockefeller Foundation (RF), was viewed as an opportunity for Atlanta to draft a climate adaptation plan that would complement the city's Climate Action Plan (CAP). However, this idea was ultimately adjusted by developing a broader focus on resilience, as an employee at the Mayor's Office of Resilience expresses:

Interviewer: So, now my question is, why did you decide for a resilience and not a climate adaptation framework?

Respondent: Um so, when I applied for "100 Resilient Cities," basically my plan, I work in climate, so my plan was to put together a climate adapta-tion plan. . . . In order to put [together] a climate adaptation plan, I need the resources. My goal was to optimize the resources for a "100 Resilient Cities". It still is, to optimize the resources that a "100 Resilient Cities" provide to put together my climate adaptation plan. So, when I honestly put together the application for a "100 Resilient Cities," my goal was to develop a climate adaptation plan. (I-22: 92f.)

In the end, the resilience efforts by the municipal government ended up becoming a much broader strategy, diverting attention away from adaptation as a specific policy field to be addressed. It was indicated that language related to extreme heat and climate change was gradually deleted from the final product, Atlanta's

Table 8.2 Atlanta's transition from shocks and stresses to vision-oriented targets

Document	Shocks (4)	Stresses (4)
Atlanta's 100 Resilient Cities Application (2013)	Infrastructure failure Terrorism Cyberattacks Rainfall flood	Unreliable transportation Aging Infrastructure Poverty and inequality Droughts
Final Resilience Strategy (2018)	**Visions (4)** Vision 1: Preserve and celebrate who we are	**Related targets (16)** (1) Leadership (2) story-sharing (3) build arts and culture sector into the fastest growing industry by 2025
	Vision 2: Enable all Metro Atlantans to prosper	(1) Prepare 100 percent of children for kindergarten (2) Create 10,000 new livable wage jobs by 2020 (3) Ensure career choice
	Vision 3: Build our future city today	(1) Improve quality, access and distribution of affordable housing (2) improve transportation access (3) Improve food security by 2025 (4) Create 500 new acres of green space by 2022 (5) sustainable and efficient energy and water infrastructure
	Vision 4: Design our systems to reflect our values	(1) Promotion of resilience planning by 2022 (2) Increase civic participation (3) Clean energy by 2035 (4) Improve public safety and community cohesion (5) Promotion Atlanta airport as resilience and workforce model

Sources: The author, based upon interviews and CoA (2017b)

Resilience Strategy. Within the policy process, the focus of attention gradually shifted from shocks to resilience visions. Those involved in the resilience planning indicated that there was a strong desire for not having another plan on paper but implementing actual action. This resulted in the articulation of 16 targets and 57 actions in the final resilience strategy. The final product is a mix of concrete and vague targets linked to visions (see Table 8.2).

There is only limited information available to what extent evaluation mechanisms were put in place as part of the strategy, to track the status of the targets and following up on the implementation of the proposed actions. The role of creating positive visions and the advantages of imaginary concepts for enabling transformative change in the context of climate change has been frequently emphasized (see, e.g., Wiek and Iwaniec 2014; McPhearson et al. 2017). According to this idea, (positive) visions can serve a critical function for large-scale transitions, as they can mobilize action. These arguments have been outlined not just in the context of sustainability transitions, but likewise in the context of climate adaptation. Visioning processes have become a key tool for sustainability and resilience planning. However, in Atlanta, this vision-oriented approach ended up being criticized for watering down concrete efforts related to adaptation and climate resilience.

When being asked what role human vulnerability plays in the setting of current policy agendas in the context of climate change, an employee at the Mayor's Office of Resilience states that it is always implicitly part of local planning but that aside from having responsibilities, certain barriers exist in actually articulating a dedicated policy:

> I mean, I think it is always there and plays a role. I think what happens is if you either have people making policies who are not as connected with humans, or you know, . . . it just ends up being something on paper. . . . I do believe that, you know, we have a great obligation.
>
> (I-19: 28)

Ultimately, resilience ended up becoming a broad umbrella term, covering very generalized societal problems without specific guidance on how to address them. Thereby, neighborhood vulnerability and broad-based societal adaptation remain unaddressed. Compared to Atlanta's initial strategy and application document to the 100 Resilient Cities program (2013), the language and thematic focus areas that had a much keener focus on climate change matter did not make their way into the final Resilience Strategy (2018). The lack of resources, political will and authority were provided as major reasons why the city of Atlanta opted for a resilience framework instead of following through with actual plans related to adaptation.

Disappointment over the final product, the "100 RC Atlanta Resilience plan," was voiced by different experts, who advise governments at different political levels. Whereas the initial draft outlined Atlanta's major challenges more clearly, the final product was criticized for its vague language and watered-down targets, aside from the almost complete disappearance of climate change-related language and content. Great dissatisfaction was also voiced over the fact that the plan was no longer focusing on vulnerability to climate change impacts, such as heat, as expressed by one expert:

Interviewer: If you look at the city level, would you state that there overall is a consciousness of human vulnerability to climate change?

Respondent: I do not see it. . . . [I]f you did a scientific assessment of vulnerability, you would find that to be the principal driver. And like I said the most recent version of the sustainability slash resilience goals in the city, I think heat is not an inch in it. They are still prioritizing things like energy conception, which is great, but it is not going to protect anyone's health. . . . [T]he very structure of municipal policy around dealing with climate related issues is changing. They have not, in the last documentation I saw, where they were listing the principal threats to the region, heat was not even listed.

(I-29: 39f)

Another interviewee was highly critical of the final plan and how the "huge grant of the Rockefeller Foundation" had been used (I-32). Further, concerns over the qualification of the person in charge of the resilience planning became loud, aside from different political interests and increasing conflicts between the Mayor's Office of Resilience and Mayor's Office of Sustainability. Political tensions in light of the mayoral elections were mentioned as further challenges impacting adaptation and resilience planning. The vision and engagement-oriented approach was criticized for its small-sectioned and inverted nature by someone working in the Mayor's Office of Resilience:

> The problem with that [community-based resilience planning], you are a poor person and they give you, they ask you for: what is your biggest problem, just decide that they gonna tell you what is their immediate problem and not what is going to happen in ten or twenty years. Nobody really cares, when you need to care for the food on the table, you do not gonna care about what is gonna happen in twenty years. Your problem is now. Does that make sense? So, when you go to do some engagement, and ask them people what is their problems, they just decided that they gonna tell you is what immediate problems they have, they just gonna tell you what is in the news, right, you know, what is more impactful. . . . And this is what is happening to the city of Atlanta.

(I-22: 112)

The Mayor's Office of Resilience and resilience planning underwent several changes due to political administrative changes in the city of Atlanta as part of the latest mayoral elections in 2017 and the Rockefeller Foundation suspending its 100 Cities Resilience program in 2019. In 2018, after 90 days in office the newly elected mayor Keisha Lance Bottoms had asked the entire cabinet for resignation letters to "enable a fresh look at our leadership" and evaluate the leadership of the administration (WSB-TV2 2018). Atlanta's Mayor Lance Bottoms ended up accepting most of these resignation letters including that of the former Sustainability Manager. The effort was part of a larger anti-corruption campaign, which ultimately resulted in a federal investigation into corruption of former mayor Kasim Reid. Many news outlets referred to this drastic step as "shake-up" of Atlanta and an effort to increase transparency and fight corruption. These political changes matter in so far for climate adaptation progress that many programs appear to be discontinued and/or lacking follow-up mechanisms. Overall, it appears that environmental governance in Atlanta underwent drastic changes in administration, thematic foci and policy priorities which may ultimately impact the effectiveness of policymaking.

NOVEMBER 2009   DECEMBER 2012

Figure 8.1 Atlanta abandoned train track 2009 and Beltline 2012.

Source: Christopher T. Martin

In addition, the Rockefeller Foundation ended the funding for its 100 Resilient Cities Program in 2019, which was considered the "largest privately funded climate adaptation initiative" in the United States (SCW 2019). The Mayor's Office of Resilience continues to exist in Atlanta, outlining several different resilience initiatives. Since its initial days, the Office appears to have transitioned from focusing on environmental matters at the beginning to (climate) resilience issues under the funding of the Rockefeller Foundation to a much broader focus as of late. It is unclear, to what extent the proposed measures of Atlanta's Resilience Plan were implemented or largely discontinued ideas. As part of the Office's new strategy "One Atlanta," the quality of housing and energy burden were identified as two key issues (Tsai et al. 2018). In addition, the broader focus includes different issues such as public health, climate policy, aging infrastructure and access to food (Lee 2019).

Neighborhood revitalization programs: The Atlanta Beltline

In light of the region's scattered growth and disconnected city pockets, the Atlanta Beltline stands out as a popular recent attempt of neighborhood revitalization (see Figure 8.1). The project grew out of a master's thesis written by then Georgia Tech student Ryan Gravel in 1999. The vision of the thesis was readjusted and realized in 2005, outlining and implementing the redevelopment of a 22-mile former railway corridor into a multiuse trail. The initial aim was to integrate green-space, employment, development and transit through increasing the number of parks, trails, transit systems and the creation of jobs across 45 in-town neighborhoods (ADA 2005; Immergluck and Balan 2017; Oakley and Greenidge 2017). One heat expert, city planner and policy advisor describes the Beltline as unintentional adaptation project:

> So, the Beltline is constructed in a very interesting and helpful way, that you have a corridor around the Beltline in which you have to meet certain standards. . . . I do not know if there are park building standards, but you have to do improvements to facilitate pedestrian movements and bicycle movements, and so this overnight district is the perfect mechanism to also assign minimum green cover standards. So, the Beltline is not being used for this now, but it could be. There are lots of things associated with the Beltline, that are helpful, they are just not intentionally helpful. So, all the new parks planning are helpful, but the parks are just not designed to reduce heat. So, and the trees that are planted along the Beltline are helpful, but they are not there for that reason. . . . [Y]ou know I often argue the Beltline is a climate adaptation project, it is just not an intentional one.
>
> (I-29: 75)

The construction of the Beltline began in 2006 and is estimated to be completed by 2030. In 2016, it became listed in the Adaptation Clearinghouse, a U.S. database that lists measures on climate change impacts and is intended to assist policymakers, resource managers and academics through providing data resources on adaptation progress. Here, the Beltline is described as one example, where underutilized space was retooled as an adaptation strategy through increasing tree canopy, enhancing green space and infrastructure. Before the Beltline was officially registered under the Adaptation Clearinghouse, some argued against referencing the Beltline as climate policy to prevent greenwashing (e.g., I-06: 20). The database promotes the Beltline as an interesting model of how cities can align adaptation efforts to "revitalize underutilized spaces" while attracting "private financing for resilience projects that increase property values" (AC a.d.). The amount of private donations was commonly mentioned as a factor that enabled the building of the Beltline, besides public support and pressure coupled with bolt political leadership (I-21, I-32). Ironically, the dominance of the private sector, strong corporate interests by select private actors were also considered to impede Atlanta's efforts related to population equity, not necessarily contributing to the resilience of certain populations.

Although the Beltline was appraised for its transformative planning persisting in its goal of integrating and cutting across socioeconomic and racial lines, conflicts of interests and a disconnect with the initial vision have overshadowed the project. The height of this struggle became apparent when Ryan Gravel and Nathaniel Smith, founder of the Partnership for Southern Equity (PSE) left the board of directors of the Atlanta Beltline Partnership in late September 2016 over inclusivity and affordability concerns.[2] Their resignation letter mentioned that the initial goal of 5,600 affordable housing units was not met and that fewer than 200 affordable units were more likely (Stafford 2016). Ryan Gravel notes the shift of focus away from community advocacy:

> Since the beginning it [the Beltline Partnership] always had this uncomfortable kind of balance between community advocacy and fundraising. . . . And you know sometimes these things are in sync and sometimes they are just not . . . with the departure of the executive director last summer, [the problem] was that, their priority was not going to be community advocacy, in the way that I think is needed. . . . You know, my resignation from the board had more to do with the board of the Beltline Partnership, [it was] not really about the project. So, I was disappointed in the way // I was disappointed in the departure of the executive director, and it just became clear that that organization was not going to focus on that kinds of things, what they should.
> (Interview with Ryan Gravel 2017)

The visionary appeal of the initial Beltline plan ultimately clashed with the on-the-ground economic and political realities (also see Oakley and Greenidge 2017). These are characterized by the lack of political enforcement mechanisms, the dominance and political power of speculative real estate as well as local political and growth regimes (also see Immergluck 2009). Critical voices describe the Atlanta Beltline as an instance of neoliberal governance, signified by a growth-first approach and select participation, which simply give an illusion of inclusion and in fact do little to address unequal power relations (e.g., see Roy 2015).

Besides affordable housing, the initial plan saw mass public transit as a core feature for enabling transformation in Atlanta (Gravel 1999). Since the construction of the Beltline, major components related to equitable access have been facing severe drawbacks. Twenty years after the initial publication of Gravel's Master's Thesis, no essential progress on public transit could be noted, which led to the formation of the Grassroots action group *BeltlineRailNow!* in 2016. Against the background of large-scale redevelopment processes and the acceleration of displacement of long-term residents, the need for better governance structures to enable equal access to mobility, its connection to affordable housing and a more equitable public transport system has been the forefront concern of the initiative (BLRN 2021).

Green infrastructure programs: floodwater management parks and permeable pavements

The city itself has come up with green infrastructure programs and published a Strategic Action Plan in 2017 as a result of a peer-exchange program with Philadelphia in 2012. This Action Plan was updated in 2018 (see previous section). The Green Infrastructure "Task Force" is run by the City's Department of Watershed Management and consists of other city partners such as the MOoR, Parks and

Figure 8.2 Old Fourth Ward flood park before.

Source: Photography by William R. Bryant, courtesy of HDR

Recreation, Planning and Community Development as well as Aviation in addition to partnerships with NGOs. One city council member lays out the core ideas:

> We have, you know, heavy rainfall events, that cause flooding that affects a very tight radius area. But it does have monumental impacts on the homes in those areas, primarily property damage at a high rate. So, what we have done to address it, one model that we use, and my district was kind of the first pilot, was a beginning of a stormwater utility. . . . So the main thing that we are doing to addressing flooding resilience, is we are investing in the infrastructure that prevents the flooding from occurring and manages the water. . . . We are investing in green infrastructure that gives us solutions to our water management, our storm water management, as well as in some cases, renewing our sewer systems.
>
> (I-33: 64)

Aside from also constituting an example of neighborhood revitalization, floodwater management parks have become the central component of Atlanta's green infrastructure approach. One flagship is the Old Fourth Ward Park (O4W), located in a historical, formerly predominantly African American district in central Atlanta, also known as home district of Martin Luther King (see Figure 8.2).

Figure 8.3 Old Fourth Ward after.

Source: Photography by Steve Carrell, courtesy of HDR

In an effort to deal with extreme rain and revitalize the neighborhood, the O4W was extensively remodeled (see Figure 8.3). The O4W is adjacent to the Beltline and Ponce City Market, another redevelopment project with an investment volume of 180 million US dollars and supposedly the largest adaptive reuse project in Atlanta's history (Sams 2014). The O4W is one of Atlanta's newest parks, which was also outlined in the Beltline Redevelopment Plan in 2005. It reopened in June 2010, after a two-year transition period, costing 23 million US dollars, which was run by the nonprofit *The Trust for Public Land*.[3] At the center of the park is a two-acre stormwater detention pond built by the Department of Watershed Management (DWM), which is said to dramatically reduce flooding in the nearby neighborhoods (H4WPC 2019).

Beforehand, local communities were increasingly confronted by stormwater runoff problems. A similar park is planned in the low-lying area of Vine City in Atlanta called Cook Park (Samual 2017), which is supposed to help address flooding issues as well. Besides increasing the quality of life by residents and addressing the city's desire for increased sustainability, recent discussions have focused on green gentrification in light of Atlanta's rapidly decreasing African-American population. The O4W is one of the most prominent examples of neighborhood revitalization programs that resulted in a relocation and rapid decline of predominantly African American residents (Taylor 2016). One member of the city council describes the nature of the problem:

> So, we kind of redesign // this is the area where people, primarily poor African Americans have lived, right? A lot of crime, drug, problems, health disparities, lack of access to education, quality healthcare, everything, right? Good jobs. Now we gonna come back with this beautiful new urbanist design for how you manage the water. Of course, when it looks pretty on a piece of paper, they will also get it and see the trade-offs, and gonna say: Oh, that would be a great place to buy. . . . There you go. Green gentrification. So, is, I mean you know, is that cool? No. So, that is one of those built-ins, we aaaaalmost get it right. So, that is the next step.
>
> (I-33: 57ff.)

There was supposed to be a non-displacement zone for indigenous local home ownership and long-term residents. Although Atlanta was the first city to offer public housing, property speculation has significantly impacted historical neighborhoods of predominantly African American citizens. No known follow-up documentation exists, and further research is needed to determine which factors stood in the way of implementing the non-displacement zone.

Besides larger green redevelopment infrastructure, Atlanta is home to the largest permeable pavement program in the United States, initiated under former Mayor Kasim Reed (also see Shamma 2015). Atlanta's DWM rolled out the program as part of the Southeast Atlanta Green Infrastructure Initiative (SAGII) and in light of a major flooding event that had occurred in 2012. The program was part of Reed's initiative of becoming "one of the top-tier sustainability cities in the nation" (CoA 2017a) aside from complying with the EPA's Clean Water Act. However, due to administration changes, it is unclear whether the program will be continued.

Altogether, the Mayor's Office of Resilience and recent resilience efforts of the City of Atlanta appear to have been focusing more strongly on addressing extreme precipitation, stormwater runoff, water scarcity and quality. Heat appears to play a more subordinate role in Atlanta's planning besides the availability of scientific evidence on Atlanta's drastic exposure to heat and urban heat islands (also see Chapters 1 and 2). In this context, educational actors have again taken on an active leadership role in combination with other non-governmental partners through setting up the six-month project UrbanHeatATL in 2021. The project is run by two higher education institutions (Georgia Tech and the Spelman College), 86 students and three NGOs and supported by the City of Atlanta. It maps Atlanta's urban heat islands (UHI) with community science through collecting data on heat extremes and how vulnerable community members are disproportionally affected, thereby operating at a nexus of climate change, racial justice and equity (UHATL n.d.).

The state of adaptation efforts in Jinhua and Zhejiang

In contrast to Georgia, the Zhejiang provincial government has initiated climate adaptation measures. Because Jinhua is a prefecture-level city, and prefectures are directly administered by the provincial government, it is useful to first examine Zhejiang's climate adaptation efforts before going into local adaptation aspirations.

Overview of recent provincial efforts

Zhejiang is the province with the highest rates of SLR. The region's historical exposure to SLR, coastal tourism, population density and economic dependence are among the reasons why the province has had a focus on undertaking coastal adaptation measures. Approximately 40 percent of northern Zhejiang is protected by seawalls (CCPF 2012). Reinforcement of the "Thousand Mile Seawall" was completed in 1997 with a traditional coastal defense structure, protecting 18 million people and one million hectares of land (CCPF 2012). The seawall is also known as "ancient seawall" and presents one of the earliest examples of ancient civil engineering during imperial China (Wei and Cunhuan 2002). Recent works have focused on how to conserve and strengthen the seawall (e.g., Wang et al. 2012). Because coastal erosion is progressing at a faster rate, building embankments, further widening and heightening the seawall are explicit adaptation measures recommended to prepare for long-term sea-level changes (Wang et al. 2018). The implementation of coastal use projects has further accelerated SLR. This has led to the control of coastal use measures through, for example, prohibiting the exploitation of sediment supplies (Wang et al. 2018). In 2017, the Zhejiang Provincial Department of Land and Resources issued a ban on land reclamation projects in order to protect the marine environment (SBNRP 2018).

Besides rural agriculture, the province is China's largest fishery and has a long aquaculture tradition. Therefore, the provincial government has been promoting adaptation efforts in the fishing industry. Altogether, marine development has traditionally been a focus of the provincial FYPs. In light of increasing temperatures, Zhejiang was chosen as a pilot province for subsidized agricultural insurance in 2006. With the intent of providing farmers with the insurance for household losses to cover a range of potential hazards, the program sought to improve disaster resilience (Zhang et al. 2008).

Despite these efforts and a strong historical consciousness, overall awareness of local governments about adaptation is still low and policy action is perceived as a responsibility of the central government (Jin and Francisco 2013). In their study of local village leaders from 21 towns located in Zhejiang province, Jin and Francisco (2012) find that most village leaders only had minimal knowledge about climate change and did not place a high priority on adaptation measures against rising sea levels. The majority of survey respondents did not see SLR adaptation as the responsibility of the local government. This was echoed by a team of government officials in Jinhua, who considered the articulation of adaptation measures a provincial responsibility (I-38).

Lack of a dedicated provincial adaptation strategy

Although China's National Adaptation Strategy (NAS), published in 2013, stipulates that each province should develop its own climate adaptation plan (also see chapter 5), Zhejiang has not done so. There is no deliberate climate adaptation strategy. In the 2013 annual update of China's Policies and Actions for Addressing Climate Change, NDRC references Zhejiang as among the few provinces that carried out their own regional strategic studies for addressing climate change (NDRC 2013). Current provincial policy plans include Zhejiang's Climate Change plans, which mentioned adaptation for the first time in 2010. The latest plan (2013–2020), published in 2014, however, focuses almost exclusively on mitigation policy, aiming to make Zhejiang a low-carbon province. Some interviewees mistakenly viewed the 2010 plan as a provincial adaptation plan. Both documents focus on broader climate policy issues, but literally translate into "Zhejiang's Response to Climate Change."

The broader context of Zhejiang's adaptation efforts is related to accelerating the construction of an ecological province and "forest Zhejiang" (ZPG 2010). Altogether, the provincial government views climate change (adaptation) primarily as a development issue and itself as a leader on modernization:

> Climate change is both an environmental issue and a development issue, but in the final analysis it is a development issue. Zhejiang is an economically developed province in China, shouldering the historic task of continuing to be at the forefront and basically realizing modernization ahead of [the national] schedule by 2020.
>
> (ZPG 2010)

Aside from aiming to construct itself as a pioneer, the need for continued development was a common reason provided for the absence of dedicated adaptation efforts. The provincial climate plan has neither a health section nor does it specifically deal with population susceptibility. The provincial government has, however, aimed to reconstruct "dilapidated houses with less than 150 percent of the minimum living standards" and promote subsistence allowances in rural areas, which is particularly relevant for exposed populations (ZPG 2010).

The 2010 plan articulates a sectoral approach, focusing on how to strengthen adaptive capacity in the agricultural, forest and wildlife sectors, and how to improve flood control in the water and coastal sectors (see Table 8.3). The plan also mentions education campaigns.

According to a 2011 report, adaptation technologies in the agricultural and forest sectors were improved, with the aim of strengthening ecosystem-based adaptation (Wu et al. 2011). Education was mentioned to popularize relevant knowledge on climate change and "actively promote a green, low-carbon, healthy and civilized lifestyle and consumption pattern so as to create a good social atmosphere for effectively coping with climate change" (ZPG 2010). To what extent these targets were followed through needs further research. However, some academics were

Table 8.3 Adaptation priorities according to Zhejiang's 2010 Plan

Focus area	Plans
Agricultural	Improved irrigation water use, agricultural insurance
Forest and wildlife	Wildlife protection and afforestation by 40 million mu by 2012, construction of wetland areas
Water	Raising drought resistance standards
Coastal	Improved monitoring and warning systems
Other	Strengthening independent innovation and public awareness

Source: The author, based on interviews and ZPG (2010)

highly critical of the environmental priorities of the provincial government, as one Hangzhou-based policy advisor expresses:

> China's economic development is still the most important. . . . Zhejiang province, takes the economic development first. Just, you know, if we have 100-yuan, Zhejiang province will spend 99 [percent] on the facility building and the urban buildings. So, it is hard to dispense other resources into this problem solving. So, I think economic development is very important. It is the key factor, to // if our // if Zhejiang province economy have developed into a certain high level, I think the government will turn to consider some measures to address the climate change influence in the vulnerable populations.
>
> (I-12:125ff.)

That GDP growth also trumps any other political concern in Jinhua's local decision-making was a dominant impression gained during fieldwork.

Municipal adaptation piloting in Zhejiang as part of central efforts

In 2016, NDRC, together with MOHURD and the Department of Building Energy Efficiency and Technology published the Urban Adaptation Climate Change Action Plan (*chengshi shiying qihou bianhua xingdong fang'an*) (NDRC 2016). The initial statement called up cities across China (prefecture level and above) to submit their recommendations of potential pilot city programs to the involved departments by September 2016 (Xu 2016). The program aimed at improving the adaptive capacity of 30 cities, with different urban characteristics, to deal with climate impacts and incorporate climate adaptation into their overall economic and social development planning (Xu 2016). In 2017, NDRC and MOHURD announced that Lishui was chosen as urban adaptation pilot (*shidian shifan* or *shidian chengshi*) in Zhejiang (NDRC 2017). According to this piloting scheme, different cities were chosen to "form a series of duplicable and transferable pilot experiences, and play a guiding and demonstrative role" (MEE 2018). In 2017, MOHURD and NDRC together with the World Bank (WB), the Asia

Development Bank (ADB) and other international organizations such as the Deutsche Gesellschaft for International Development Cooperation (GIZ) held an International Workshop on Planning Climate Adaptive Cities in prefecture-level city Lishui, north of Jinhua. Supported by the GIZ, the first results of two adaptation pilots were presented.

The pilot cities are expected to improve the urban ability to adapt by integrating climate change-related indicators into their urban and rural planning by 2020 (CCCIN 2016). One target includes the promotion of green building standards by 50 percent. Construction standards and industrial development plans shall be adjusted to enable the "cities' ability to deal with water-logging, drought and water scarcity, high temperature and heat waves, strong winds and freezing disasters" (MOHURD 2016).

MOHURD launched another pilot series on green infrastructure planning, focusing on urban design (*chengshi sheji xi shidian*) (MOHURD 2017). In 2017, plans for 37 cities were announced. Four of these cities – Ningbo, Hangzhou, Taizhou and Yiwu – are based in Zhejiang. Taizhou, for example, has been promoting the construction of "passive houses" (*beidong fang*), characterized by zero energy consumption and built to withstand hot summers and cold winters. The concept was first introduced as "Hamburg House" at the Shanghai Expo in 2010. In 2011, a memorandum of understanding on technical cooperation for passive houses was signed by the Ministry of Housing and Urban-Rural Development and the German Federal Ministry of Transport, Construction and Urban Development (Gu 2016). Ningbo disclosed plans to improve public infrastructure, waterways and green spaces "for a more livable environment" (China Daily 2017). Ningbo had also served as a pilot city for the World Bank Climate Resilient Cities Program. The outcome was a local Climate Resilience Action Plan, published in 2011 (World Bank 2011). The elderly population (one million) and the growing floating population (3.94 million) are mentioned as vulnerable groups, due to less protective resources and limited social services, respectively. The growing urban–rural income gap is also mentioned; however, no further links nor policy recommendations are explored aside from suggestions on improved insurance, further research and targeted information programs for different parts of society.

These piloting schemes are examples of parallel policy development with overlapping content. The distinctive political pathway is characterized by the launch of a new policy program through the guidance of a central governmental actor. It is commonly known that pilots are chosen on the basis of pre-existent efforts in order to be able to reach the announced political targets. Further research is needed on this matter.

Regional-local quasi-adaptation efforts: Jinhua as sponge city

Although Jinhua is receiving much of its guidance from the provincial level, the city has not come up with dedicated adaptation policy strategies and is not part of current adaptation piloting at the provincial or central government levels.

The preexisiting site (2011)   Before (2011)   After (2014)

Figure 8.4 Floodwater landscape in Jinhua, before and after.

Source: Kongjian Yu, Copyright Turenscape, Yanweizhou Sponge City Park in Jinhua

Several measures were announced to constitute quasi-adaptation policies. The most prominent ones are policy documents related to the ecological civilization and Jinhua's sponge city project (SCP) (see Figures 8.3 and 8.4). When asked how human vulnerability to climate change is addressed, the emergency management plans were only mentioned on the side, if at all (I-38, I-39G).

Due to its long history of flooding episodes, Jinhua was among the first to become a sponge city (*haimian chengshi*). The Yanweizhou Flood Park[4] is Jinhua's most recent completed infrastructure adaptation project, a water resilient landscape. At the heart of the project is a bridge that is elevated above the 200-year flood level, which connects west and northern Jinhua. The JMPG assigned Turenscape, an architecture bureau with the task of designing the floodwater park. The planning was rolled out in August 2010 and completed in May 2014, when Jinhua's flood adaptive landscape was formally opened. This was one year before the Central Government and City Council published their technological guide on building sponge cities. A team of researchers on SCPs referred to Jinhua as a city governing by example (Dai et al. 2018). Andrew Buck of the Chinese landscape architecture firm Turenscape, based in Beijing, has referred to the flood-resistant topography as a civil engineering approach to water management to enable "flood control in an ecological way" (see Figures 8.4–8.6).

In 2015, the Chinese government formally introduced the SCP. As part of the initiative, the State Council selected 30 cities to act as pilots. The Jinhua case is an example of testing and rolling out policy efforts before launching a larger pilot

Figure 8.5 Flood-resilient green infrastructure landscape in central Jinhua.
Source: Kongjian Yu, Copyright Turenscape, Yanweizhou Sponge City Park in Jinhua

Figure 8.6 Flood-resilient green infrastructure landscape in central Jinhua, flooded.
Source: Kongjian Yu, Copyright Turenscape, Yanweizhou Sponge City Park in Jinhua

scheme based on the successful outcome of the foregone case. Direct guidance was offered by the Ministry of Housing and Urban-Rural Development, the Ministry of Finance (MoF) and the Ministry of Water Resources (MWR). The implementation of China's SCP is at the core of China's response to urban flooding and the problem of water scarcity. The sponge city construction is also promoted under the framework of urban adaptation pilots, which aims to:

> Vigorously build the urban 'sponge' such as green roofs, rainwater gardens, water storage ponds, miniature wetlands, sunken green spaces, vegetation grasses and biological retention facilities to enhance the urban sponge capacity. Rainwater tanks, storage tanks and other rainwater harvesting facilities should be built according to local conditions so that the rainwater can be collected and utilized on the spot and the utilization efficiency of rainwater resources can be increased.
>
> (NDRC 2016)

The most recent policy documentation outlines Zhejiang as a provincial pilot area for sponge cities. Lanxi city is supposed to take the role of providing first-hand experience through scientific exploration. In Jinhua, five major sponge city demonstration zones are about to be rolled out in order to accelerate sponge city construction (JMPG 2018). Other counties, cities, and districts are instructed to likewise promote the "demonstration of constructing sponge-type parks and green space, sponge-type residential areas, sponge-type roads and sponge-type small towns and form a contiguous demonstration effect as soon as possible" (JMPG 2018). Jinhua's sponge city park is considered a transformational approach to urban resilience and flood management (Perepichka and Katsy 2016). Government officials outlined how it is being used by approximately 40,000 visitors each day and has helped to create a new cultural identity (I-38, I-39G).

Sponge cities and ecological corridors

Yanweizhou Park is one of China's larger policy efforts to create ecological corridors (*shengtai langdao*). It was initiated by the Jinhua Municipal Party Committee (*shiwei*) and the Municipal Government (*shi zhengfu*). It constitutes one of Jinhua's most recent developments regarding urban planning and the reconstruction of the city center according to ecological standards. The corridor presents a holistic approach by envisioning the ecosystem as consisting of economic, social and natural subsystems.

The Jinhua City Master Plan (2000–2020), (*chengshi zongti guihua*) lays out the development of Jinhua's ecological environment and is grounded on the "Zhejiang Urbanization Development Program" of 1999. According to the plan, "Jinhua should become the central city and transportation hub of the central and western regions of Zhejiang Province" (JMPG 2009). The inner city along the six banks of the three rivers "shall be taken as the core, aligning along the traffic axis and shall be consistent with wedge-shaped green space consisting of scenic spots, suburban recreational green land and natural woodland" (JMPG 2009). According to the Master Plan, this central area is viewed as suitable for the future economic and social development of Jinhua City. Jinhua's Sponge City Construction is to be a long-term effort, with a great amount of detailed preparatory work on the part of municipal and provincial governments. The 2015 annual work report of the Jinhua government speaks of the city's rapid advancement in urban construction and further strengthening of the core city (JMPG 2015). Aside from perceiving different concerns as more immediate tasks deserving of governmental attentions, decision-makers at the municipal level pointed to these policy efforts as quasi adaptation endeavors:

> So, in Jinhua, although there is an absence of such climate adaptation documents or opinions, the ecological corridor is actually a very useful exploration and attempt to change the climate and to adapt to climate change.
>
> (I-38: 73)

Corridors and large-scale constructions of infrastructure are principles of China's urbanization strategy. Wu (2009) describes how urbanism as "a new way of life" has become China's primary design principle and accumulation strategy, thereby breaking with historical trajectories of "under urbanization" (lack of urbanism) in the Mao era. Consequentially, Chinese cities have become more diverse and nowadays present elements of both, advanced capitalist economies and developing countries (Wu 2009). The diversity is said to have substantiated the urban scale of capital accumulation and the juxtaposition of rural villages, migrant enclaves and exotically transplanted Western-style gated communities (Wu 2009). Lan'xi and Pan'an, the most vulnerable study units of the SVA can be considered such instances of juxtapositioned rural sides in contrast to the more affluent, urbanized centers of Jinhua. The design principle of simulated landscapes has been referred to as neo-urbanism, which is in the making of China's market transition (Wu 2009). Others explain how the beautification of urban city centers under the ecological civilization is a form of urban entrepreneurialism paired with environmentalism (Xu 2017).

Jinhua 12th FYP on meteorological development

In 2012, the municipal government published the 12th FYP for Meteorological Development in Jinhua City (*qixiang shiye fazhan shi'erwu guihua*). The document can be understood as an irregular policy measure, based on existing provincial and municipal plans: the Zhejiang Meteorological Development Plan (2011–2015) and Jinhua's 12th FYP on National Economic and Social Development. Meteorological services are seen as increasingly urgent, due to the impact weather climate and climate change will have on the economy and social development (JMPG 2012). The plan outlines Jinhua's specific vulnerability in geographic, economic and social terms:

> Jinhua is located in the eastern part of the Jinyu Basin. Its special natural geographical environment and economic and social characteristics make it highly vulnerable to the adverse effects of climate change. At the same time,

Jinhua's light, heat, precipitation and other climate resources are abundant, and wind power, solar energy, air cloud water resources, mountain climate resources, etc. have reasonable development and utilization value. Mitigating the impact of climate change and building livable cities and ecological markets, there is an urgent need to strengthen climate change, monitoring and evaluation of climate resources, and enhance scientific and technological support for climate change.

(JMPG 2012)

It is the first document in which heat and climate vulnerability are mentioned. Surprisingly, the plan is outspoken on the weaknesses in current meteorological forecasting and the inability to meet the needs of disaster prevention and mitigation, a problem which is especially acute at the rural and grassroots levels (JMPG 2012). The Jinhua Municipal Government notes that the public meteorological services are insufficient to meet the needs of agricultural services, the economy and social development: "The scientific and technological support ability to cope with climate change and climate resource development and utilization is relatively weak" (JMPG 2012). It becomes clear that coping with climate change is understood as a technological task of improving warning systems. The aim of the five-year period is to speed up meteorological development, improve meteorological services and enhance Jinhua's sustainable capacity (JMPG 2012). The plan remains largely mitigation-focused aside from mentioning adaptation technologies such as improved early warning services.

Most of the policy efforts are grounded in provincial and/or central regulations but articulate local recommendations. Further, there is a strong cross-linking of different policy efforts. The environmental emergency management plan, for example, is based on the "Environmental Protection Law of the People's Republic of China," the "Safety Production Law of the People's Republic of China," "Regulations on the Safety Management of Dangerous Chemicals" and "Emergency Plan for National Sudden Environmental Incidents" and "Overall Emergency Plan for Public Emergencies in Zhejiang Province." The cornerstones of Jinhua's quasi-adaptation planning are policies concerning Sponge City construction and reform, which are related to the planning and implementation of constructing an ecological civilization, which in turn are connected with Jinhua's building of the Ecological Corridor. The policy documents that were commonly mentioned as quasi-adaptation efforts are listed in Table 8.4.

Five-water co-governance

Zhejiang province is at the center of the iconic water policy "Five Water Treatment" (*wushui gongzhi*) whose implementation began in 2014. The idea evolves around the coherent treatment of five waters in light of Jinhua's and Yiwu's longstanding challenges with industrial pollution (I-14). Among officials and state practitioners, the issue and importance of "water safety" (*shui anquan*), as part of the provincial government directive of Five Water Treatment (FWT), also known

Table 8.4 Overview of relevant policy efforts – Jinhua

Year, Name	Main purpose	Issuing agency
2018: Implementation Plan for the Reform of Ecological Civilization System	Identifies 42 priorities in the reform of the ecological civilization system, establishment of the "Blue Sky Defence Office," annual target of 10% growth in the ecological protection and environmental management industry was reached ahead of target	Jinhua Municipal Government (JMPG)
2017–2018: Jinhua Ecological Civilization Construction Planning and Jinhua City Ecological Civilization Demonstration Creation Action Implementation Plan	Build environmental and ecological protection knowledge	Provincial and municipal EPD, JMPG, Planning Bureaus
2016–2030: Jinhua sponge city special planning and Implementation Opinions on Promoting the Construction of Sponge City	Maximize natural accumulation; infiltration and purification of rainwater in urban areas, accelerate the construction of a healthy and perfect urban water ecosystem.	Municipal Bureau of Construction, Municipal Planning Bureau
2012, 12th FYP for Meteorological Development in Jinhua City	Outlines current deficiencies to improve meteorological services, speed up meteorological development, enhance sustainable development capacity	JMPG
2015: Emergency Plan for Environmental Emergency in Jinhua City	Improve the government's ability to respond to sudden environmental incidents involving public crises	JMPG
2013: Jinhua flood control and drought emergency plan	Deals with natural disasters and water conservancy risks caused by heavy rain, floods, typhoons, droughts	JMPG

Sources: JMPG (2012), JMEPB (2018), JMPG (2018)

as Five Water Co-Governance were emphasized. The notion of water safety implies five different elements, as one policy practitioner of the Five Water Treatment Department in Jinhua lays out:

> It includes sewage treatment (*zhi wushui*), flood prevention (*fanghong shui*), water drainage (*pailao shui*), assurance of water supply (*bao gongshui*) and water conservation (*zhuajie shui*).

> (I-39G, respondent a: 7)

In the recurring context of "water safety," the exposure to floods is considered a matter related to *being safe from* water, aside from holistically aiming to treat the different elements of exposure related *to* water. Most of the documents on climate-induced or weather-related stress in this context interlink successful coping with the issue of water safety. Within the province, Jinhua is considered to have been more proactive in its water treatment. In line with local governance approaches, Pujiang, a county to the North was chosen as an FWT pilot. Currently, several other pilots are on the way, also feeding into governmental plans to construct eco-villages. When local officials were asked about their motivations, Xia Baolong's directive of "three reforms and one demolition policy" (*san gai yi chai*) at the 18th National People's Congress was referred to.

Aside from the FWT, the provincial and municipal governments have taken up several measures to boost water conservancy, build water reservoirs and establish water cooperation with, for example, Dayang and Yiwu City in the eastern part of the province. The province and Jinhua in particular also became known for their concept of river chiefs.

Embeddedness in the Ecological Civilization

The most recent efforts related to the city's construction and implementation of an Ecological Civilization Construction and demonstration of implementation plans aim at building environmental and ecological protection knowledge by adding three provincial-level ecological civilization education bases, 20 provincial-level green schools, 26 provincial-level green households, 25 municipal-level green schools, and 190 municipal-level green households. In addition, a green environmental protection bus line is to be launched (JMEPB 2018).

Jinhua is a recent example of China's shift between hard engineering solutions and soft nature-based adaptation solutions. Meng and Dabrowski (2016), for example, show how the deeply rooted top-down planning culture has affected how different departments and the local government deal with urban adaptation in light of flooding in the background of a historical Fengshui-based philosophy that emphasizes the coexistence with water. Different accounts have examined the cultural connotations of human culture and nature prevalent in Daoist thought (Girardot et al. 2001) and command-and-control relationships during imperial China (Elvin 1998), as part of Confucian ideals (Kassiola 2013), or massive geo-engineering projects for flood control purposes during and after the time of Chairman Mao Zedong (Economy 2010). Though hard adaptation measures continue to be common practice, the political discourse on the "ecological civilization" and recent adaptation planning efforts have been emphasizing a shift to nature-based and ecosystem-based adaptation concepts. Research on the future of sustainable urban construction, resilience and adaptation thrives in China in the context of the CG's Sponge City Program and the historical significance of water management despite the acknowledgment of the increased danger of flooding (Li et al. 2017; Dai et al. 2018; Hermaputi and Hua 2017). Accordingly, revolutions in green infrastructure planning and low impact development mark a succinct change from earlier approaches to water management and urban planning (Dai et al. 2018). Aside from reducing air pollution, the overarching problem themes are related to green development of the greater Jinhua region, industry 4.0, green agricultural development and green intelligence.

No understanding of adaptation as specific policy arena

Altogether, climate adaptation was not really abstracted from mitigation or treated as an own policy arena. Despite the fact that the term climate adaptation (*shiying qihou bianhua*) was largely unknown and preparing for climate impacts (*zhunbei qihou yinxiang*) was not understood, as it is commonly conceptualized in (Western) policy discourses around adaptation, there was some awareness about matters related to adaptation. When asked about the increasing human exposure of flooding in Lanxi in light of climate change, an employee at the Five Water Treatment Department, who is nestled at the Environmental Protection Bureau reiterated:

> Why should it be the climate? That is not very specific. So, climate pollution involves the atmosphere and many other aspects. First, vegetation // [the proportion of] green vegetation is very important. Then, speaking from the drain sewage from the industry, there are many factories, then the heat may be more, the climate will change. I think these two points are still the most important. In terms of the nature of our work, for the construction of dams, the first is that we have done a good job in water control and cleaning the water. So, the climate will certainly be fine. Second, the air and haze was reduced, the air is more clear now. Accordingly, the climate is more livable.
>
> (I-39G, respondent c: 169)

This constituted a very common response from government officials at different levels, policy advisors and practitioners. Instead of talking about how to prepare communities for unavoidable climate impacts, interviewees rather mentioned environmental protection, water management, and efforts related to mitigation such as China's new energy cities, low-carbon pilots, international climate change action. Very soon, conversations turned to how to manage local carbon emissions in the realm of green development. It became obvious that adaptation was not understood in terms of adapting to and preparing for climate change, but the adaptation to circumstances. When asked about the aspect of human health or how to support communities to adapt to a changing climate, the conversation would then focus on how factors related to environmental pollution lead to a diminished quality of life. When attempting to probe social vulnerability to and/or the human health impacts of climate change, most officials focused on uneven exposure to air pollution. The majority of responses were signified by low adaptation sensitivity.

When trying to grasp human vulnerability to events that had already occurred, such as the major and historical floods during the summer months of 2017, members of the municipal standing committee pointed to management frameworks and well-experienced governmental institutions:

> The way we have done it has reduced the losses (*sunshi*). [In this year's flood] there were no dead and no wounded people, everything went smoothly. This is related to the construction of ecological corridors (*shengtai lang dao*). And the construction of ecological corridors is actually related to water safety (*shui anquan*) // the foundation of the ecological corridor is to ensure water safety. It precisely means that I am safe after a flood has come.
>
> (I-38: 46)

This demonstrates the point that human vulnerability was understood as an all-or-nothing principle rather than a precondition to prepare for, cope with and recover from growing climate impacts. Zhejiang was perceived as having low population susceptibility in light of the provincial economic development, as pointed out by one academic and policy advisor:

> Well, as you know that the poverty population in Zhejiang province is not as much as in Ningxia province. So, our effort on this problem is not very much. But as I remember, there is a project carried [out] in Taijun City and on poverty alleviation.
>
> (I-12: 79)

This suggests that having greater economic resources was perceived to automatically correlate with greater resilience and thereby lowered the risk awareness. Aside from the acknowledged need of addressing SLR, due to population density and a coastal concentration of economic activities, and agricultural sensitivity due to food security, adaptation was seen as a major concern of developing regions.

Summary: Accidental and infrastructure-focused adaptation

In Atlanta and Jinhua, interviewed experts marked the lack of dedicated adaptation policy strategies but spoke of unintentional or quasi-adaptation efforts. In both cases, floodwater management parks and the increase of green cover were retroactively labelled as adaptation efforts, but it was unclear, to what extent the local population and their vulnerability had been considered in the planning. In Jinhua, there appears to exist only little sensitivity about the human dimension of vulnerability, and how different parts of the population are unevenly affected. Overall however, and in contrast to the state government of Georgia, multifold adaptation efforts have been rolled out by the provincial government of Zhejiang. Some of these governmental efforts are signified by high degrees of cooperation between different government departments, especially in the water sector. This might enable more holistic forms of governance. Yet, political efforts are also signified by ambivalence when it comes to the way political priorities are being set and continue to focus on urban development and accumulation at the local and provincial level.

Atlanta and Georgia are signified by the lack of a coherent climate change adaptation strategy. In comparison to other states in the United States, Georgia's resilience planning was considered immature. For most part, Atlanta's accidental adaptation efforts are characterized by their reactive nature, and only incremental decision-making when it comes to addressing the uneven aspects of human vulnerability to climate change. This was especially imminent. The municipal climate policy was strongly mitigation focused only including adaptation as co-benefit without further specifications. The resilience and climate policy efforts seem to aim for status quo reproduction. Here, floodwater management and green infrastructure programs were coupled with neighborhood revitalization programs and urban efforts for redevelopment. The examined cases of accidental adaptation present powerful examples of green gentrification: the revitalization of certain neighborhoods through floodwater retention programs and building of green infrastructure led to an increase in property value and as a result increase in housing prices, causing an affordability crisis and an acceleration of relocating predominantly African American populations. Thus, although planned efforts are intended to increase resiliency toward climate impacts in the context of addressing the social impacts of uneven population exposure, they present forms of maladaptation.

Public inclusion in decision-making processes was not standard procedure as part of the city's green infrastructure and neighborhood revitalization programs but became an important cornerstone in setting up Atlanta's resilience strategy. Nevertheless, initial findings suggest that the quality of participation was rather consultation-based, characterized by exchanging information and appeasement rather than having actual ownership and power in the decision-making process. The lack of political leadership on climate adaptation at municipal, state and federal levels paired with the lack of federal guidance was provided as a major reason for the absence of dedicated adaptation efforts. Climate adaptation illiteracy, i.e., the lack of knowledge on complex matters related to adaptation of decision-makers paired with the lack of training and lack of resources were mentioned. By adopting a lock-in perspective, this book argues that these are surface explanations and that the root causes are found in path-dependent lock-ins at different levels (also see Chapters 6 and 7). Political activism on climate mitigation on the other hand was attributed to the desire of becoming a pioneer within the conservative South. Training and exchange opportunities at the international level were provided as primary motivators besides the support received by initiatives such as ICLEI or the Rockefeller Foundation.

In contrast, more deliberate adaptation efforts exist at the provincial level in China, in Zhejiang province but local governmental levels continue to have very low adaptation awareness. Preliminary findings suggest that local policy motivation was low with regard to addressing the uneven nature of human vulnerability to climate change in the rural pockets of the city. Further research is needed

Table 8.5 Differences in problem recognition, Atlanta and Jinhua

City	Recognition	
Problem	*Human vulnerability to climate change*	*Climate change adaptation*
Atlanta	More mature but fragmented	Nonexistent, redevelopment constituting maladaptation
Jinhua	Little sensitivity	More local knowledge but not adaptation-intentional
Overall problem recognition	Immature	Immature

to solidify these findings. In the United States, a greater focus on human health and limited participation mechanisms stands out, while no deliberate adaptation efforts were taken at the state and local level (see Table 8.5). Current efforts are enabled through the back doors of other, existing policies. In both cases, the problem stream is relatively immature. This is particularly the case regarding the human dimension in policy planning in Jinhua and the climate adaptation component in Atlanta.

The prevalent notion throughout China was that a lot of items could be attributed to climate change adaptation, "but people do not talk about that much" (I-37: 29). Aside from different forms of localized knowledge, the low problem recognition in Jinhua about climate adaptation as a distinctive policy was also traced to the small size and political insignificance of Jinhua, with only "an area of ten thousand square kilometers" (I-38: 91).

Aside from not perceiving adaptation as a local but rather provincial responsibility, there was an overall lack of knowledge about what adaptation actually is. Very often, to cope with climate change (*yingdui qihou bianhua*) was holistically understood, but also interchangeably used with the terms related to adapting to the unavoidable impacts of climate change (*shiying qihou bianhua*). Most conversations revealed how climate adaptation was blurred within the broader field of climate policy. Meanwhile existing environmental and flood-risk planning were referred to as informal adaptation efforts.

Lock-ins of knowledge and politics related to adaptation

The following section discusses the main finding of the foregone chapter, which relates to low problem recognition and sensitivity about the social dimension of adaptation, as well as the predominance of accidental forms of adaptation. Accidental adaptation is characterized by the retroactive labeling of policy efforts that were not per se planned as deliberate adaptation ambitions.[5] Further, both cases exhibit a strong focus on creating green infrastructure and redevelopment projects. Social vulnerability was not a major concern and redevelopment projects in the city of Atlanta even contributed to the vulnerability of local populations (see Figure 8.7). This chapter explores the lock-in characteristics. These

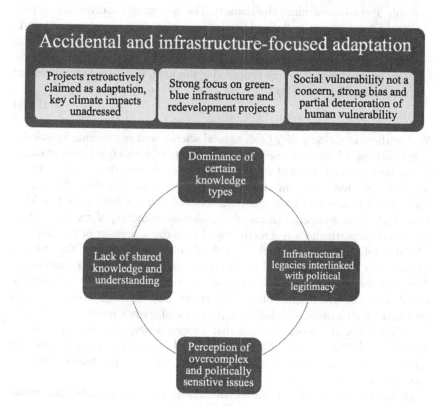

Figure 8.7 Lock-ins related to accidental and infrastructure-focused adaptation
Source: the author

become evident at the intersection of path-dependent patterns of knowledge construction, politically motivated knowledge patterns and the lack of a shared understanding.

Dominance of certain knowledge types

Natural-science focused knowledge paths in China and legitimacy legacies

In Jinhua, the lack of knowledge and little awareness about the interrelational aspects of adaptation and human vulnerability to climate change was a core finding. The lack was explained with the continued absence of social science studies as core characteristic of the Chinese political system. The tradition of science and technology research from an engineering point of view was pointed to by most experts interviewed in the academic sector. One academic and adaptation expert from Beijing expresses:

> Social vulnerability, I think, we know, it is very important. But for China [the] academic problem is, there are not enough social science studies. It is

only about [researching] the impacts. The government has realized to look at the social part as well. But I think this is still a process and ongoing work. In the past, the government did not [have] so many expertise on human health. The national adaptation report noticed human health as an important part. . . . more research is needed in this field. Some was being done, but more is needed.

(I-35: 21)

Within the adaptation policy field, natural science and engineering knowledge paths are coupled with the engineering tradition of China's floodwater management. Treating the politics of water as a mirror for the way the Chinese government operates has grown into a solid explanatory angle that also reflects upon the cultural factors and historical trajectories of political legitimacy (see Mertha 2008). During imperial China, the inadequate handling of natural hazards and floods in particular was directly linked with the legitimacy of the emperor. The storytelling goes as far back as to the founder of the earliest Xia Dynasty, Emperor Da Yu (also translated as: "Yu the Great"), who became known as the "tamer of the flood."[6] Not managing a flood correctly, could result in losing the mandate of heaven. Besides historical legacies, and although China has been an emerging scientific actor globally, China's social science research is still lacking behind. Social science research that informs adaptation planning remains likewise weak (Nadin et al. 2016). The limited influence of certain actors on adaptation processes, including social scientists, has been pointed out early on (also see Li 2013).

Good water treatment and disaster handling have also grown to become important political motivators in light of China's evolving mandatory target system. Good government is now being measured through, for example, the cleanliness of water (Yang and Zhao 2015). The River Chief System (RCS) is one mechanism, which was put into place to increase governmental efficiency and achieve environmental targets. It is also another example of policy development that evolved from local levels to become a national mechanism (Huang and Xu 2019). Anhui, Jiangsu and Zhejiang were among the first provinces which saw experimentation with river chiefs at the local level until the State Council declared it a national agenda in 2013 and appointed 61,000 river chiefs at various levels of government across the county. RCS grew from being a local policy effort that was largely driven by local emergencies of drastically polluted rivers and urgent need for the protection of water resources, into a national political will, which was implemented across all 31 provinces and autonomous regions (Huang and Xu 2019). "River chiefs" or "river leaders" in form of civil servants are in charge of monitoring and improving water quality across their assigned rivers, streams and lakes. Zhejiang was the first province to introduce the law on river chiefs, which came into effect in 2017 and connects with the province's status of being an eco-province, which has set a particular focus on water co-governance.

The submission of political targets and governmental accountability are core items of the RCS. In fact, some consider the RCS a mechanism that has effectively put in place the responsibility for governing waters (Wang and Chen 2019). Others argue that the national RCS mechanism values more political control than governance efficiency (Huang and Xu 2019). The particularly political task of water governance is provided as a major reason. The greater context has been referred to as sustainable legitimacy (see, e.g., Wang 2013), or old sources of path-dependent legitimacy (see e.g., Holbig 2006; Noesselt 2015).

Knowledge production in the adaptation area has been traditionally nestled at research departments with a focus on agriculture, hydraulics and development. The Chinese Academy of Agricultural Sciences (CAAS), based in Beijing, is the most prominent research institution on adaptation. Compared to the West, research institutions in China are more closely affiliated with the government and usually directed through important political figures. Several interviewees suggested, this enables research to be more directly influenced and shaped by specific political agendas (I-35, I-37, I-40). One adaptation expert implied that the choice of Guangzhou as pilot city under the ACCC (see Chapter 5) and its focus on health were largely personality-driven coupled with a stronghold of health-focused research institutions and expertise in that region. When asked how they would explain the little emphasis on humans as part of adaptation planning, one vulnerability scholar expresses:

> [Very long pause] It is difficult. Maybe it is because they [the government] think the investment will deal with it, [that investment into] technology and other sectors will address human health and vulnerability as a side effect.
>
> (I-35: 44f.)

Against the background of a dominant focus on technological solutions, human exposure has become a sidelined concern that is addressed across specific policy paths, not necessarily deliberate, but like adaptation, as a co-benefit.

Politically sensitive knowledge paths

In China, it became clear once again that there is a strong political bias toward social-scientific perspectives and that human vulnerability is a politically sensitive knowledge path, even within the adaptation policy field. Direct human vulnerability to flooding as well as the indirect health effects of, for instance, polluted rivers are considered political weaknesses and viewed as a proxy of illegitimate government. Admitting to having vulnerable populations to climate change was also equated with economic weakness and potentially in contradiction with submitted political targets, as part of the mandatory target system (I-40). In China, fewer government officials and policy practitioners perceived human exposure as a standalone problem that needs addressing in the context of climate change but had overall higher global conviction about growing climate impacts. First observations suggest that a typology of awareness existed, which needs to be looked at more cautiously in a follow-up study.

There is sporadic engagement with adaptation through a health perspective, but this has not entered the mainstream yet, nor did mainstreaming of vulnerability occur into political processes from a perspective of broader human exposure. In Jinhua, flooding was considered primarily as a problem that needs more innovative governmental and infrastructural problem solutions and approaches of civil engineering, in terms of organizational management, strong governmental and military guidance. The issue of social vulnerability was considered a sensitive topic that is difficult to address, particularly as a political decision-maker. Furthermore, the subtle impression was gained that although certain knowledge existed, there was a deliberate choice involved to not-knowing and not deciding upon matters, due to the perceived politically sensitive nature. One academic reflects upon local governance and how government officials may choose to keep an observant position:

> In China, I think, it has to come from the very top . . . the higher-level environment. I think the local government, they are pretty responsive, I mean reactive. If something happens, they take care of it. But if there is no urgent need, they think they are less likely to make a mistake.
>
> (I-15: 81)

In Atlanta, knowledge paths that were related to climate change and segregation-based patterns of vulnerability were viewed as politically sensitive by some decision-makers. Further, there was a strong differentiation of different climate change beliefs among decision-makers. It appeared that decision-makers generated the importance of climate policy from different levels, e.g., those who are globally convinced and build strong links to the global level through, for instance, partaking in city-to-city competitions such as that of the Rockefeller Foundation. Then there were those, who appeared to be more focused on the local level and building up community work on issues related to resilience. Different types of skepticism were also hinted at to impact decision-making on climate matters at the municipal level. The literature distinguished those who are skeptical of humans have caused the issue (anthropogenic skeptics) and those who are skeptic that the climate is changing altogether (climate skeptics) (also see Heimann 2019). To what extent adaptation decision-making in Atlanta was impacted by these different beliefs could not be systematically assessed. However, it appeared that all these different beliefs coexist side by side, informing adaptation action but also impacting why certain aspects of knowledge are considered more sensitive than others (I-27). What also became obvious is climate skepticism as a dominant southern legacy, which has strongly impacted patterns of knowledge creation, dissemination and discourses. These factors too are also closely linked to questions of political legitimacy. Some interviewees indicated that climate-related language and following this knowledge path too closely could potentially hamper their career.

These insights demonstrate the close intertwinement of dominant knowledge patterns, path-dependent legacies and political legitimacy. These do not only become present in terms of water being historically prominent issues based upon old storytelling of emperors who lost their "mandate of heaven" they also relate to career opportunities, (the lack of) pressure from the top to explore certain topics, and issues that are considered politically sensitive. In Atlanta, the need for depoliticizing climate change policy was reiterated to be able to act, as this policy practitioner notes:

> It is slowly getting there, so people can talk about climate change, and you have to divorce from the discussion of climate change, human-induced climate change, whether humans are causing it or not, has to become a non-issue to talk about the whole bigger picture. Things are changing, and they do go through cycles, but unfortunately, what is hard to convince a lot of people of, yes, things cycle, but the earth cycles are somewhat longer than humanity's cycle that, while you think "well this is just a little cycle", well that might be a cycle of last five thousand years, we just do not know. And so . . . it is not something I am doing.
>
> (I-27: 66)

In both cases, selectively choosing topics that enter the political agenda were also driven by issues that decision-makers considered politically sensitive. In the United States, one core element seemed to be how to communicate climate change in a state with a partisan climate agenda. Employees within and outside government institutions in Atlanta emphasized the need to be very careful and cautious on matters related to climate change in general. Actors working for federal agencies in Atlanta and Georgia explained how the politically sensitive climate had led to adjusted patterns of communication. Another stakeholder expressed:

> They [the municipal government] are more willing to deal with the resiliency as a concept, and not putting the word climate change in front of it, just talking about resiliency.
>
> (I-31: 133)

It became clear that decision-makers and planners had adjusted the way they are approaching climate-related issues: "[In Georgia,] We are careful and cautious" (I-11: 32). This presents a particularly worthwhile research endeavor, to what extent decision-makers, who are engaged in the issue of climate change but working in climate skeptic environments, adopt their practices and language to still design political strategies (also see Teebken fc.).

Hard infrastructural legacies of flood management and narrow focus on human health in Atlanta

Decision-makers in both Jinhua and Atlanta seemed to have long traditions of addressing climate-induced weather events and natural hazards with a traditional understanding of infrastructure. Knowledge related to water management was more deeply rooted than, for instance, knowledge related to heat or social vulnerability (also see following sections). In Atlanta, this was reflected at the city and county levels, where elected officials talked about flood as an issue of stormwater control and an improvement of existing physical structures. The county levels admitted that their awareness about issues such as flooding is low overall, because they consider this a job of the emergency management bureaus and the municipalities. A policy practitioner within the Fulton County Government put it this way:

> But we have not // I do not know that we have a flooding issue. Nor do I know that we have populations that are vulnerable to the // be impacted by that. And that is just something we have not had a lot of attention on as a county.
>
> (I-24: 46)

What could be observed in the example of Atlanta was the strong decentralization of responsibilities to different levels of government. Here, stormwater management is nestled at the city level and receives some guidance from the state level. However, the lack of a clear link to climate-related knowledge was always explicit in these supposedly "traditional" lines of work. When looking at political institutions that were considered to be more responsible for these "new types" of policy challenges, such as the Mayor's Office of Resilience, their endeavors were criticized for their broad resilience understanding. It was also implied that climate change related language had been deliberately deleted from formal presentations and reports, to not influence the audience and "be spontaneous" (I-22, I-31). One employee at the MOoR expressed this aspect of the 100RC organization needs urgent review, as "it is flawed in the protocol honestly" (I-22: 128). A majority of interviewees outlined how the human vulnerability aspect is addressed in a narrow sense as part of human health in the resilience strategy, if at all. This was mirrored by one policy practitioner at the metro level, who indicated a technology-based understanding of human vulnerability in terms of emission reductions, and status quo reproduction:

> I would not say [human vulnerability as a policy concern] has [played a role]. I do not know that it has in [the United States of] America much at all. Like, other than the CDC's work and some of the states that are more focused on the health impacts of climate change, I do not know that there has been a lot of focus on human vulnerability. I think it has mostly been dealt with through like, a technological lens of, you know, what can we do to reduce emissions? And then, what can we do to make sure, you know, bridge x or y will not flood? In the spirit of keeping the system activated but not really looking at how that impacts individuals. Other than the fact that like, you know if you cannot drive to the hospital because the one bridge out of your area is flooded, then you are // that is a public health problem.
>
> (I-31: 150f.)

Lack of shared knowledge and understanding

The problem of effective communication about climate change adaptation science, a disconnected science–policy interface and the lack of a sound evidence-base to inform policy decisions constitute a set of interconnected factors that are interlinked with how the political systems work in both contexts. These, it will be argued lead to a lack of shared knowledge and understanding to drive political action in this regard.

Science policy public gap in Atlanta

The overall lack of climate adaptation efforts and awareness was explained with a disconnect of science and policy and a disconnect of experts and local people (I-27, I-29, I-35, I-37). Explanatory attempts focused on the way knowledge is made available; limited governmental capacity to process knowledge and the sensitive political climate under climate-skeptic governments at state and federal levels. This is in line with research, which stresses low science literacy based on isolated and late production of knowledge; politically motivated misinformation, inherent biases and cognitive dissonances (see, e.g., Stover 2019). Further, the lack of knowledge in Atlanta was also noticed:

> The people who are charged with these issues in Atlanta, do not have an extensive background in these kinds of issues and so are struggling to prioritize.
> (I-29: 13)

The need to incorporate climate literacy into multiple areas of the school curriculums to develop well-educated citizens has been argued for (Bhattacharya et al. 2018). This strand of research seeks to explain, why climate literacy continues to be relatively low despite voluminous amounts of research and scientific consensus. It also coincides with practical efforts guided by the Climate and Natural Resources Working Group for the Council of Climate Preparedness and Resilience for "Building Climate Literacy and Capabilities Among Federal Natural Resource Agencies" (2016) under Executive Order 13653. Besides a continued disconnect of available science, awareness and the capacity to act, unawareness in the United States is politically motivated. To fill the void, not using "climate change" has become a "communication strategy" (Stover 2019).

More advanced shared knowledge in Jinhua

Jinhua seems to be more advanced regarding its cross-departmental and long-term floodwater management efforts, which appears to be based on a more holistic form of cooperation. The Five Water Treatment Department and related governmental efforts are an example of multi-sectoral cooperation that involves the expertise of different government departments, who work together from their angles on issues related to water. This integration has structured the municipal planning accordingly.

The Sponge City Program has been a pioneer of floodwater management and attracted great attention from different levels of government and city planners across the world. Although it is seen as a transformational approach to holistic flood management, it is also an example of urban construction that aims for economic and rapid advancement of the core city, commonly neglecting suburban and rural offshoots such as Lan'xi and Pan'an. The neglected focus on proactively preparing communities with less access ahead of time became obvious. Institutions that enable people to take own precautionary measures were largely absent and it seemed that a strong dependence on punctual government intervention has become the norm. Shared knowledge appears to be significantly more advanced on issues related to the environment, but not as it relates to humans.

Communication and governmental learning about knowledge

In Atlanta, the lack of a shared understanding of knowledge was more pronounced and appears to be driven by different political agendas and political partisanship. Further, government officials claimed, the knowledge had not traveled to them, whereas experts were critical of their interest in receiving information related to climate change. Government officials at the municipal level, who were interested in working on human vulnerability in the context of climate adaptation talked about the selective way of choosing not to know, due to the politically sensitive nature of the topic. The availability of and fallback on best practices, knowledge exchange programs, and training to guide local policy processes has played a major role in Atlanta's climate policy process. Due to the absence of guidance from other government levels, Atlanta's municipal government seemed to depend on guidance through other non-state actors. Federal protocols such as NOAA's guidance on vulnerable communities were mentioned to provide important background knowledge nevertheless. In the context of adaptation, one government official at a federal office expressed that it will "take steady, continuing effort of education for local governments, departments and state agencies" (I-11: 59).

In addition to the science–policy–public gap, and the way different types of knowledge (did not) travel, the lack of a shared understanding was also attributed to the way complex science is communicated. In both cases, adaptation experts were indicating that the scientific work they had done "for a long time is not new," but that the problem rather was "communicating the science" (I-35: 34). Communicating effectively with different stakeholders was pointed out as a lengthy process in Georgia. Here, the southern way of communicating was emphasized as a particular challenge (I-19). Making information relatable and not just focusing on record-breaking temperatures was referred to as core difficulty.

Perception of overcomplexity

Both cases are characterized by very selective awareness about unavoidable and growing climate impacts and the simultaneous need to address human vulnerability. Whereas some politicians seemed to deliberately underestimate the risk and did not see climate adaptation as a necessary policy priority in light of strategic political concerns, others seemed to lack knowledge on the matter overall. It was often understood as an overtly complex policy issue to be addressed. Aside from an often politically motivated bias, the following section looks at several factors that

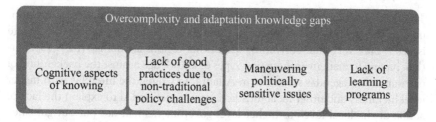

Figure 8.8 Factors leading to the perception of overcomplexity

seem to contribute to the perception of adaptation and social vulnerability as an overcomplex issue (see Figure 8.8).

Cognitive aspects of knowing

In both cases, low-risk awareness appeared to correlate with the ubiquity of extreme events such as storms and flooding. Decision-makers in Atlanta emphasized that the Southeastern United States receives a great number of different storms such as hurricanes and tornadoes and has always coped with that (I-10, I-23, I-27). The long tradition of the need to address flooding and well-working governance mechanisms were noted in China (GI-01, I-13, I-15, I-16, I-38, I-39G). At the cognitive level, those who have experienced vulnerability and/or were affected previously, were considered to have greater capacity in being more attuned to these types of issues. This coincides with an observation which is: municipalities at the cost that are significantly more progressed in their daily experience with climate change, especially sea-level rise, are more advanced with regard to their adaptation planning. The example of Tybee Island stands out in this context and was often referred to by different interviewees. Tybee Island is Georgia's most densely populated small barrier Ireland in the northern stretch of the state. The island and mainland are connected by one highway route. The street is flooded several times a year, cutting off the island multiple times per year. Here, the problem of climate change has become especially visible, making Tybee Island adopt the first sea-level rise adaptation plan in Georgia.

This is in line with adaptation scholarship, which points to the close proximity to an event and cultural as well as environmental experience as important triggers for change. The experience with the environment is considered to support a holistic way of thinking that is not just based on abstract knowledge but a relationship with the environment experienced by the immediate context (Lammel et al. 2013). Recent research found that culture and cognition and the lack of shared knowledge play a distinctive role for effective adaptation to climate change, with no real climatic experience leading to more cognitive vulnerability (Lammel et al. 2013). This connects with what was elaborated earlier: Vulnerability is a deeply relational concept and experience.

Lack of good practices in light of nontraditional or sensitive policy issues

The lack of action on social vulnerability and adaptation was also often attributed to the lack of good practices in this area, as "Atlanta's responses to climate issues . . . are very much driven by what other regions are doing" (I-29: 13). Further, having to maneuver these new types of knowledge against the background of partisan policy agendas stood out in Atlanta. In China, the interplay with higher level hierarchy and other local jurisdictions was mentioned to explain the lack of action at the local level, as higher level governments had not put out guidance documents on the issue of climate adaptation and social vulnerability. One policy advisor implicated that it is important to get the ball rolling at higher government level: "This is like the snowball effect, once it gets noticed, then the snowball gets mucher and mucher" (I-37: 82).

In Atlanta, the lack of good practices was also explained by the fact that heat and social vulnerability constitute nontraditional policy challenges. Water on the other hand was presented as a common policy topic and traditional type of knowledge for political decision-makers. The same can be argued for Jinhua, where water has been a traditional policy concern and knowledge capacity in this area is especially pronounced.

Little political capacity for learning how to address complex problems

The findings indicate that local decision-makers had rather low problem recognition and a traditional infrastructure focus based upon past-knowledge patterns that are coupled with concerns over political legitimacy. The psycho-social level and how actors in political institutions learn also appear as core aspects in this regard. Although it is impossible to detect, it appeared that some complex problems, such as socially vulnerable populations to climate change were deliberately neglected. Local decision-makers in both political systems did not seem to be equipped in order to address the larger political problems that are interlocked at different levels of society and politics.

How political institutions learn and adjust to political volatile atmospheres appeared as a relevant factor that is also closely related to the cognitive level and psycho-social processes of *experiencing* a related problem. When asked, what unwritten rules underly proactive policy responses, this decision-maker and academic expresses:

> This is a cultural phenomenon to me. There is a great quote from Mark Twain: "Travels is the only cure of ignorance". A different one goes: "The world is a book and those who don't travel have only read the first chapter." If government officials only exist in their only little corner of the world, there are less likely to be open-minded. The biggest impact on the public service commission was an organized solar trip panel to Germany in 2013. Those were very Southern elected officials, who didn't believe in any solar power, who went to Germany for a week, toured the entire country, came back and wanted to do

this. One of them was McDonald, who is a big renewable energy supporter, who believes in the power of solar, but is a Trump supporter.

(I-09: 25f.)

Interviewees, who were experts on both cases, spoke of the importance of dialogue and cooperation programs to advance a shared understanding and knowledge about matters to adaptation and that this was currently lacking (I-10, I-29, I-35, I-37). The lack of political learning opportunities in terms of knowledge-exchange programs and learning how to deal with complex challenges was pointed to. It was argued again and again that first-hand experience and seeing something with your own eyes will help to get decision-makers sensitive for certain issues (I-09). The field of urban planning was pointed to as a good practice example, where dialogue was "changing the way elected officials are looking" (I-29: 13). Although policy experimentation is a dominant practice in China, the parameters are often set from the top, with a strong selection of stakeholders and issues that can be addressed. Future research could examine how to strengthen the social science perspective as part of governmental knowledge exchange and learning programs.

Summary

This chapter investigated how adaptation practices and recognition unfold across two different political jurisdictions. It was found that adaptation is often occurring accidentally, with policy efforts being retroactively labelled "adaptation." In both cases, creating green-blue infrastructure was the dominant practice, often following a (re)development paradigm. The social component of adaptation was at best incrementally addressed as a co-benefit without being specified. In Atlanta, neighborhood revitalization programs in the form of floodwater parks or greening the environment, actively co-created vulnerability due to a rise in property values, which triggered a housing affordability crisis. Thereby, Atlanta's accidental adaptation measures in the area of floodwater management have become a powerful example of maladaptation, demonstrating, how one adaptation target (floodwater management) may stand in conflict with another (addressing social vulnerability).

The main findings, accidental and green infrastructure-focused adaptation, were then explained with lock-ins that sit at the intersection of knowledge and politics. Three facets were discussed in greater detail: 1) the dominance of certain knowledge paths, 2) the lack of shared knowledge and 3) the perception of over-complexity. Natural science-focused knowledge paths have a long-standing political legacy in China. In line with this, social-science informed adaptation is absent. Issues such as highly uneven vulnerability to climate change vulnerability are also considered sensitive political issues. Although adaptation processes in the United States are informed through social science studies, politically sensitive issues could also be detected here. These are uneven human vulnerability as it relates to climate change and path-dependent patterns of segregation of African American populations. There was a strong partisan divide on these matters. Those working for predominantly Republican decision-makers actively adjusted their language and only talked about it reluctantly. Further, the uneven distribution of vulnerability in the context of heat were considered nontraditional issues in Atlanta, where good practices are often lacking, but are needed to inform local government. In contrast to Atlanta, the findings point to a more advanced understanding and forms of shared knowledge on climate- and water-related issues in Jinhua. That being said, sensitivity about the social dimension of adaptation was largely absent among decision-makers in Jinhua. Greater sensitivity existed in Atlanta, but also appeared divided across party lines. Lastly, the lack of cognitive and environmental experience coupled with the lack of good practices and little political capacity to learn about complex political matters such as unequal vulnerability to climate change, together contribute to this being perceived as an overtly complex challenge. The normative implications of the findings are that local governments across both political systems have only low adaptive capacity of governmental actors to deal with complex issues.

Notes

1   See Mayor Keisha Lance Bottoms in the full hearing of the Senate Democrats' Special Committee on the Climate Crisis, published via Senate Democrats July 17, 2019.
2   The Atlanta Beltline Partnership is a private nonprofit that claims to enable the vision of the Atlanta Beltline project. For more information see: "About the Atlanta Beltline Partnership", https://beltline.org/about/atlanta-beltline-partnership/.
3   Since its foundation in 1972, the organization has mainly focused on conservation projects across the United States that seek to "work with communities to ensure that development happens for them, and not to them," see: "How we work," available at www.tpl.org/how-we-work, last accessed July 27, 2019.
4   For more information, please find a project description with pictures at the Landeszine architectural platform available at: www.landezine.com/index.php/2015/03/a-resilient-landscape-yanweizhou-park-in-jinhua-city-by-turenscape/, accessed April 14, 2019.
5   It is beyond the scope of this research to examine, to what extent the "accidental" adaptation efforts constitute their own form of cultural knowledge, or are differently understood by policy practitioners and may as well be named differently. Further research could also study whether the adaptation concept is rather referring to the academic practice and scholastic tradition of adaptation research, but poses certain limitations when examining adaptation actions on the ground.
6   See "Da Yu, Chinese mythological hero," *Encyclopedia Britannica Online*, accessed July 25, 2019: www.britannica.com/topic/Da-Yu.

References

Adaptation Clearinghouse (AC). "Case Study of the Atlanta BeltLine – Adaptation Aspects." www.adaptationclearinghouse.org/resources/case-study-of-the-atlanta-beltline-adaptation-aspects.html, last accessed June 14, 2019.
Atlanta Development Authority (ADA). "Atlanta BeltLine Redevelopment Plan." Atlanta Development Authority Report, November 2005.

Aton, Adam. "Ga. Races Could Shape U.S. Climate Policy for a Generation." *E&E News*, January 4, 2021. www.eenews.net/stories/1063721619, last accessed July, 13, 2021.

BeltLine Rail Now (BLRN). "Atlanta Beltline Rail: A Blueprint for Transit Funding." January 2021. https://static1.squarespace.com/static/5b6799645b409b9876b2a389/t/603275a4e6476567183babc7/1613919654798/2021+BeltLine+Rail+Now+-+White+Paper+-+Blueprint+for+Transit+Funding.pdf, last accessed July 12, 2021.

Bhattacharya, Devarati, Cory T. Forbes, Mark A. Chandler, K. Carroll Steward, and A. McKinzie Sutter. "Climate Literacy: Insights from Research on K-16 Climate Education." *Green School Catal. Q.* 5, no. 4 (2018): 26–37.

China Climate Change Info Net (CCCIN). "The National Development and Reform Commission and the Ministry of Housing and Urban-Rural Development issued the Action Plan for Urban Climate Change Adaptation" (Original: 发改委, 住建部发布 "城市适应气候变化行动方案"). February 2016. www.ccchina.org.cn/Detail.aspx?newsId=58941&TId=57, last accessed April 29, 2019.

China Climate Change Partnership Framework (CCCPF), Smith, Edward Clarence and Catherine Wong. *China Climate Change Partnership Framework*. United Nations Spain, MDG Achievement Fund, published online, 2012. https://www.sdgfund.org/sites/default/files/ENV_DISCUSION%20PAPER_China_%20Climate%20Change%20Partnership%20Framework.pdf.

China Daily. "Ningbo Selected as Pilot City for Urban Design." March 30, 2017. http://subsites.chinadaily.com.cn/ningbo/2017-03/30/c_73617.htm, last accessed April 29, 2019.

City of Atlanta (CoA). "Renewal Through Creative Stormwater Management: Rodney Cook, Sr. Park." Department of Watershed Management, GAWP Fall Conference, 2017a.

CoA. "Resilient Atlanta – Actions to Build an Equitable Future." 2017b.

CoA. "City of Atlanta Green Infrastructure Strategic Action Plan." City of Atlanta, Department of Watershed Management, 2018.

Cronin, David. "Is Georgia a Tipping Point for New US Climate Policy?" *Carbon Tracker*, January 13, 2021. https://carbontracker.org/is-georgia-a-tipping-point-for-new-us-climate-policy/, last accessed July 12 2021.

Dai, Liping Helena F. M. W. van Rijswick, Peter P. J. Driessen, and Andrea M. Keessen. "Governance of the Sponge City Programme in China with Wuhan as a Case Study." *International Journal of Water Resources Development* 34, no. 4 (2018): 578–596. Doi: 10.1080/07900627.2017.1373637.

Economy, Elizabeth C. *The River Runs Black: The Environmental Challenge to China's Future*. Ithaca and London: Cornell University Press, 2010.

Elvin, Mark. "The Environmental Legacy of Imperial China." *The China Quarterly* 156, Special Issue: China's Environment (1998): 733–756. www.jstor.org/stable/656123.

Georgia Climate Project (GCP). "Our Work." www.georgiaclimateproject.org/our-work/, last accessed July 13, 2021.

Girardot, N. J., James Miller, and Liu Xiaogan. *Daosim and Ecology. Ways within a Cosmic Landscape*. Cambridge, MA: Harvard University Press, 2001.

Gravel, Ryan Austin. "Belt Line – Atlanta, Design of Infrastructure as a Reflection of Public Policy." Master Thesis, College of Architecture the Georgia Institute of Technology, December 1999.

Gu, Shanghui. "So Amazing! The Houses They Built Were Warm in the Summer and Cool in the Summer, Almost no Electricity!" (Original: "这么神奇！他们造的房子冬暖夏凉 几乎不用电!"). *Zhejiang Daily Newspaper*, October 20, 2016. https://zj.zjol.com.cn/news.html?id=466441, last accessed April 29, 2019.

Heimann, Thorsten. *Culture, Space and Climate Change. Vulnerability and Resilience in European Coastal Areas*. Series: Routledge Advances in Climate Change Research. London and New York: Routledge, 2019.

Hermaputi, Roosmayri L., and Chen Hua. "Creating Urban Water Resilience: Review of China's Development Strategies 'Sponge City' Concept and Practices." *The Indonesian Journal of Planning and Development* 2, no. 1 (2017): 1–10. Doi: 10.14710/ijpd.2.1.1-10.

Historic Fourth Ward Park Conservancy (H4WPC). "Clear Creek Basin." 2019. www.h4wpc.org/clear-creek-basin/, last accessed June 27, 2021.

Holbig, Heike. "Ideological Reform and Political Legitimacy in China: Challenges in the Post-Jiang Era." *GIGA Research Program: Legitimacy and Efficiency of Political Systems* no. 18, 2006.

Huang, Qidong, and Jiajun Xu. "Rethinking Environmental Bureaucracies in River Chiefs System (RCS) in China: A Critical Literature Study." *Sustainability* 11, no. 6(1608) (2019). https://doi.org/10.3390/su11061608.

Immergluck, Dan. "Large Redevelopment Initiatives, Housing Values and Gentrification: The Case of the Atlanta Beltline." *Urban Studies* 46, no. 8 (July 2009): 1723–1745. Doi: 10.1177/0042098009105500.

Immergluck, Dan, and Tharunya Balan. "Sustainable for Whom? Green Urban Development, Environmental Gentrification, and The Atlanta Beltline." *Urban Geography* 18, no. 4 (2017): 546–572. https://doi.org/10.1080/02723638.2017.1360041

Jin, Jianjun, and Hermi Francisco. "Sea-Level Rise Adaptation Measures in Local Communities of Zhejiang Province, China." *Ocean & Coastal Management* 71 (2013): 187–194. Doi: 10.1016/j.ocecoaman.2012.10.020.

Jinhua Municipal Environmental Protection Bureau (JMEPB). "Jinhua City Ecological Civilization Construction Demonstration City Planning (2017–2025) passed expert review." (Original: "金华市生态文明建设示范市规划 (2017–2025) 通过专家评审". September 9, 2018. https://m.huanbao-world.com/view.php?aid=40947, last accessed April 29, 2019.

Jinhua Municipal People's Government (JMPG). "Jinhua City Master Plan 2000–2020." (Original: "金华市城市总体规划"). 2009. www.cityup.org/case/general/20090429/47932.shtml, last accessed January 17, 2020.

JMPG. "12th Five-Year Plan for Meteorological Development." (Original: 气象事业发展十二五计划). 2012.

JMPG. "2015 Jinhua Government Annual Working Report." (Original: "2015 华市政府工作报告 – 金华市政府"). 2015.

JMPG. "Jinhua City Ecological Civilization Demonstration Creation Implementation Plan." (Original: "金华市生态文明示范创建实施方案"). December 27, 2018. www.jinhua.gov.cn/11330700002592599F/02/szfwj/201812/t20181229_3620931_2.html, last accessed April 29, 2019.

Kassiola, Joel J. "China's Environmental Crisis and Confucianism: Proposing a Confucian Green Theory to Save the Environment." In *Chinese Environmental Governance. Environmental Politics and Theory*, edited by Ren B. and Shou H. New York: Palgrave Macmillan, 2013.

Kingdon, John W. *Agendas, Alternatives and Public Policies*. Boston: Little, Brown, 1984.

Kingdon, John W. *Agendas, Alternatives and Public Policies* (Updated 2nd ed.). Boston: Longman Classics in Political Science, 2011.

Lammel, Annamaria, Elisa Gutiérrez, Emilie Dugas, and Frank Jamet. "Cultural and Environmental Changes: Cognitive Adaptation to Global Warming." *Steering the cultural dynamics: Selected papers from the 2010 Congress of the International Association for Cross-Cultural Psychology*, 2013.

Lee, Maggie. "Atlanta's Resilience Chief Departing." Saporta Report, June 19, 2019. https://saportareport.com/atlantas-resilience-chief-departing/sections/reports/maggie/, last accessed June 30, 2021.

Li, Bingqin. "Governing Urban Climate Change Adaptation in China." *Environment and Urbanization* 25, no. 2 (2013): 413–427. https://doi.org/10.1177/0956247813490907.

Li, Hui, Liuqian Ding, Minglei Ren, Changzhi Li, and Hong Wang. "Sponge City Construction in China: A Survey of the Challenges and Opportunities." *Water* 9, no. 9 (2017): 1–27. Doi: 10.3390/w9090594.

McPhearson, Timon, David Iwaniec, and Xuemei Bai. "Positive Visions for Guiding Urban Transformations Toward Sustainable Futures." *Current Opinion in Environmental Sustainability* 22 (2017): 1–8. Doi: 10.1016/j.cosust.2017.04.004.

Meng, Meng, and Marcin Dabrowski. "The Governance of Flood Risk Planning in Guangzhou, China: Using the Past to Study the Present." 17th IPHS Conference, Delft 2016.

Mertha, Andrew C. *China's Water Warriors: Citizen Action and Policy Change*. Ithaca and London: Cornell University Press, 2008.

Ministry of Environment and Ecology (MEE). "China's Policies and Actions for Addressing Climate Change." November 2018.

Ministry of Housing and Urban-Rural Development (MOHURD). "Cities Adapt to Climate Change Action Plan." (Original: "城市适应气候变化 行动方案"). 2016. www.mohurd.gov.cn/wjfb/201602/W020160224041125.doc, last accessed August 31, 2019.

MOHURD. "Ministry of Housing and Urban-Rural Development Announces New Pilot List of Urban Design." (Original: "住房城乡建设部公布城市设计新试点名单"). 2017. www.mohurd.gov.cn/zxydt/201707/t20170725_232719.html, last accessed May 1, 2019.

Nadin, Rebecca, Sarah Opitz-Stapleton, and Xu Yinlong (eds.). *Climate Risk and Resilience in China*. London: Routledge, 2016.

National Development and Reform Commission (NDRC). "China's Policies and Actions for Addressing Climate Change (2013)." 2013. https://en.ndrc.gov.cn/newsrelease_8232/201311/P020191101481854882171.pdf, last accessed January 17, 2020.

NDRC. "The People's Republic of China First Biennial Update Report on Climate Change." 2016. http://qhs.ndrc.gov.cn/dtjj/201701/W020170123346264208002.pdf, last accessed November 2, 2019a.

NDRC. "Notice on Printing and Launching Pilot Work for Climate-Adapted City Construction." (Original: "关于印发气候适应型城市建设试点工作的通知"). 2017. www.ndrc.gov.cn/zcfb/zcfbtz/201702/t20170224_839212.html, last accessed May 1, 2019b.

Noesselt, Nele. "Relegitimizing the Chinese Party-State: 'Old' Sources of Modern Chinese Party." *Power ASIA* 69, no. 1 (2015): 213–233. Doi: 10.1515/asia-2015-0014.

Oakley, Deirdre, and George Greenidge Jr. "The Contradictory Logics of Public-Private Place-making and Spatial Justice: The Case of Atlanta's Beltline." *City & Community* 16, no. 4 (2017): 355–358. https://doi.org/10.1111/cico.12264.

Perepichka, Anzhela, and Iulia Katsy. "How Landscape Infrastructures Can Be More Resilient? Positive Practice of Wetland Urban Adaptation to Stormwater Extreme Events in China." Research Paper, 2016.

Roy, Parama. "Collaborative Planning – A Neoliberal Strategy? A Study of the Atlanta Belt Line." *Cities* 43 (2015): 59–68. Doi: 10.1016/j.cities.2014.11.010.

Russel, Duncan. "Enabling Conditions for the Mainstreaming of Adaptation Policy and Practice." In *Research Handbook on Climate Adaptation Policy*, edited by E. C. H. Keskitalo and B. L. Preson. Cheltenham, UK and Northhampton: Edward Elgar Publishing, 2019.

Sams, Douglas. "Ponce City Market Secures Largest Atlanta Construction Loan Since Recession." *Atlanta Business Chronicle*, April 28, 2014. www.bizjournals.com/atlanta/real_talk/2014/04/ponce-city-market-secures-largest-atlanta.html, last accessed January 17, 2020.

Samual, Molly. "Park Planned to Bring Vine City a Place to Play, Flood Relief." *WABE*, February 13, 2017. http://news.wabe.org/post/park-planned-bring-vine-city-place-play-flood-relief, last accessed June 27, 20221.

Shamma, Tasnim. "Atlanta Is Home to Largest Permeable Pavers Project in US." *WABE*, November 1, 2015. www.wabe.org/atlanta-largest-pavers-project/, last accessed April 29, 2019.

Smart Cities World (SCW). "Rockefeller Foundation Winds Down Funding for 100 Resilient Cities." 2019. www.smartcitiesworld.net/news/news/rockefeller-foundation-winds-down-funding-for-100-resilient-cities-4037, last accessed June 4, 2021.

Songyang Bureau of Natural Resources and Planning (SBNRP). "Notice of the Ministry of Land and Resources of Zhejiang Province on the Catalogue of Administrative Documents of the Provincial Department of Land and Resources, which continues to be valid, declared invalid, and abolished." (Original: "废止的行政规范性文件目录的通知"). 2018.

Stafford, Leon. "Home Affordability Concerns Prompt BeltLine Leader to Resign." *Atlanta Journal Constitution*, September 28, 2016. www.ajc.com/news/local-govt – politics/home-affordability-concerns-prompt-beltline-leader-resign/5W6x7eanw2zSd3h7uk8glK/, last accessed January 10, 2020.

Stover, Dawn. "Marshall Shepherd: Connecting Atmospheric Science and Society." *Bulletin of the Atomic Scientists* 75, no. 4 (2019): 205–209. https://doi.org/10.1080/0096340 2.2019.1628517.

Taylor, Josey. "When Fourth Ward Grew Old: A Discussion on Gentrification Focusing on One of Atlanta's Most Famous Historic Neighborhoods." *Taylor Josey Portfolio, Just another Georgia State University site* (blog), April 29, 2016. http://sites.gsu.edu/tjosey3/2016/04/29/when-fourth-ward-grew-old-a-discussion-on-gentrification-focusing-on-one-of-atlantas-most-famous-historic-neighborhoods/, last accessed July 27, 2019.

Teebken, Julia. "Playing Hide and Seek – Adapting Climate Cultures in Troubled Political Waters in Georgia, United States." In *Climate Cultures in Europe and North America*, edited by Thorsten Heimann, Jamie Sommer, Margarethe Kusenbach, and Gabriela Christmann. Routledge, forthcoming.

Tsai, Ethan, Haeun Kim, Nobuhiro Arai, Rishi Chakraborty, RobertPaton, and Yu Ann Tan. "Atlanta Rising: An Analysis on the Climate Change Impacts and Resilience-Building in the City of Atlanta." Columbia School of International and Public Affairs Report, 2018.

UrbanHeatATL (UHATL). "About." https://urbanheatatl.org/about/, last accessed July 13, 2021.

Wang, Yahua, and Xiangning Chen. "River Chief System as a Collaborative Water Governance Approach in China." *International Journal of Water Resources Development* 36, no. 4 (2019): 610–630. Doi: 10.1080/07900627.2019.1680351.

Wang, Lizhong, Yonggui Xie, Youxia Wu, Zhen Guo, Yuanqiang Cai, Youcheng Xu, and Xibing Zhu. "Failure Mechanism and Conservation of the Ancient Seawall Structure along Hangzhou Bay, China." *Journal of Coastal Research* 28, no. 6 (2012): 1393–1403.

Wang, Qiu-Shun, Cun-Hong Pan, and Guang-Zhi Zhang. "Impact of and Adaptation Strategies for Sea-Level Rise on Yangtze River Delta." *Advances in Climate Change Research* 9, no. 2 (2018): 154–160. https://doi.org/10.1016/j.accre.2018.05.005.

Wang, Qiu-Shun, Cun-Hong Pan, and Guang-Zhi ZhangWang, Alex L. "The Search for Sustainable Legitimacy: Environmental Law and Bureaucracy in China." *Harvard Environmental Law Review* 37 (2013): 365–440.

Wei, Jiang, and Tao Cunhuan. "The Seawall in Qiantang Estuary." In *Engineered Coasts. Coastal Systems and Continental Margins*, edited by J. Chen et al., Vol. 6. Dordrecht: Springer, 2002.

Wiek, Arnim, and David Iwaniec. "Quality Criteria for Visions and Visioning in Sustainability Science." *Sustainability Science* 9, no. 4 (2014): 497–512. https://doi.org/10.1007/s11625-013-0208-6.

World Bank. *Climate Resilient Ningbo Project Local Resilience Action Plan*. World Bank, 2011. https://openknowledge.worldbank.org/bitstream/handle/10986/12823/702050v10ESW0P00Ningbo0Final0Report.pdf?sequence=1&isAllowed=y, last accessed April 14, 2019.

WSB-TV-2. "Atlanta Mayor on Mass Resignations: 'We Needed a Fresh Look at Our Leadership'." Published with WSBTV. 2018. www.wsbtv.com/news/local/atlanta/atlanta-mayor-keisha-lance-bottoms-to-discuss-mass-forced-resignations/730128104/, last accessed June 4, 2021.

Wu, Fulong. "Neo-Urbanism in the Making Under China's Market Transition." *City* 13, no. 4 (2009): 418–431. Doi: 10.1080/13604810903298474.

Wu, Hongmei, Qiwei Zheng, Heng He, and Jiezhen Wu. "Zhejiang Province's Response to Climate Change Research." (Original: "浙江省适应气候变化政策"). *Development and Planning Research* 55 (2011). www.zdpri.cn/sanji.asp?id=222629.

Xu, Jiang. "Bargaining for Nature: Treating the Environment in China's Urban Planning Practice." *Urban Geography* 38, no. 5 (2017): 687–707. Doi: 10.1080/02723638.2016.1139414.

Xu, Weixing. "Zhejiang Environmental News Agency: Two Ministries and Commissions Publish 'the Plan to Climate Adaptation Model for City Construction pilot scheme operating plan'." (Original: "浙江环保新闻网. 两部委印发 '气候适应型城市建设试点工作方案' 推动试点城市适应气候变化"), November 8, 2016. http://epmap.zjol.com.cn/system/2016/08/11/021261988.shtml, last accessed February 15, 2017.

Yang, Hongxing, and Dingxin Zhao. "Performance Legitimacy, State Autonomy and China's Economic Miracle." *Journal of Contemporary China* 24, no. 91 (2015): 64–82. Doi: 10.1080/10670564.2014.918403.

Zhang, Linxiu, Renfu Luo, Hongmei Yi, and Stephen Tyler. *Climate Adaptation in Asia: Knowledge Gaps and Research Issues in China, the Full Report of the China Team*. Chinese Academy of Sciences Institute of Geographic Sciences and Natural Resources Research (IGSNRR) Centre for Chinese Agricultural Policy (CCAP), published online 2008.

Zhejiang Provincial Government (ZPG). *Zhejiang's Response to Climate Change* (Original: 浙江省应对气候变化方案). ZPG, published online 2010, updated version from 2012: http://www.zj.gov.cn/art/2012/8/17/art_1229591319_64009.html.

Cited interviews

GI-01, July 5, 2016

I-06, October 28, 2016

I-09, October 31, 2016

I-10, November 2, 2016

I-11, November 4, 2016

I-12, March 7, 2017

I-13, March 7, 2017

I-14, March 7, 2017
I-15, March 9, 2017
I-16, March 14, 2017
I-19, April 11, 2017
I-21, May 2, 2017
I-22, May 3, 2017
I-23, May 3, 2017
I-24, May 10, 2017
I-27, May 12, 2017
I-29, May 16, 2017
I-31, May 17, 2017
I-32, May 18, 2017
I-33, May 19, 2017
I-35, August 10, 2017
I-37, October 20, 2017
I-38, November 20, 2017
I-39G, November 21, 2017
I-40, April 27, 2018

9 Lock-ins of political epistemology across different political systems

The following chapter will discuss the main empirical findings on vulnerability and adaptation lock-ins in greater depth. First, vulnerability as an inherent characteristic of the two different political systems of China and the United States will be presented. Next, the analytical framework of lock-ins is revisited. Because the empirical cases both identified lock-ins at the interface of knowledge and political institutions, this chapter discusses the findings in the newly emerging field of research on political epistemology. The way knowledge is produced, accessed and distributed in combination with the way political institutions operate are core elements of the inherent vulnerability of both countries. Drawing from recent studies in the field of political epistemology, the chapter will discuss three core facets of this type of lock-in. The last part reflects on knowledge as an indicator for the manifestation of a class-based society. In both cases, access to knowledge and education became visible as class-based phenomena, which link to future economic and political opportunities.

Vulnerability as an inherent characteristic of both political systems

Adaptation research explains the lack of adaptive capacity in developing countries vis-à-vis developed countries (see, e.g., Parks and Roberts 2006; Adger et al. 2009a; Fankhauser and Dermott 2013; Nur and Shestra 2017). More nuanced analysis rejects the notion of higher adaptive capacity of richer countries in light of widely occurring climatic events in Europe, the United States and Australia and an observed lack of efficient adaptation responses here as well (see, e.g., Wolf 2011).

Another strand of literature has examined the coming of environmental authoritarianism and relative capability of authoritarian regimes to deal with complex environmental challenges (e.g., Beeson 2010). Democracy scholars have argued for some time now that we are confronted by an era of democratic recession, signified by a low quality of democratic institutions (Diamond 2004), especially in the context of climate change (see the latest e.g., Fiorino 2018). This literature has spread serious doubt about the question whether democratic systems are well-equipped to deal with a challenge of the magnitude of climate change. It is

DOI: 10.4324/9781003183259-9

more rarely investigated, which democratic principles correspond with improved climate performance (see, e.g., Hanusch 2017).

In democratic theory, a positive relationship between federalism and enhanced democracy is assumed on the basis of the fact that government is more responsive when closer to the people (see, e.g., Diamond 2004). Greater governmental accountability and social concern are commonly considered a result of legitimate actions of a democratic government (see, e.g., DeLauro 2017). Against the background of these strands of literature, the underlying sentiment has increasingly been that quasi authoritarian governments, like China, are more capable of dealing with the challenge due to their unitary political system and centralized decision-making opportunities. Yet, the ambiguities of the "authoritarian alternative" in implementing environmental governance and driving mitigation are more frequently pointed out (e.g., Eaton and Kostka 2014; Beeson 2018; Engels 2018).

The findings of this study show that social vulnerability is co-created by and anchored within both political systems. The findings also demonstrate the deeply entangled nature of large-scale adaptation deficits at the local level. For this reason, the book challenges the argument that China's environmental authoritarian structures are better equipped in dealing with matters related to climate change (adaptation). For the same reason, it challenges the argument that liberal democracies are better suited in dealing with social justice issues, as is commonly assumed. What became visible as part of this research is that being able to effectively address patterns of inequality is a key aspect for adaptation policymaking. The two chosen regions shed light on the deep nature of path-dependent processes, which are interlocked at multiple origins such as infrastructure, actors, knowledge and political institutions. These, the book argues, are anchored in the political ecology (see Chapters 1 and 3). At the current stage, both political systems appear to have low adaptive capacity in dealing with problems of this complex and interrelated magnitude. The analysis of codependent lock-ins brought forward the different dimensions of political marginality as major characteristics found in, both China and the United States. This political marginality drastically impacts vulnerability and manifests adaptation deficits further. Therefore, this book also questions the widespread assumption of many adaptation scholars that developed countries like the United States have higher adaptive capacity than developing or emerging countries like China.

Overview of interrelated lock-in dimensions

The analytical framework of Siebenhüner et al. (2017) proposes four different dimensions of adaptation lock-in (see Figure 3.1 in Chapter 3). In this book, their intertwinement was analyzed and the descriptive dimensions of knowledge, physical infrastructure, institutions and actors conceptualized through a political ecology lens (see Figure 3.2 in Chapter 3). Due to scope, not all of the dimensions could be examined in detail. What Siebenhüner et al. (2017) refer to as lock-ins of knowledge, information and expertise, in this research became apparent as lock-ins of political epistemology, which sit at the intersection of how political institutions and routines operate in combination with knowledge and actors. The following section will briefly reflect upon some of the side findings and then take a deeper look at political-epistemological lock-ins.

Lock-ins related to infrastructure

Physical infrastructure

Due to scope, the analysis of biophysical and geographic factors could not be discussed in detail (see A1.1). These factors nonetheless play a distinctive role for the affectedness and scope of action for both of the inland located municipalities. For instance, Atlanta has endless room to grow further due to its landscape environment. This will affect urban growth, land transition, the environmental conditions and put greater pressure on critical infrastructure, such as access to affordable housing. In contrast, Jinhua is surrounded by a mountainous landscape, which is rich in forests and rivers and limits Jinhua's room to expand and urbanize further. It also became obvious how these biophysical and climatological factors more drastically interrelate with lock-ins related to critical infrastructure, some of which were examined in Chapter 6. The water resources of Atlanta are scarce due to the current societal demand for water and dependence on the Apalachicola-Chattahoochee-Flint River Basin, which is also claimed by Alabama and Florida. This conflict is reflected in the ongoing political struggle of the tri-state "Water Wars." Against the background of urbanization and intensifying periods of more frequent extreme heat events, Atlanta's needs for resources will likely deteriorate.

Critical infrastructure

In addition to water, urbanization and housing, the transport sector has been a major challenge, due to Atlanta's insufficient and highly segregation-based public transport system (MARTA) and the cultural dependence on cars, which have been overwhelming the limited roadways for decades. This pressure too will exacerbate, based on urbanization and the absence of cultural shifts related to the use of cars and pedestrian culture. Atlanta is a nightmare for any urban planner when it comes to retrofitting and making the city more livable for its population. Private sector interests and the conflict over better public transport and affordable housing in light of Atlanta's biggest neighborhood revitalization project, the Beltline, have elucidated this problem. Transportation, education, housing, health and access to fresh resources correlate with opportunities for a better life and demonstrate the deep patterns of inequality.

Although the pressure through urbanization is not as drastic in Jinhua, the city and especially Zhejiang province have been coping with substantial in-migration. Jinhua is confronted by an urban–rural divide, which is also reflected in reverse trends of out-migration in rural pockets of the city, such as Lanxi and Pan'an. Here, limited access to education and health are the dominant challenges aside from a subtle notion of cultural devaluation. These factors combined determine patterns

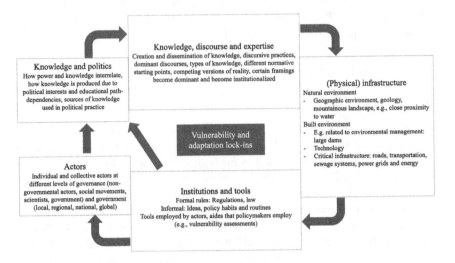

Figure 9.1 Political epistemological lock-ins

Source: the author

of social vulnerability to climate change. Interestingly, psycho-social lock-ins and cognitive components play out somewhat differently across the two cities. In Jinhua, the omnipresence of flooding, typhoons and governmental support appears to have resulted in greater dependence on governmental intervention and lessened means for personal provision. In Atlanta, interviewees commonly reported that the direct effects are not as clearly visible but will strongly impact how people and local governmental agencies choose to respond in the future. The underlying sentiment was that more extreme events are needed in order to trigger action. These preliminary insights on some of the infrastructural aspects (and how they too relate to questions of power) also hint at matters that are deeply rooted in knowledge patterns and dominant ideas of how to build a city. Further research is needed to explore this connection more systematically.

Main finding: Political epistemological lock-in

The empirical chapters revealed that the persistence of uneven human vulnerability to climate change is deeply connected to questions of power and knowledge. Across these two cases, the main finding was that knowledge, access to education and the way decision-makers in institutions learn crucially impact the maintenance of adaptation deficits. These are considered political epistemological lock-ins (see Figure 9.1). It was found that segregated access to knowledge affects interest representation, which appeared to be very selective and defined by a dominant class in both Atlanta and Jinhua. (The interests of) vulnerable populations appear to have lower interest representation than (the interests of) people with greater access to education. Access to education seems to positively correspond with greater access to decision-making. Path-dependent patterns and cultural forms of vulnerability conceptions as well as adaptation knowledge further impact the way knowledge is used by decision-makers, often resulting in political bias and stigmatization. How actors within institutions learn and use (selective types of) knowledge appears as core aspect, which suggests low adaptive capacity of local governments in Atlanta and Jinhua. The political knowledge economy and education were the key dimensions of the identified lock-ins.

In the examined cases, it was found that power is carried by those who 1) get to define vulnerable populations and thereby influence the dominant discourses on vulnerable populations, 2) those who decide over who gets to qualify for being recognized vulnerable in the political process and 3) those who are able to set political (adaptation) agendas and priorities in this regard. In both political contexts, power was commonly interconnected with the question: "Are you in a position to know?".

Political epistemological lock-ins interrelate with institutional lock-ins at the level of interorganizational learning. Top-down tools such as assessments and political target setting seem to be important, as they regulate local political priorities, but to date were not tailored toward vulnerability or adaptation in political practice. The adaptive capacity of both municipal governments appeared to be rather low, based on little room for self-reflexive and experience-based learning about how to address complex challenges such as adaptation. Chapter 7 discovered some of the challenges local decision-makers mentioned. In Atlanta, this includes mitigation-focused learning exchange programs with other government officials, which resulted in putting mitigation and renewables onto the agenda of local officials. Such learning exchange programs, coupled with pressure from local educational institutions and the broader public and financial resources of non-state actors, such as the Rockefeller Foundation resulted in Atlanta taking major steps regarding climate mitigation. However, in the adaptation field, similar learning programs appear to be absent in addition to pressure more broadly. In both local political contexts, motivation to act politically is still triggered through the visibility of short-term results.

The politically unfavorable environment of Georgia has influenced policymaking in Atlanta to the extent that local decision-makers maneuver in predetermined waters of international protocols. Altogether, room for taking courageous action and new measures seems to be absent. Although the reasons for Jinhua may be different, a similar pattern can be observed here as well: interviewed decision-makers indicated that they were not willing to take a risk as long as it was not mandated from higher government levels. The reliance on top-down protocols, which all still appear to be non-adaptation-related and embedded in rather reactive disaster risk reduction frameworks, coupled with little career incentives and little pressure from very few non-state actors have shaped the motivation of local decision-makers to not put the issue of climate adaptation onto the agenda. This, Chapter 8 argued, is also an extension of educational legacies of natural-science-focused knowledge paths in China. Further, political scientists have pointed out that local Chinese officials know well how to maneuver with unspecified government directives and leave enough room for their own interpretations and implementation of policies (see, e.g., Zhong 2003). Adaptation is simply an undesirable policy field for local decision-makers in Jinhua. The lack of social science studies to inform adaptation processes has grown out of a long, historic path dependence.

Therefore, explanations that focus on the lack of provincial pressure in Jinhua and limited political support from state and federal governments in Atlanta are too short-sighted. The degree of reflexivity and the way political institutions are set up, how they learn and respond to complex challenges, also by drawing from different forms of knowledge appear to be more important.

Facets of political epistemology lock-ins

The field of epistemology is concerned with the study of knowledge, especially regarding its methods, validity and scope, and how knowledge is brought about through political institutions. It is interested in learning about sufficient conditions of knowledge, its sources, structure, and limits.[1] Political epistemology speaks to the political dimension of epistemology by highlighting the intrinsic political dimension of knowledge production, knowledge exchange and routines of meaning-making.[2] It is interested in understanding how political, economic and social conditions shape normative assumptions under which knowledge is historically generated and enacted (MPIWG n.d.). Political ideas and knowledge are variables that affect political action and simultaneously are products of political behavior and institutions (Althaus et al. 2014). Therefore, the findings will be treated as political epistemological lock-ins. Lock-ins of political epistemology can be understood as trajectories of political practices related to sense-making, knowledge production and political action.

The findings of Chapters 7 and 8 discover that political epistemological lock-ins play a major role in the maintenance of social vulnerability and adaptation deficits. Political epistemology evolves around the study of knowledge and ideas and how they affect political action (Althaus et al. 2014) (also see Chapter 6.3). The view of "political epistemology" helps to critically explore relationships between theory, policy and society (Straßheim 2015). It precisely uncovers the entanglement of politics and knowledge and engages in a critical reflection of the classical dictum "knowledge is power" (Omodeo 2019). In this book, it became obvious that knowledge is power for those who can access it. Knowledge and ideas in turn influence political behavior and are intertwined with questions such as: are you in a position to know and as a result shape discourses? Are you (culturally) in a position to access decision-makers?

One-sided epistemic agency

Amongst other things, political epistemology investigates the extent to which politically marginalized groups are in a position of epistemic privileges (Hannon and Edenberg 2020). The examined cases reveal that the production of vulnerability and adaptation knowledge is problematic, as it reflects a form of one-sided epistemic agency. It became obvious that access to certain forms of knowledge fundamentally matters to be in a position to make (in)vulnerability claims and as a result having the ability to shape discourse agendas and/or decide over policy priorities related to adaptation. The findings scratch at the fundamentals of vulnerability assessments and adaptation practices by critically questioning the way knowledge about climate impacts, human vulnerability and policymaking is generated, how it operates, and is used. These observations make this a particularly tricky form of adaptation lock-in as it fundamentally relates to the way society is organized, who has access to what kind of knowledge type, discursive power, discourse sovereignty and ultimately epistemic agency.

The issue of vulnerable groups as contested phenomenon demonstrates how the practice of assessing vulnerability is a deeply politicized act, which can have detrimental impacts on how governmental interventions in this area are thought about and approached. In Atlanta and Jinhua, the designation of vulnerable populations carried a strong connotation of incapability, thought of as "those that cannot take care of themselves," and are dependable. This is an especially interesting finding, as this perception feeds the problem of disempowering and marginalizing certain populations further by stigmatizing the vulnerable. At the same time, it points to a phenomenon that political ecologists have emphasized, which is knowledge as an embodied practice. Meaning is externally ascribed onto humans. The practice of vulnerability assessments and acknowledgment of vulnerable populations speaks to the role of social practices in truth regimes, which impact, how authorities and institutions manage, rule and control social life (Valdivia 2015). One-sided epistemic agency appeared to be driven by a very segregated access to knowledge and education.

The populations that were outlined as particularly vulnerable in Atlanta and Jinhua by the social vulnerability assessments and climate justice studies (see Chapter 7), are also those, who are politically particularly disenfranchised due to their lack of access to knowledge and limited opportunities in producing knowledge and advising decision-makers. Besides striking efforts in poverty reduction, which marks a strong contrast to the United States, the post-reform era of relaxed political control has been quite vulnerable to pressure of interest groups in the context of decentralization and marketization in China (Yang and Fang 2000; Ngok and Zhu 2007). Education marks one of the key tenets of discriminatory policies, which is based on the affirmative treatment of urban residents, who continue to receive major welfare privileges.

Historical processes and (in)access to engineering-focused education in China

The experiences and consequences of the Great Proletarian Cultural Revolution (GPCR), which lasted from 1966 to 1969, had a significant effect on the content of political education and access to knowledge (Goodman 1989) long after the Mao era. The educational system was one of the primary targets of Mao's leadership in his efforts to "engineer the socialist transformation" (Wang 1975) and try to prevent the spreading of capitalist ideology and bourgeois ideas. Formal education ceased to exist, besides some rural schools continuing to function during this period (Wang 1975). Most universities and schools remained closed until 1972 with the direct impact on popular education being felt until the 1980s. Although measures initiated during the GPCR had negative effects on the literacy rate and the quality of higher education of urban elites, educational opportunities for rural children increased significantly (Andreas 2009). The goal of redistributing access to education yielded impressive results regarding the rapid expansion of basic education (Andreas 2009). As a result, tertiary education presented a dilemma for Mao's educational revolution as it was closely linked to elite status. This resulted in a compromise of Mao's egalitarian vision by focusing on colleges of science and engineering (see Andreas 2009: 167f.). Engineering training was significantly strengthened and the number of engineering university students grew from 23,035 to 292,680 between 1947 and 1965. Reaching through the Reform and Opening Up period of Deng Xiaoping in the early 1990s, "engineering training was associated with the outlook and aptitude most appropriate for leadership" (Andreas 2009: 243). China's contemporary educational system is characterized by an uneven distribution of educational opportunities and an ostensibly meritocratic nature (Ma and Adams 2014). The egalitarian attempts of redistributing educational opportunities under Mao backfired in the opposite direction: today holders of urban household registrations (hukou) are systematically advantaged in contrast to rural residents and migrant children for whom it is nearly impossible to be accepted into high-quality educational institutions (Ma and Adams 2014).

China's education developed from a universal education system to an elite education system, where the state provides advanced education for the urban population (Yang and Fang 2000). The elite education system was strengthened particularly after the 1980s and has become increasingly marketed in the 1990s toward paid education with university fees getting as high as the average annual income of an urban employee (Yang and Fang 2000). "This system increasingly helped to force prospective students from poorer backgrounds to give up education" (Yang and Fang 2000: 17). Access to (higher) education is nowadays denied to students from poor families (Ngok and Zhu 2007). The lack of economic opportunity for some populations has stretched from rural to urban–rural levels, as one policy advisor expresses:

> And for the urban, the poor urban groups, they have low incomes and which their income is dependent on their salary, which is very low and depends on their unemployment insurance which is provided by the government. And as you can see, the southern part of China is developed better than the northern part. So, the poor, the urban poor group people are more in the north than the south. And as for the rural poor group people, and in the Chairman Mao period, we have an insurance policy which can provide some financial support to the very low-class people. We have the minimum wages.
>
> (I-12: 53)

Kwong and Qui (2003) examine the role of social security, which changed significantly in the post-Mao era.[3] The provision of social services now depends on workers contributing to their own insurance premiums. The new financial burden the Chinese population has been facing regarding the basic service sectors of education, healthcare and housing became reflected in the phrase of "new three mountains" (*xin sanzuo dashan*) (Ngok and Zhu 2007).

Historical segregation of education in the United States

Until today, the public education system in Atlanta is highly segregated. Atlanta is a particularly drastic example of school zones that are divided upon a stark racial and ethnic composition. Despite the need and agreement of school system reform, there is a drastic lack of consensus on the basic issues of urban education (Stone et al. 2001). Stone considers the obstacles to urban school reform part of a larger problem that is political in nature based on powerful interest groups (Stone et al. 2001). When talking about the education system of Atlanta, one policy practitioner explains the problem of breaking with past legacy regarding segregation:

> It is very segregated still. I think once you get to higher education it is less segregated than our key through twelve model for educating our students. But at that point it is a matter. So, here is the trick, right, the big gotcha, is that we train children from the time that they are five, when they enter kindergarten, till the time that they are in the twelfth grade, that is K through twelve, that is our elementary medal in high school education, that you should be with your own. And when they are in Highschool we say, here fill out application and go some // anywhere in the country that you want to go, go anywhere in the world that you want to go. Well, if we have been telling them, to stick with your own from when they were five years old, do you think that they are going to choose, when they are in the twelfth grade, to go to a college that is integrated to be around people that do not look like them, do not think like them, do not have the same car as them, you know they are not going to choose that. After seventeen years of you telling them, that is not the way to be. It is not a choice. By that time, you have taught them the social behavior that you do not want them to adopt when they grow into society. . . . [I]f there is any institution that we ought to target for difference, it needs to be that.
>
> (I-24: 98)

What became obvious was that the lack of educational opportunities often translated into lack of economic opportunity, which resulted in a lack of political opportunity (I-04G, I-24, I-33). It was also noted that this divide has emerged over a long time, that it is racial and geographic and decision-makers are doing "their best to not perpetuate that. . . . But it exists, you know it is kind of there, and it is hard to do that" (I-24: 68f.). To conclude, access to knowledge and education appeared as the key tenets that significantly determine the political disenfranchisement and one-sided epistemic agency of local populations in China and the United States.

Ideology in science

Another observation that was made is the problem of ideology as exhibited by related scientific practices. The way social vulnerability assessments are conducted points to severe methodological shortcomings, which persisted in intransparent assumptions. Very rarely, the studies disclose the underlying assumptions that would explain why, for example, a lower socioeconomic status corresponds with higher vulnerability to climate change. The underlying causal problems remain likewise unexplored and detected vulnerability indicators unverified. In the Chinese case, one explanatory attempt looks at the causality of declining family size and the country's one-child policy. Although the policy was abandoned in 2016, its effects will continue long into the century. However, this study also fails to bring forward the underlying assumptions why these characteristics necessarily correspond with higher vulnerability. At the same time, most of these assessments reproduce a distinctive status quo and norm of development. Furthermore, forms of self-governance which may act as buffer that can increase resilience are not explored. Alternative cultural practices, habits and means of community resilience as well as indigenous knowledge are also not reflected upon. Whereas researchers are confronted by severe data availability and coherence in China, data restrictions regarding temperature records mark an important limitation in the United States as well (also see Stone 2012). This finding relates with the demand for self-critical adaptation studies (see Chapter 10).

The reconstitution of a class-based society

The findings of this book suggest that the locked vulnerability is a result of larger political developments, which are found to varying degrees in both political systems: the manifestation and reconstitution of a class-based society. There seemed to be general agreement on class formation as a political development in both cases based on a limited access to public goods. Limited access of some groups to certain public goods is an underlying political pattern, which crucially conditions the adaptive capacity of local populations and governments alike. The political-epistemological dimension of lock-ins detects inaccess to knowledge, also as enabled through access to education, as a key determinant for vulnerability. In China, those with rural hukou have only limited access to education. In the United States, it is the historical and ethnic minorities of African Americans and Latin Americans, who have limited access to education. As the empirical chapter pointed out, Georgia and Atlanta have a long history of race-based politics, ethnic division and income segregation. Here, state-maintained segregation acts as a vulnerability determinant. Compared to the long history of the American South, this is a relatively new phenomenon in Jinhua and China. The following section will briefly provide some preliminary insights on the manifestation of classes, as voiced by interviewees. Although this was not the main focus of the study, it was a recurring red thread upon which further research can be built.

Empirical findings

In both political contexts, interviewed policy planners and practitioners observe the reconstitution of different classes. The problem of limited access to public goods based on hukou seems to have also expanded to the urban sphere and is nowadays a problem of urban-to-urban inequality. A policy planner sums up the situation:

> I guess I would say, a pretty severe problem right now is sort of the . . . formation of different classes within China, which has always been implicit. But I think it has become more and more explicit as a result of the last thirty years of stellar, astronomical economic growth, so you are seeing kind of the elites in the urban areas. It used to be that there was mostly urban and rural, you know inequality. But now there is a lot more urban to urban inequality.
>
> (I-36: 78)

In Atlanta, the issue of poverty and racial segregation were pointed out as inherent characteristics ("legacy") of the political system. When being asked about the causes of vulnerability, this political official at the county level states:

> My official answer would be the underlying cause of these risk factors is probably legacy, right? It is probably [that] over the years, we have somehow divided our community in the haves and the have nots. And that has // the fallout of that is, that it has been a racial divide, right? And so, you know, economic opportunity has kind of landed where the haves are and the have nots are not. It has landed where, folks who are Caucasian are and folks who are Afro-American are not. And so, it is just, it is perhaps a consequence of our legacy of systems.
>
> (I-24: 85)

Government officials were reluctant to talk about segregation and racism, but some referred to the unequal access and distribution of resources, as an instance of segregated geographical development (I-24). With a long history of race-based politics and ethnic division, Georgia and Atlanta specifically sit at the heart of different political tensions. Certain contradictions inherent in sustainability and resilience governance became apparent as one employee within the Mayor's Office of Resilience states:

> We cannot just keep ignoring that, while, like, you know, the haves keep getting more and the have-not populations keeps on growing and . . . and we call ourselves a "world-class-city".
>
> (I-19: 58)

Georgia, Atlanta and Fulton County all experience widespread income segregation. Whereas the north of Fulton County is inhabited by a white upper middle class, it is largely impoverished Afro-American and Hispanic households, who live in the southern part of the county. The previous chapter already indicated that segregation was commonly pointed out as a deeply interwoven characteristic of Atlanta, which is particularly apparent in the transport sector. Specific historical, social, political and economic factors are evident and have a great influence on population susceptibility to natural hazards. In China, the household registration system (hukou), the one-child policy, the aging population and rapid urbanization processes can be seen as some of the China-specific policies, which are the baseline of social and spatial inequality in Jinhua and beyond. One China-focused scholar elaborated on social stratification and the Chinese societal division into lower and upper classes:

> I think your study needs to be discussed from the aspect of social structure, that means from a different level, different social status. For example, that is according to the German theory of Max Weber, social stratification – the theory of social classes. So, I think, regarding this vulnerable population, in the context of climate change, you are talking about the bottom level of our society, the problem of the lowest class people.
>
> (I-12: 45f.)

Explanations in the literature: education as key determinant for classes

The restoration and reconstitution of class power have become apparent. To interpret class phenomena, scholars concentrate on wealth, the accumulation of capital in the hands of a few (Harvey 2018), income and race (e.g., Stone 2008; Reardon and Bischoff 2011). Tomba (2004) and (Yang 2017) concentrate on resources and power coupled with a culture-based urban–rural divide where urban elites exhibit cultural hegemony over migrant workers as a result of state-facilitated urban development in China. There has been some consensus that viewing political power as a matter of an elite group exercising power in some observable form is only a narrow depiction of power (e.g., Gaventa 1980). Power has a systemic character, which is characterized by a system of stratification that involves economic, associational, social and political dimensions (also see Stone 2008).

Sociologist Reckwitz (2019) conceptualizes classes along three dimensions: cultural, resource-based and political dimensions. The cultural dimension refers to shared cultural forms of life and practices such as work, family and relationship practices. The resource-based dimension speaks to the endowment of resources, which enable and disable life. The political dimension entails cultural struggles about status and influence. Based upon these three dimensions, Reckwitz (2019) draws a new model of a class-based society. According to this model, a new middle class developed for which access to higher education and therewith related cultural status is a key door opener for a good life.

Although income is an important component when studying inequality patterns, Reckwitz pays specific attention to the cultural and social dimensions of equality, such as lifestyle and the way of living, reflected in everyday practices. The culture of classes has an own, complex reality, for which the condition of material resources is only one factor amongst many (Reckwitz 2019: 65). Accordingly, different lifestyles are not equal in terms of their cultural acknowledgment but are concerned with cultural prestige and characterized by different opportunities and attitudes toward life. Reckwitz postulates a revolution of social structure by looking at the expansion of the educational sector. The academization of the postindustrial knowledge economy has occurred in which the middle class is no longer carried by specialized or low-skilled workers but an educated elite that enjoys specific social prestige (Reckwitz 2019).

The new lower class has lower qualifications and nowadays works in the service sector. In a time of a steadily rising service sector, Reckwitz views industrial workers as a dying species – a cultural development that is accompanied by a loss of cultural significance of industrial workers and an increase in low-paid workers. Meanwhile, educational professions have expanded. It is against this educational-historical background, Reckwitz states, that we have to understand the academization as a driving revolutionary change. Education has initiated a cultural devaluation process. The expansion of academized professions has resulted in a lowered status and prestige of other educational achievements. In addition to the new lower class, which consists predominantly of nonacademic and former industrialized workers, a precarious class developed which partially operates outside the labor market and belongs to the educationally disadvantaged ("Bildungsverlierer").

Despite the cultural dimension, Reckwitz studies the nature of work and takes the expansion of the tertiary sector as one characteristic for describing the new division of classes. In the United States, the service sector expanded from 47 percent (1960) to 73 percent (2015) (Reckwitz 2019: 79). In China, the service sector has become the bottleneck of structural change and is considered the major driver for growth (Zenglein 2016). Despite problems of access to reliable data on China's service sector (Heilmann 2015), the replacement of the agricultural and industrial sectors with tertiary employment is clearly visible (see, e.g., Zenglein 2016). These major labor market changes are considered "the great employment transformation" in China (Majid 2015).

If the development of the tertiary sector and the expansion of education are taken as indicators for the development of a new middle class and a descent of the old, industrial middle class, though less advanced, we can argue the same for China. In Atlanta and Jinhua, the core resource, which the powerful appear to have acquired is education. Having a higher educational background seems to positively correlate with access to decision-making, cultural status and resources. This pattern was even reflected in the process of ascribing vulnerability. Those who were qualified to determine who gets to be vulnerable and those who are in a position to recognize vulnerability are part of the upper middle class.

Summary: knowledge as power for those who can access

The chapter explored the main findings on vulnerability and adaptation lock-ins in greater depth. One main observation is that social vulnerability is co-created by and anchored within both political systems. Against this light, the first chapter section discussed vulnerability as an inherent characteristic of the two different political systems of China and the United States. Literature on democracy as it relates to climate policy as well as scholarship on environmental authoritarianism both have had an underlying sentiment that quasi authoritarian governments, like China, are more capable of dealing with the challenge due to their unitary political system and centralized decision-making opportunities. The core findings of this research however demonstrate the deeply entangled nature of large-scale adaptation deficits at the local level. For this reason, the book challenges the argument that China's environmental authoritarian structures are better equipped in dealing with matters related to climate change (adaptation). For the same reason, it challenges the argument that liberal democracies are better suited in dealing with social justice issues, as is commonly assumed. Being able to effectively address social inequality is a key aspect for adaptation policymaking.

The two chosen regions shed light on the deep nature of path-dependent processes, which are interlocked at multiple origins such as physical and critical infrastructure, actors, knowledge and political institutions. These, the book argued, are anchored in the political ecology (see Chapters 1 and 3). The way knowledge is produced, accessed and distributed in combination with the way political institutions operate are core elements of the inherent vulnerability of both countries. The chapter revisited how these manifests as lock-ins and conceptualized them in the current literature on political epistemology. Three core facets of political epistemological lock-ins are: 1) one-sided epistemic agency when determining vulnerability, 2) path-dependent political legacies of educational opportunities and 3) ideology in science. These facets impact adaptation processes and are core reasons for the maintenance of uneven vulnerability to climate change. The last chapter section went a step further, discussing knowledge as a (new) indicator for the manifestation of a class-based society.

Notes

1   In accordance with the Stanford Encyclopedia of Philosophy, see "Epistemology". First published December 14, 2005, https://plato.stanford.edu/entries/epistemology/, last accessed July 4, 2019.

2   Although the term "political epistemology" is fairly new and research in this field has been flourishing as of late, scholars have long been interested in the intersection of political philosophy and epistemology, see Hannon and Edenberg 2020.

3   For a more detailed discussion on the regional political economy of work in a post-socialist era, see William Hurst, *The Chinese Worker after Socialism*. Cambridge: Cambridge University Press, 2009.

References

Adger, Neil W., Irene Lorenzoni, and Karen L. O'Brien. *Adapting to Climate Change. Thresholds, Values, Governance.* Cambridge: Cambridge University Press, 2009a.

Althaus, Scott, Mark Bevir, Jeffrey Friedman, Hélène Landemore, Rogers Smith, and Susan Stokes. "Roundtable on Political Epistemology." *Critical Review* 26, no. 1–2 (2014): 1–32. https://doi.org/10.1080/08913811.2014.907026.

Andreas, Joel. *Rise of the Red Engineers: The Cultural Revolution and the Origins of China's New Class.* Stanford: Stanford University Press, 2009.

Beeson, Mark. "The Coming of Environmental Authoritarianism." *Environmental Politics* 19, no. 2 (2010): 276–294. Doi: 10.1080/09644010903576918.

Beeson, Mark. "Coming to Terms with the Authoritarian Alternative: The Implications and Motivations of China's Environmental Policies." *Asia & the Pacific Policy Studies* 5, no. 1 (2018): 34–46. Doi: 10.1002/app5.217.

DeLauro, Rosa L. *The Least Among Us Waging the Battle for the Vulnerable.* New York, NY: The New Press, 2017.

Diamond, Larry. "Why Decentralize Power in A Democracy?" Paper presented to the Conference on Fiscal and Administrative Decentralization, Baghdad, February 12, 2004.

Eaton, Sarah, and Genia Kostka. "Authoritarian Environmentalism Undermined? Local Leaders' Time Horizons and Environmental Policy Implementation." *The China Quarterly* 218 (2014): 359–380.

Engels, Anita. "Understanding How China is Championing Climate Change Mitigation." *Palgrave Communications* 4, no. 101 (2018). Doi: 10.1057/s41599-018-0150-4.

Fiorino, Daniel J. *Can Democracy Handle Climate Change?* Oxford: Polity Press, 2018.

Fankhauser, Samuel, and Thomas K. J. McDermott. "Understanding the Adaptation Deficit: Why are Poor Countries more Vulnerable to Climate Events than Rich Countries?" Centre for Climate Change Economics and Policy Working Paper No. 150, Grantham Research Institute on Climate Change and the Environment Working Paper No. 134, September 2013.

Gaventa, John. *Power and Powerlessness. Quiescence and Rebellion in an Appalachian Valley.* Urbana: University of Illinois Press, 1980.

Goodman, David S. G. *China's Regional Development.* Location unspecified: Routledge, 1989.

Hannon, Michael, and Elizabeth Edenberg. "Political Epistemology." Oxford Bibliographies, July 29, 2020. Doi: 10.1093/obo/9780195396577-0408.

Hanusch, Frederic. *Democracy and Climate Change.* Routledge Global Cooperation Series. London: Routledge and Taylor & Francis Group, 2017.

Harvey, David. *The Limits to Capital.* London, UK and New York, US: Verso Books, 2018.

Heilmann, Sebastian. "Unter Abwärtsdruck: Wie reagiert Chinas Regierung auf die Krisenanzeichen der Wirtschaft?" *Internationale Politik* 6 (2015): 84–91.

Kwong, Julia, and Yulin Qui. "China's Social Security Reforms under Market Socialism." *Public Administration Quarterly* 27, no. 1–2 (2003): 188–209.

Ma, Damien, and William Adams. *Introduction to in Line Behind a Billion People: How Scarcity Will Define China's Ascent in the Next Decade.* Upper Saddle River, New Jersey: FT Press, 2014.

Majid, Nomaan. "The Great Employment Transformation in China." In *International Labour Office, Employment Policy Department.* Employment Working Paper, no. 195. Geneva: ILO, 2015.

Max Planck Institute for the History of Science (MPIWG). "Political Epistemology New Approaches, Methods and Topics in the History of Science Workshop Series 2016–17." www.mpiwg-berlin.mpg.de/page/political-epistemology, last accessed July 14, 2021.

Ngok, Kinglun, and Guobin Zhu. "Marketization, Globalization and Administrative Reform in China: A Zigzag Road to a Promising Future." *International Review of Administrative Sciences* 73, no. 2 (2007): 217–233. Doi: 10.1177/0020852307077972.

Nur, Ismawaty, and Krishna K. Shestra. "An Integrative Perspective on Community Vulnerability to Flooding in Cities of Developing Countries." *Procedia Engineering* 198 (2017): 958–967. https://doi.org/10.1016/j.proeng.2017.07.141.

Omodeo, Pietro Daniel. "Towards a Political Epistemology: Positioning Science Studies." In *Political Epistemology*, edited by Omodeo, 13–49. Springer, 2019. https://doi.org/10.1007/978-3-030-23120-0_2.

Parks, Bradley C., and J. Timmons Roberts. "Globalization, Vulnerability to Climate Change, and Perceived Injustice." *Society and Natural Resources* 19, no. 4 (2006): 337–355. Doi: 10.1080/08941920500519255.

Reardon, Sean F., and Kendra Bischoff. "Income Inequality and Income Segregation." *American Journal of Sociology* 116, no. 6 (2011): 1934–1981.

Reckwitz, Andreas. *Das Ende der Illusionen. Politik, Ökonomie und Kultur in der Spätmoderne.* Berlin: Suhrkamp, 2019.

Siebenhüner, Bernd, Torsten Grothmann, Dave Huitema, Angela Oels, Tim Rayner, and John Turnpenny. "Lock-ins in Climate Adaptation Governance. Conceptual and Empirical Approaches." Conference Paper 2017, unpublished.

Stone, Brian. *The City and the Coming Climate: Climate Change in the Places We Live.* Cambridge and New York: Cambridge University Press, 2012.

Stone, Clarence N. "Civic Capacity and Urban Education." *Urban Affairs Review* 36, no. 5 (2001): 595–619. Doi: 10.1177/10780870122185019.

Stone, Clarence N. "Social Stratification, Nondecision-Making, and the Study of Community Power." In *Power in the City: Clarence Stone and the Age of Inequality*, edited by Marion Orr and Valeria C. Johnson, 55–75. Lawrence: University Press of Kansas, 2008.

Straßheim, Holger. "Politics and Policy Expertise: Towards a Political Epistemology." In *Handbook of Critical Policy Studies*, edited by Frank Fischer, Douglas Torgerson, Anna Durnová, and Michael Orsini, 319–340. Edward Elgar, 2015. Doi: 10.4337/9781783472352.00026.

Tomba, Luigi. "Creating an Urban Middle Class: Social Engineering in Beijing." *The China Journal* 51 (2004): 1–26. Doi: 10.2307/3182144.

Valdivia, Gabriela. "Eco-Governmentality." In *The Routledge Handbook of Political Ecology*, edited by Tom Perreault, Gavin Bridge, and James McCarthy, 467–478. Abingdon: Routledge, 2015.

Wang, Robert S. "Educational Reforms and the Cultural Revolution: The Chinese Evaluation Process." *Asian Survey* 15, no. 9 (1975): 758–774. www.jstor.org/stable/2643172, last accessed July 27, 2019.

Wolf, Johanna. "Climate Change Adaptation as a Social Process." In *Climate Change Adaptation in Developed Nations*, edited by James D. Ford and Lea Berrang-Ford, 21–32. *Advances in Global Change Research.* Dordrecht: Springer, 2011. https://doi.org/10.1007/978-94-007-0567-8_2.

Yang, Dennis Tao, and Cai Fang. "The Political Economy of China's Rural-Urban Divide." In *Stanford Center for International Development*. Working Paper No. 62. 2000. https://siepr.stanford.edu/sites/default/files/publications/62wp.pdf, last accessed January 17, 2020.

Yang, Qinran. "Interpreting Gentrification in China: The Rising Consumer Society and Inequality in The State-Facilitated Redevelopment of the Central City of Chengdu." Dissertation Submitted for the Degree of Doctor of Philosophy at The University of British Columbia, Vancouver, 2017.

Zenglein, Max. "China's Overrated Service Sector Missing Dynamics Threaten Growth Target." *Merics China Monitor Series*, October 13, 2016.

Zhong, Yang. *Local Government and Politics in China. Challenges from Below.* Armonk, NY: M. E. Sharpe, 2003.

Cited interviews

I-12, March 7, 2017
I-04G, October 25, 2016
I-12, March 7, 2017
I-19, April 11, 2017
I-21, May 2, 2017
I-24, May 10, 2017
I-33, May 19, 2017
I-36, August 29, 2017

10 Adaptation policy and transformation?

The foregone chapters explored lock-in dynamics related to actors, infrastructure, institutions as well as knowledge and their various intertwinements. Lock-ins were conceptualized as anchored in the cultural ecology and political economy – they are a result of the way humans interact with the environment and are affected by it. The close examination of vulnerability manifestation as part of local decision-making in the two cases Atlanta and Jinhua revealed that vulnerability and adaptation lock-ins are fundamentally connected to questions of social justice. Political epistemological lock-ins were identified, which sit at this intersection of knowledge, power and institutions. This regards the way knowledge is produced, disseminated, perceived and used in decision-making concerned with adaptation and vulnerability. Access to education shapes knowledge patterns and dominant discourses and was found to be very unevenly distributed. Further, political bias toward using certain types of knowledge as well as selectively recognizing only certain populations were main findings of the empirical chapters. This chapter discusses the implications for discourses on transformative adaptation.

Before the chapter delves into the role policymaking and the state can play regarding transformative adaptation, the first chapter section lays out why climate change adaptation is an inherently political concept and why greater attention needs to be devoted to questions of power and conflict. This also regards the need for more self-critical adaptation research. In the following, knowledge is presented as a fundamental aspect that matters for transformative adaptation (chapter section "Knowledge, education and transformative adaptation"). This section also discusses some of the implications for dominant scientific practices and what science can do differently. This section briefly revisits some of the dominant discourses to place the argument. The subsequent chapter section then explores the role of the state for transformative adaptation (section "The role of adaptation policy for transformative adaptation"). It critically asks, if the policy field of adaptation is at all equipped to deal with grand challenges, such as social justice or whether vulnerability governance in the sense of avoiding its roots causes is the golden limit of political decision-making. Is transformative adaptation a utopia? The chapter ends with a couple of low- and high-hanging policy fruits in an attempt to approach the identified lock-ins.

DOI: 10.4324/9781003183259-10

Adaptation as a political concept

What this book more broadly emphasizes is the political nature of adaptation deficits and vulnerability maintenance. This is not a new perspective and has been emphasized by authors from different backgrounds, who agree that climate adaptation is not a predominantly technical or economic aspect but is deeply ethical and political in nature (e.g., Ribot 2011; Javeline 2014; Mikulewicz 2018). In this realm, researchers have called for a (re)politicization of climate change adaptation and vulnerability (e.g., Mikulewicz 2018).

The need to pay attention to the political dimension of human vulnerability

Political institutions do play a role in maintaining and in some cases worsening population vulnerability to climate change. In the examined cases, this became evident in different patterns of vulnerability recognition (see Chapter 7). It was found that the idealized vulnerability conceptions are disconnected from the political practices in place, which are characterized by a very different understanding and awareness about vulnerable groups. Whereas some groups appeared to be considered politically sensitive, other groups were stigmatized for being weak and not self-sufficient. Outside the immediate awareness and recognition practices of local politicians, vulnerability as maintained by state institutions also manifested in more fundamental ways, like limited access to knowledge and shaping dominant discourses about vulnerable populations (and adaptation). Those empowered by their political offices and access to knowledge could choose not to divert attention to the issue of vulnerable populations, because it is simply "an overtly complex matter" or was not politically recommended.

With regard to adaptation awareness and practices, there was only little sensitivity about the social dimension of adaptation in terms of formulating policy responses that ameliorate the uneven distribution of climate impacts. This was explained by dominant knowledge patterns related to path-dependent knowledge traditions in engineering on the one hand, and low political capacity to deal with complex problems such as human vulnerability to climate change on the other. At the same time, certain quasi-adaptation practices in the form of floodwater management parks even led to a deterioration of vulnerability in the case of Atlanta. Here, programs that were supposed to ameliorate the resilience to extreme precipitation led to green gentrification (see Chapters 8 and 9).

The need for self-critical adaptation studies

The findings build upon critical adaptation scholars such as Ribot (2014), Taylor (2015) and Morchain (2018): the way adaptation and resilience are commonly understood by a majority of decision-makers and policy advisors engages in a problematic reproduction of a development paradigm, which fails to introduce transformational thinking on how to reproduce ourselves differently (Taylor 2015;

Morchain 2018). Social vulnerability assessments and adaptation studies often do not explore the depth of vulnerability: why is it, for instance, economic characteristics and certain employment categories that lead to greater vulnerability? Instead, we should ask: why have we become so dependent on certain economic means? How do we want to live in the future, when conditions become even more challenging as a result of depleting natural resources and increasing climate change? How do we plan to address political marginality and social inequality? These questions have significant ethical and moral dimensions. Oels (2019) argues that a more radical disruption of the status quo may be a more effective way of pushing for change, while Keskitalo and Preston (2019) suggest that this may also be the reason why transformational changes may not be perceived as an inherent good. Radically questioning the status quo as a more drastic form of transformational change touches upon cultural aspects of daily life, such as the way we consume, work and live in and with our contextual environment and ultimately also giving up some of the privileges.

Understanding the way, we want to live and structure our lives, whether out of deliberate choices or social routines, or as part of decision-making is fundamental for recognizing some of the values with which we alter and respond to our environment. When studying climate impacts from a sectoral perspective or multidimensional lens on how sectors interdepend, we continue to lack a profound understanding of how our cultural and personal values impact decisions related to our physical and social environments and vice versa. Marcus Taylor (2015) calls this "the way we collectively produce ourselves" (p. 193). Questioning the naturalness of adaptation not just relates to political practices but also to a science that is self-critical. In reaction to an article that explored the potentials of social-scientific adaptation research, Winterfeld (2012) reacted by calling for critical and self-reflective climate adaptation studies. Accordingly, the task of critical social science research on adaptation would be to also question the prevailing funding practice with regard to its motivation (p. 168). This chapter section would like to go a step further by pointing to a self-critical research that explores the deficiencies of how the research was conducted (see following sections).

Shifting the focus to questions of justice

The lack of critical enquiry in this "rush to marry climate change adaptation and development" is questioned by looking at the problematic inherent dichotomy between climate and society, and through analyzing the predominant managerial discourses that present adaptation as a self-evident concept (Taylor 2015: xif.). Taylor argues "that the seeming naturalness of adaptation stands as a considerable barrier to critical thinking about climatic change and social transformation" (Taylor 2015: xii). Within the adaptation and overwhelming resilience literature, and particularly within the growing adaptation barrier field, too little research investigates the ambiguous notions of "adaptation" and related political practices. The transformation envisioned is oftentimes closely entangled in the idea of making climate change governable based on the premises of continued development and consumptive practices. Despite the need of recalibrating the political and social institutions, there is an urgent need to politicize adaptation, ideas of development and related inequality. Certain types of resilience thinking within the development field likewise depoliticize poverty through focusing too narrowly on access to markets, entrepreneurship and self-exploitation (Walsh-Dilley and Wolford 2015).

> Without attention to power and conflict, resilience thinking necessarily fails to address the structural conditions of poverty and the limitations of relying on individuals who face extreme vulnerability, insecurity and violence; indeed, a depoliticized framing of resilience threatens to exacerbate such problems.
>
> (Walsh-Dilley and Wolford 2015: 175)

Morchain (2018) calls for rethinking the framing of climate change adaptation and demands radical changes to adaptation in order to foster social justice. "Failure to do so risks promoting a paradigm whereby groups with little power in the Global South are constrained to play the role of the helpless, while the Global North and Southern enclaves of powerful elites and unrepresentative governments recognise themselves as rightful providers of adaptation solutions" (p. 55). As much as Morchain emphasizes the need for a framework that supports a more equitable framing of climate change adaptation in the context of global divides, as much is there a need for a more equitable framing at the local and regional levels, which consider cultural legacies and sensitivities.

Knowledge, education and transformative adaptation

Within the discourses on transformative adaptation, no consensus exists on what constitutes "good" adaptation practices, yet there is a growing acceptance of the need for transformative adaptation (see, e.g., Pelling 2011; O'Brien 2012; Kates et al. 2012; Godfrey-Wood and Naess 2016; Moser et al. 2017; Schlosberg et al. 2017). Adaptation actions of the past, such as infrastructure and planning, i.e., remodifying built landscapes retroactively, are considered to be no longer sufficient to cope with the magnitude of climate change underway (Biagini et al. 2014).

Transformation as a contested knowledge process

Besides the multitude of existing notions, "definitions of what constitutes transformational adaptation remain elusive" (Keskitalo and Preston 2019: 11). Some papers have reviewed the plurality of ideas and debates about adaptation and transformation (O'Brien 2012; Bahadur and Tanner 2013; Fazey et al. 2017; Few et al. 2017). Here too, O'Brien (2012) has pointed to the different processes of meaning-making and why the difference in definition persists: "Transformation means different things to different people or groups, and it is not always clear, what exactly needs to be transformed and why, and whose interest these transformations serve, and what will be the consequences" (O'Brien 2012: 670). As much as actionable efforts related to adaptation and vulnerability are contested processes, as much does the often-implicit critique that there is no unified definition of transformative adaptation, neglect different epistemic traditions, which can involve different actors, objectives and ultimately conflicts of interests.

To echo, what has been said earlier, and in line with O'Brien and colleagues (2007), different interpretations matter also regarding the various aspects transformative adaptation can take (also see chapter 2). Yet, caution is also appropriate, on what is labelled transformative, as some (transformative) adaptation measures related to flooding and infrastructure may lead to worsening of population vulnerability, and in this sense constitute maladaptation, as the example of neighborhood vitalization programs in Atlanta have demonstrated (see chapter 8). Exploring conflicting goals and negative side effects of planned adaptation efforts is therefore key when designing adaptation interventions.

Dominant notions of transformative adaptation: participation

Research on transformative adaptation has focused on process-oriented aspects of inclusive political decision-making (see, e.g., Ribot 2014; Kaswan 2014; Schlosberg et al. 2017). Discussing procedural elements of participation reflects a traditional concern of the environmental justice movement. Not surprisingly, equitable adaptation has become a common demand, not just by organizations in the practical adaptation field, such as the National Association for the Advancement of Colored People, but also by legal scholars and policy planners alike (see, e.g., Kaswan 2008, 2014; GCC 2019). The importance of different forms of broad-based public participation appears to have become an element in which the political governance of adaptation is being approached. This is not necessarily the case for China, where there is only little information available on public participation in decision-making related to adaptation. Few existing research-into-practice adaptation projects in Ningxia and Inner Mongolia acknowledge the need for broader stakeholder engagement to identify, prioritize and implement different adaptation options (Nadin et al. 2016). However, to what extent this is common practice in China, could not be examined.

When approaching vulnerability in political practice, enabling participation of different types of groups, including vulnerable populations, appears to have become the lower hanging fruit, in contrast to addressing the root causes of vulnerability production. The quality of participation and outcome of participatory practices is more rarely reflected upon. In their study on regional governance of a climate adaptation project in Northern Germany, Garrelts and colleagues (2018) deliver one of the rare examples, which critically reflects upon the way participants were chosen in local adaptation processes. They criticize that anticipatory structures turned out to be characterized by a highly unequal representation of actor types.

In the examined cases in Atlanta, local employees at the city level considered broad-based community participation in resilience planning a novel policy approach. However, the involved planner did not critically reflect upon the quality of participation. City and resilience planners also missed the critical evaluation of the political feasibility of the envisioned resilience targets and how they are supposed to be implemented. Municipal actors and a big consortium of non-state actors (see chapter 8) spent large amounts of time designing participatory processes. However, they failed to set follow-up mechanisms to actually implement the community-designed targets.

Critique of participation

Other researchers have become increasingly skeptical of participatory adaptation approaches (see, e.g., Oels 2007, 2019, O'Brien 2016; Garrelts et al. 2018). They argue participation does not necessarily lead to transformative adaptation or sustainability as most participation processes tend to reproduce the status quo and rarely lead to the change needed (Oels 2010, 2019; Taylor 2015). Oels (2019) has effectively discussed participatory processes as neoliberal governmentality in the adaptation context. This lens holds that government-led citizen participation is a "technology of government seeking to steer citizen behavior at a stance" (p. 146). The problem that Oels indicates may best be understood by looking at Arnstein's ladder of participation (1969) and participation as a form of alibi act. Arnstein (1969) developed a typology of participation with eight rungs that correspond with the increasing degrees of "decisionmaking clout" (p. 217). The ladder presents fundamental differences in participation between citizens. Some forms of participation at the lower rungs, such as therapy and manipulation actually constitute nonparticipation and perform an alibi function to citizen power (Arnstein 1969). One may also consider "adaptation information workshops" an alibi act of participation, as the emphasis persists on a one-way flow of information.

Power and knowledge

In the context of this book, adaptation and vulnerability lock-ins were identified to relate fundamentally to power in terms of knowledge. These were framed as "political epistemological lock-ins." They manifest amongst other things in form of epistemic injustice, i.e., the power not to frame your own vulnerability as well as questionable patterns of select and politically motivated vulnerability recognition. In some cases, groups were being stigmatized due to their supposed "low adaptive capacity" and dependence on state resources. In the adaptation context, lock-ins also sit at the interaction of power and knowledge to choose preferred types of adaptation. In both Atlanta and Jinhua, political practice had a strong focus on green infrastructure-focused adaptation, which in the case of Atlanta even deteriorated the vulnerability of certain communities' vulnerability (see previous section). The low sensitivity of the social dimension of adaptation was partially explained with the legacy of dominant educational systems in China: the strong focus on engineering education and lack of social science studies (see Chapter 8). In Atlanta, interviewees spoke more broadly about the uneducated black teenager being absent in formal decision-making processes – the lack of vulnerability experience in this sense was provided as one reason to explain matters of protracted vulnerability.

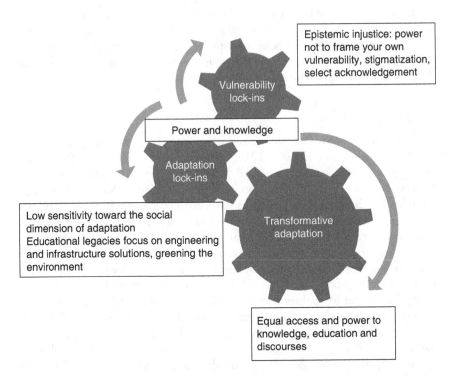

Figure 10.1 Overview of key vulnerability and adaption lock-ins preventing transformative adaptation

Knowledge as a key aspect to achieve transformative adaptation

What this book argues more broadly is that there is a necessity to look at the deeper levels and systemic nature of path dependencies, such as long-established forms of inequality, which are reflected at different levels such as access to education and knowledge (also see Chapters 7 to 9) or transportation (Chapter 6). Addressing political epistemological lock-ins to adaptation would mean to reform the educational system and imply deeper political reform regarding processes of sense-making and knowledge dissemination as part of political decision-making. In China, this would certainly include reforms related to the Gaokao System, the public examination system, which regulates the access to universities and accessing higher education. It is currently still largely based on hukou, and signified by strong competition of best higher education institutions that are located in few provinces and municipalities and characterized by different quotas, reserving places to local residents (Donzuso 2015). Accessing these higher education systems, however, is considered an entry ticket for future job opportunities. The notion of transformative adaptation implied in this book is one of enabling a more equal form of access to different types of knowledge, education and power to shape discourses (see Figure 10.1).

Ideology and knowledge production

The findings also revealed that ideology is embedded in scientific practices of conceptualizing social vulnerability by equating certain characteristics with a higher propensity to be affected by climate change without disclosing the underlying assumptions. In his book on political epistemology, Omodeo (2019) points to the way in which cultural-political agendas have constantly shaped academic discourses on science. In China, social sciences are still a vastly underfunded field of research and for that matter practice. Throughout the years, the social science sections in libraries have decreased in size. Without the educationally supported production of social scientific insights, the perspective of social vulnerability of climate change, a deeply social scientific matter, will likely to not reach the forefront of local decision-makers but will continue a niche existence that is likely to be facilitated by educational exchange programs and the transfer of knowledge through scholastic activities.

When going back to the empirical cases and local problem recognition, it became visible that concepts related to social vulnerability presented a type of scientized knowledge that was not readily accessible for most decision-makers. Instead, they would choose the perspectives they accept or are politically more permissible. As part of the assessments, the supposedly vulnerable populations appeared to not have the power of producing their own type of knowledge (such as declaring who is vulnerable and why), which seemed to be the widely accepted method within everyday political practices, governmental consultancy and scientific assessments. This problematic practice has been outlined previously based on the argument that labeling a population as "vulnerable" is itself an exhibition of power (O'Brien et al. 2007b; Ribot 2011). Exhibiting knowledge has become a distinct political mechanism, which also serves the function of reproducing vulnerability. The argument that climate change adaptation is an intrinsically political process (Taylor 2015), can be extended to vulnerability ascription.

Here too, political epistemology has relevant insights to offer on the centrality of agency through knowledge. Knowledge creation is a contested field of political action and often a matter of expert elites (Omodeo 2019). Epistemology cannot be avoided; however, knowledge production can make transparent issues of agency, limitations of methodology and political influence on knowledge processes. This is in line with earlier demands of self-critical adaptation studies that are often co-funded by governmental institutions.

Improving social vulnerability assessments through systematic validation

Research is growing that validates social vulnerability indices (see, e.g., Fekete 2009; Zahran et al. 2008; Khunwishit et al. 2012; Ignacio et al. 2015; Liu et al. 2016). Existing assessments provide an ambivalent picture. Some researchers find that social vulnerability had a statistically positive effect on disaster impact in the case of Hurricane Katrina (Khunwishit et al. 2012) and was a decisive factor that determined higher risk to river floods in three federal states in Germany (Fekete

2009). Other empirical findings suggest that exposure rather than social vulnerability plays a greater role in determining the magnitude of losses and damages experienced at a local scale (Ignacio et al. 2015; Liu et al. 2016). This research suggests that differential vulnerability may disappear in extreme hazard contexts overall. One study focused on ethical tensions in light of the 2010 Haiti earthquake and emphasized that protracted power differentials mattered more significantly (Durocher et al. 2017). The matter of access to information and recovery resources and/or assistance was also pointed out in the context of the 2013 Boulder floods (Adams et al. 2013) and extreme climatic events in the Brazilian Amazon (Parry et al. 2018). Fekete (2019) reviews the usefulness of spatial indicators and critically reflects upon the shortcomings of social vulnerability assessments using an empirical case study of an earlier SVA. The article is an important contribution to a diversifying field of social vulnerability studies, which often overlooks validation, methodological and conceptual shortcomings.

Responsible research and discursive framing

Power of method in terms of being critically self-reflexive is not just an important aspect of self-determination, but in line with Oels (2019)'s argument: "To stimulate self-government, people need to be constituted as agents capable of self-government and guided towards governmental goals" (p. 146). According to this understanding, social learning as part of deliberation processes is the ability of self-organizing. More research is needed on the conditions under which participation-oriented adaptation planning is successful in reaching its targets and what potential downsides participation-heavy processes can have, including hindering effective adaptation action. More research is needed on factors that enable participation-based evaluations and assessments of vulnerability and adaptation. Development scholar Robert Chambers has extensively studied participatory methods for conducting responsible research and evaluation and is one source to be drawn from.

The analysis of social vulnerability falls short by not looking at *how* the state distributes resilience. In line with the path-dependent perspective of lock-ins, future studies should ask: what is the lack of adaptive capacity of certain people a greater phenomenon off? What is the constitution of our political system and economy, and how is society equipped in dealing with climate extremes (and not making them as powerful) in the future? Often people that have the least advantage are assumed to have less resilience. Thereby, we engage in a political and discursive construction of vulnerability by referring to them as "vulnerable". We fail to ask: why is it that certain resources and privileges are limited to only some parts of society?

Therefore, this book views institutions not just as adaptation enablers, but likewise as being responsible for upholding certain structures that disable peoples' access. When the need for causal explanations is emphasized, there is an analytical gap that speaks to the problem of how certain social vulnerability factors such as economic status or rural and urban factors correspond to what political dimension. There is a clear research need to unveil which political and economic institutions and past policy decisions impact uneven vulnerability in the here and now.

Giving weight to political feasibility of vulnerability categories

This stunning literature gap on political feasibility of proposed concepts is of course not exclusive to transformational adaptation concepts. It is remarkable that most scholars do not research how to operationalize SVA into local policy processes, thereby decoupling the problem from the political practice further. Adaptation programs that target "socially vulnerable groups" are at risk of falling into the same trap like other population targeted policy programs by judging the socio-demographic as well as economic characteristics of certain people, in order for them to qualify for recognition, thereby building and upholding a certain type of cultural stigma. Idealized vulnerability conceptions are drastically disconnected from the political practice in place, which is often characterized by a very different understanding and awareness about certain groups. In the examined cases the political perception of "the vulnerable" differed strongly from the groups defined by the social vulnerability literature.

Disclosing and verifying main assumptions of vulnerability categories

Other methodological shortcomings that are commonly outlined include data availability and coherence. Eriksen and Kelly (2007), for example, argue that the anticipation of external factors is difficult and determining how certain vulnerability factors interact is neglected. They criticize the subjective component of vulnerability assessments:

> Significantly, the indicators are largely descriptive measures aggregating population characteristics. Vulnerability, like happiness, is a human state or condition that cannot be measured directly in any objective fashion. . . . We consider that, from this perspective alone, the policy relevance of existing studies, in that they largely focus on population characteristics rather than causal processes, is limited. Moreover, the lack of a process-based framework results in a static view of vulnerability, which, in reality, is a highly dynamic state.
>
> (Eriksen and Kelly 2007: 500ff.)

In the adaptation context, it is questionable how some factors were chosen to be representative of vulnerability. Even though adaptation scholars themselves acknowledge the problem of subjectively choosing factors that supposedly lead to vulnerability, they continue to identify and subjectively weigh indicators to call out the "most vulnerable" countries and people (e.g., Brooks et al. 2005). Here, vulnerability is defined as a function and an internal property of a social system besides being context-dependent. The naming of "generic developmental factors" such as poverty, health status and economic status are commonly equated with lower adaptive capacity and higher vulnerability (Brooks et al. 2005). In line with Füssel's critique of existent vulnerability indices (2010), social vulnerability assessments are oftentimes opaque, hiding problematic normative assumptions about social factors that are equated with greater community or individual vulnerability. The idea of resilience carries a notion of being self-sufficient. Within public health, caretakers have actively pushed for a community of self-reliance. Here, those who lack resilience and adaptive capacity are considered to be autarchic. This has profound implications on how vulnerable groups see and place themselves within society and how they are being perceived by the supposedly self-sufficient.

Shifting the focus: who has privilege, why?

Framing of adaptation and vulnerability influences how we think about solutions. In line with critical feminist legal scholar Martha Fineman, rather than looking at who is vulnerable and in what way, vulnerability assessments need to start looking at who has what privileges, and why this is the case. In light of climate change and the lack of transformative adaptation strategies, this book argues for an exhibited political vulnerability. In different words, the largely unaddressed adaptation deficits, which are especially prominent at the social level, expose weak political systems. As a result, we need to rechannel attention to the political institutions that act as enablers of resilience. Governments are partially responsible for mediating vulnerability instead of just engaging with a narrow-minded recognition of differential vulnerability.

At its core, political vulnerability is characterized by interdependent institutions operating in a political economy and cultural ecology that constantly redesign uneven vulnerability. The presented urban attempts in Atlanta and Jinhua also constitute examples of governmental failure and the vulnerability of political institutions that govern. The reasons for these failures need to be reflected upon. Those involved in adaptation planning spend too much time on defining passive adaptation frameworks, which are reactive rather than anticipatory and proactive. They reactively try to reduce vulnerability without foreseeing and touching upon the root causes of vulnerability. Because this is an all-government challenge, climate adaptation practitioners rely on the efforts of all government sectors.

Schlosberg et al. (2017) take a capabilities approach that goes beyond distributional and procedural inequity by focusing on the provision of basic needs necessary for a functioning life. Their "just transformation" climate adaptation approach is characterized by "progressive pragmatism" of governments and enhancing "community capability" (Schlosberg et al. 2017). The authors seek to move beyond the prevailing narrow and pragmatic perspective of governmental adaptation approaches in the context of Australia, through moving to community-oriented concerns, calling attention to vulnerable populations (Schlosberg et al. 2017). There is a greater need for these justice-oriented adaptation lenses to critically question the notion of "vulnerable populations" and consider how the corresponding categories are methodologically designed and how they will actually apply to local political contexts.

The role of adaptation policy for transformative adaptation

A greater and more frequent engagement with adaptation by political scientists is needed. The necessity for not making adaptation a small subfield of environmental politics, but rather a growing super-field that connects existing fields of political science has been emphasized (Javeline 2014). Adaptation scholar Susan Moser contrasts this by questioning the illusionary concept of "perfect governance approaches that promise perfect adaptation" (Adger et al. 2009: 17; Moser 2009). Going back to the findings of Ford et al. (2018), who point to the superficial treatment institutions and politics have received, rarely investigating actual implementation processes of proposed recommendations, this chapter section briefly aims to lay out some of the lower hanging policy fruits, which it is hoped, can be more easily implemented. As has been pointed out by Biesbroek and colleagues (2015), decision-making is a highly dynamic process, which is often neglected by empirical vulnerability researchers. This research hopes to have reflected at least on some of the dynamics – in terms of (the lack of) shared knowledge patterns and looking at the origins of political bias toward certain issues.

Early inclusion and co-creation of adaptation responses

The quality of adaptation can be substantially advanced by adhering to the Habermasian principles of the ideal speech situation (Oels 2019). The current situation of vulnerability ascription and recognition practices is, to paraphrase Habermas, far from a nonhierarchical discourse. Several criteria, which Habermas identifies for an ideal-speech situation, are drastically absent in local political practices related to vulnerability recognition. In the cases examined, vulnerable people and vulnerability researchers and practitioners were not equal communication partners. People did not have equal opportunities to express themselves. In both political contexts, access to knowledge, education and decision-making is severely limited and divided unequally. Before more complex problems such as a more egalitarian or Rawlsian access to education is the reality, we must rely on other means. Other means regard an adjusted adaptation cycle (see following sections), which does not predefine existing vulnerabilities but have people define their own experience of climate change (also see Liverman 2015).

When looking at vulnerable populations, a dominant perception has been that "these populations" are in need of external support and depend on people in power leveraging for them (e.g., see DeLauro 2017). Despite the pressing lack of research on how better social safety nets can enhance social welfare and contribute meaningfully to adaptation (and addressing vulnerability), symptomatically waging the battle for the vulnerable, thereby potentially disempowering people will not make it work. There is a greater need for inclusion in the actual policy process when defining and framing the challenges instead of consulting populations once the assessment of vulnerability is done. Examining, for instance, how a fair deliberative process regarding socially just adaptation planning and framing would look like in Atlanta and Jinhua and how this could be politically implemented would provide an insightful follow-up study to this book. Further instruments include participatory vulnerability assessments.

Increasing ownership over predefining vulnerability in the adaptation policy cycle

Climate adaptation is often defined as a process of different systems (such as climatological, economic, ecologic, or human systems), to prepare for, adjust to and recover from already occurring or expected climate change and its effects (IPCC 2014). This research aimed to show that adaptation options depend not only on the envisioned system but also on the way vulnerability is thought about and conceptualized. A contingent adaptation policy cycle usually starts by asking "vulnerability of whom to what"? (also see Chapter 3). After clarifying the unit of analysis and related concepts, an impact and vulnerability assessment is usually conducted. Although more research is needed to verify this observation, the first step (determining the unit of analysis) is often undertaken by decision-makers and policy planners, thereby potentially neglecting forms of bias.

Adaptation researchers emphasize that adaptation planning is a continuous improvement process, involving multiple tasks and iterations (e.g., Vogel and Henstra 2015; Street et al. 2016). There are different approaches regarding where to include stakeholders and the public in the policy cycle. Some integrate a broader audience only at the latter stages of the policy process when evaluating adaptation measure (Giordano et al. 2011: 12ff.), others add stakeholders midway (Vogel and Henstra 2015), or right at the beginning and at multiple points throughout the process (Nadin et al. 2016). Understandings of what constitutes inclusive and participatory adaptation planning differ strongly.

An ideal adaptation policy cycle would spend greater time predefining vulnerabilities, framing the issue and enabling processes of co-owning vulnerability assessments with local populations. This would be a strong procedural readjustment. Inclusive adaptation planning would be understood as peoples' means to define who and what is vulnerable and how this vulnerability is manifested. This would ultimately increase the ownership of local populations over and potentially increase the success of adaptation options.

Need for constant policy iteration

Policy iteration is needed on the basis of the validation of existing social vulnerability studies and pairing them with potential cultural practices. In a regular policy cycle, based upon the definition of vulnerability, adaptation priorities are identified, and plans outlined. These steps are followed by looking at adaptation

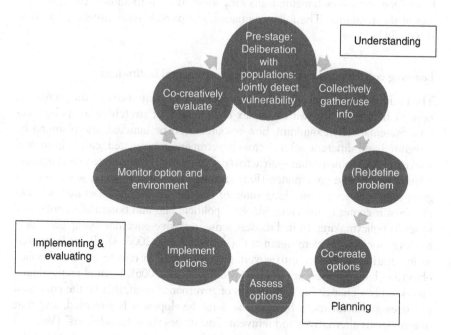

Figure 10.2 Ideal-type phases and substages of the adaptation planning process.
(Source: The author, adjusted from Moser and Boykoff 2013 adapted from Ekstrom et al. 2011)

implementation opportunities. The last step of a more ideal policy cycle includes monitoring and evaluation of activities (for policy integration, see Candel and Biesbroek 2016). Adaptation policy seeks to build and/or increase the capacity of an affected system (see Chapter 3). A more localized version of an adaptation policy cycle spends greater time defining key terms and concepts and including local communities in defining adaptation priorities.

Moser and Boykoff (2013) have outlined the ideal-type stages of the adaptation decision-making process. According to their analysis, three general phases exist: (1) increasing the understanding of the challenge and gathering information, (2) planning through developing, assessing and selecting options (3) managing adaptation through implementing, monitoring and evaluating the selected adaptation options (see Figure 10.2). In the understanding stage, it is important to ask: *who* detects the problem, *who* gathers and uses the information, and *who* is involved in (re)defining the problem? If one were to clearly outline these items, some of the problems related to political epistemology identified in the empirical chapters could potentially be addressed.

In the planning stage, iteration seems to have already received some consideration through first developing, assessing and then selecting options. In both contexts, the most pressing lock-in detected, occurred at a political epistemological level. One of the low-hanging fruits for policymaking is to address the epistemo-logical shortcomings. The following paragraphs provide some insights as to how this could be done.

Learning capacity and coordination in governmental institutions

The findings brought to light that governments are confronted by the increasing need of having to deal with complex challenges that stretch across policy sec-tors. Systematically examining how resources can be bundled, the planning be integrated and different actors across government coordinated, can help to deal with complex problems that span across policy sectors. At the same time, concepts related to reflexive governance (RG) and organizational learning seem to be of great relevance for approaching some of the identified political-epistemic lock-ins when it comes to unveiling ideology, political bias and potentially conflicting logics in policymaking. In its broadest sense, reflexive governance explores "new modes of societal problem treatment" (Voß and Kemp 2006: 4).[1] RG is interested in integrating reflexivity into governance processes to enable a shift of policy objectives based on new knowledge (Voß and Kemp 2006). Related political pro-cesses imply a change in "foundations of governance itself, that is, the concepts, practices and institutions by which societal development is governed, and that one envisions alternatives and reinvents and shapes those foundations" (Voß and Kemp 2006: 4). Learning capacity has also been an important aspect within the Adaptive Capacity Wheel (ACW), a methodological tool that looks at the adap-tive capacity of political institutions more systematically (also see Gupta et al. 2010). According to the ACW, learning capacity is based upon factors such as trust, single- and double-loop learning, the opportunity to discuss doubts and institutional memory (cf. Gupta et al. 2010; Grecksch 2012). The governmental ability to learn corresponds with the above-mentioned need of self-reflected prac-tices of knowledge production and knowledge iteration, which is important for any planning process, and especially important when it comes to dealing with une-ven and dynamic patterns of vulnerability (reproduction) that constantly change. Making transparent the different understandings of vulnerability, their origins, as well as changes in knowing and explaining and self-critically examining how the own institutions are (not) equipped might be one step in the right direction. Other ACW indicators include aspects such as resources (human, financial, authority), leadership, room for autonomous change, diversity and aspects of fair governance (cf. Greksch et al. 2012). The ACW has also been readjusted to put a stronger focus on psychological components such as adaptation motivation and adapta-tion belief of policy practitioners and actors involved in adaptation strategizing (Grothmann et al. 2013).

The way political actors learn, approach complex challenges and apply existing knowledge in the context of adaptation is important for unlocking the socio-cognitive processes which may constrain adaptation action. Similar arguments have been made by adaptation scholars on the aspects of organiza-tional and transformative learning (Boyd and Osbahr 2010; Moser et al. 2017) and self-organizing by means of access to education (Boyd et al. 2008). As has been made sufficiently clear, the cognitive factors in adaptation, such as risk perception and perceived adaptive capacity, are decisive factors that impact the judgment and governmental responses to climatic stimuli as well as human decision-making under uncertain conditions (Grothmann and Patt 2005). Socio-cognitive and cultural factors likewise impact what type and degree of inequality are considered legitimate. Due to scope, this research was unable to reflect adequately on this important and growing literature of reflexive governance and transformative learning within political institutions. Follow-up research is urgently needed on these aspects and not just in the context of local adaptation governance dynamics but also in consideration of governance dynamics related to social inequality.

Examples of other policy tools

How political institutions and their organization are themselves equipped to deal with uneven population vulnerability has not been assessed in systemized way as part of this study. Self-critically examining the adaptive capacity to deal with complex challenges, such as uneven population vulnerability is a starting point for the facilitation of adaptation. The adaptation literature has produced analytical tools for understanding different facets of adaptation (also see Purdon and Thornton 2019). The ACW is one example for an analytical tool and methodology which helps to systematically examine the adaptive capacity of political institutions to adapt to climate change (see previous section).

In their case study on adaptive capacity under a changing institutional landscape in two municipalities in the southwest of Burkina Faso, Brockhaus and Kambiré (2009) initially assume that in light of political reform processes and decentralization, the transfer of resources, competencies and planning authorities to the local level will have a positive effect on the municipalities' adaptive capacity. However, their findings suggest that successful adaptation is driven by two key factors "(1) individual understandings; and (2) institutional flexibility in governance structures, ensured by strong direct connections ('short distances') to local realities" (Brockhaus and Kambiré 2009: 414). Although decentralization can be advantageous for adaptation in light of its institutional flexibility, and an opportunity for selective planning, it is not per se a guarantee for successful adaptation (Brockhaus and Kambiré 2009). In line with the aforementioned findings, low adaptive capacity was explained with the lack of knowledge and shared-learning structures coupled with biased perceptions and passive participation mechanisms in form of consultation. These factors led to a disconnect between higher levels of government and local realities. The lack of institutional flexibility prevalent in the distance to local realities (and populations' vulnerabilities) and the way governmental decision-makers learn and/or generate the knowledge also appear as key factors, which explain adaptation deficits in Atlanta and Jinhua.

Participatory vulnerability assessments

Some jurisdictions in the United States address vulnerability more specifically as part of their adaptation policy planning. Examples include but are not limited to: Austin, Baltimore, Boston, Chicago, Los Angeles, New York and Washington DC.[2] A systematic review of these efforts was not possible but the initial impression suggests that these socially sensitive adaptation plans focus on three primary tools: (1) the inclusion of social vulnerability, health risk assessments and hazard mitigation plans into adaptation planning processes, (2) opportunities and engagement of a broader stakeholder community to partake in local adaptation planning workshops and (3) communication tools and dissemination of information regarding adaptation.

There appears to be great development across local jurisdictions in the United States on issues related to vulnerable populations and climate change. One example is the Headwaters Economics nonprofit research group, which helps people and organizations to develop strategies on issues local communities face.[3] Aside from collaborating with local governmental institutions, this too is a non-state initiative. Although social vulnerability assessments and mapping are evolving in China, and research has flourished examining economic, environmental and social factors in climate change processes, on the ground, social vulnerability concerns are rare and if they exist continue to be embedded in health planning initiatives (e.g., in Guangdong). Herders' adaptation practices in Inner Mongolia are one of the first studies, where regional climate projections were integrated with participatory assessment research on social vulnerability and adaptation planning (see Hang et al. 2016).

Summary: the importance of unveiling bias

The foregone chapter aimed to discuss the main findings of this book in the context of discourses on transformative adaptation. The first section briefly laid out, why adaptation is a political concept and that political institutions do play a role in maintaining and in some cases worsening population vulnerability. This section reflected upon the need for self-critical adaptation studies. The end of the section argues that there is a greater need to research, how adaptation relates to questions of social justice and what can be done about it.

The subsequent section presented access to knowledge and education as two key components addressing political epistemological lock-ins. This argument was placed in dominant perspectives on transformative adaptation, which often focus on participation. Because knowledge production is such an important aspect for breaking with path-dependent vulnerability patterns, this chapter also reflected upon the methodological shortcomings of existing (social) vulnerability studies and what they can do differently to cope with the limitations. In line with the earlier demand, self-critical adaptation studies could, for instance, disclose the underlying assumptions that they operate with certain vulnerability categories and spend more time on exploring and communicating the limitations of related methods. Further, the need for a systemic validation of vulnerability assessments has been pointed out before, together with the need of exploring social adaptation practices, and how they might actually mitigate some of the deeply rooted vulnerabilities.

Against the background of adaptation being deeply rooted in matters of social justice, the next section critically examined the role the adaptation policy field can play in this regard. Higher hanging fruits include educational reform and enabling access to education and production to different kinds of knowledge. In Atlanta, attention must be paid to a segregated school system that is in big parts racially divided and significantly interrelates with access to public transportation. In China, this relates to Gaokao and hukou reform and enabling migrant worker children to access schools and job opportunities. This does not necessarily imply that this book pushes an egalitarian understanding of social justice, signified by equal access, or on the contrary pushing for affirmative action to the benefit of certain profiles. This discussion is to be left to those who have written monumental volumes on this and know the respective contexts of migrant workers and rural hukou holders in China and (racially) segregated African American population in southeastern Georgia.

Some lower hanging fruits were also reflected upon by arguing for the early and constant inclusion of vulnerable populations in the adaptation policy cycle. Aside from self-validating who is vulnerable in a prolonged definition phase in the adaptation cycle, studies need to cross-validate population vulnerability. Through providing iterative elements within the policy cycle a constant integration of generated knowledge could be enabled, which would offer greater justice to epistemological problems. This requires the state to enable the participation of those, who cannot afford to be absent from work. Further, letting people frame themselves and their own vulnerability, through, for instance, participatory vulnerability assessments might be other lower hanging fruits decision-makers can implement. The last identified opportunity regards intra-organizational learning. The chapter poses the argument that in order for some lock-ins to be addressed, a reform of the way political administrations and organizations work may become necessary, as these current institutions appear to have limited capacity in dealing with issues of greater complexity, such as social vulnerability. Experience-based learning formats and exchange programs with other decision-makers and relevant actors outside the governmental sphere with on the ground-knowledge on the matter may diminish some of the cognitive barriers, which hamper an emotional engagement with certain issues. Setting up good practice partnerships with other actors and allowing decision-makers to use and embrace a plurality of methods is another complex need in this context, as not every method is fit for every context. Below the line, the need to be critically self-reflected, unveiling political bias across different contexts (politics, science) and embracing not just complexity but also vulnerability alike is what the chapter set out to argue for.

Notes

1 For a recent review of different reflexive governance conceptualizations, see Feindt and Weiland. "Reflexive Governance: Exploring the Concept and Assessing its Critical Potential for Sustainable Development. Introduction to the Special Issue." *Journal of Environmental Policy & Planning* 20, no. 6 (2019): 661–674. Doi: 10.1080/1523908X.2018.1532562.

2 See the Adaptation Clearinghouse for the different initiatives addressing vulnerable population across different policy sectors related to adaptation, www.adaptationclearinghouse. org/, last accessed December 3, 2019. Also see a recent article published in the Guardian "Killer heat: US cities' plans for coming heatwaves fail to protect vulnerable," published December 6, 2019, www.theguardian.com/cities/2019/dec/06/killer-heat-us-cities-plans-for-coming-heatwaves-fail-to-protect-vulnerable, last accessed December 18, 2019.

3 See "Research" and "Tools" of Headwaters Economics, https://headwaterseconomics. org/about/, last accessed December 18, 2019.

References

Adams, Melanie, Melanie Ferraro, Patrick Gantert, Taryn Lee, Cameron May, Abigail Peters, Michael Salka, and Abby Hickcox. "Differential Vulnerability and Environmental Justice: A Preliminary Report on the 2013 Boulder Floods." Honors 4000 Paper: Environmental Justice, December 16, 2013.

Adger, Neil W., Irene Lorenzoni, and Karen L. O'Brien (eds.). *Adapting to Climate Change. Thresholds, Values, Governance*. Cambridge: Cambridge University Press, 2009.

Arnstein, Sherry R. "A Ladder of Citizen Participation." *Journal of the American Planning Association* 35, no. 4 (1969): 216–224. https://doi.org/10.1080/01944366908977225.

Bahadur, Aditya, and Thomas Tanner. "Distilling the Characteristics of Transformational Change in a Changing Climate." International Conference Proceedings of Transformation in a Changing Climate, University of Oslo, Oslo, Norway, 19–21 June 2013.

Biagini, Bonizella, Rosina Bierbaum, Missy Stults, Saliha Dobardzic, and Shannon M. McNeeley. "A Typology Of Adaptation Actions: A Global Look at Climate Adaptation Actions Financed Through the Global Environment Facility." *Global Environmental Change* 25 (2014): 97–108. https://doi.org/10.1016/j.gloenvcha.2014.01.003.

Biesbroek Rorbert, Johann Dupuis, Andrew Jordan, Adam Wellstead, Michael Howlett, Paul Cairney et al. "Opening Up the Black Box of Adaptation Decision-Making." *Nature Climate Change* 5 (2015): 493–494. https://doi.org/10.1038/nclimate2615.

Boyd, Emily, and Henny Osbahr. "Responses to Climate Change: Exploring Organisational Learning Across Internationally Networked Organisations for Development." *Environmental Education Research* 16, no. 5–6 (2010): 629–643. Doi: 10.1080/13504622.2010.505444.

Boyd, Emily, Henny Osbahr, Polly J Ericksen, Emma L Tompkins, Maria Carmen Lemos, and Fiona Miller. "Resilience and 'Climatizing' Development: Examples and Policy Implications." *Development* 51, no. 3 (2008): 390–396. https://doi.org/10.1057/dev.2008.32.

Brockhaus, Maria, and Hermann Kambiré. "25 Decentralization: A Window of Opportunity for Successful Adaptation to Climate Change." In *Adapting to Climate Change Thresholds, Values, Governance*, edited by W. Neil Adger, Irene Lorenzoni, and Karen L. O'Brien, 399–416. Cambridge: Cambridge University Press, 2009. Doi: 10.1017/CBO9780511596667.026.

Brooks, Nick, W. Neil Adger, and P. Mick Kelly. "The Determinants of Vulnerability and Adaptive Capacity at the National Level and the Implications for Adaptation." *Global Environmental Change* 15, no. 2 (2005): 151–163. https://doi.org/10.1016/j.gloenvcha.2004.12.006.

Candel, Jeroen J. L., and Robbert Biesbroek. "Toward a Processual Understanding pf Policy Integration." *Policy Sciences* 49, no. 1 (2016): 211–231. Doi: 10.1007/s11077-016-9248-y.

DeLauro, Rosa L. *The Least Among Us Waging the Battle for the Vulnerable.* New York, NY: The New Press, 2017.

Donzuso, Nunzio Nazareno. "'Equality of Opportunities' in Education for Migrant Children in China." *Global Social Welfare* 2 (2015): 9–13. https://doi.org/10.1007/s40609-014-0012-y.

Durocher, Evelyne, Ryoa Chung, Christiane Rochon, Jean-Hugues Henrys, Catherine Oliver, and Matthew Hunt. "Ethical Questions Identified in a Study of Local and Expatriate Responders' Perspectives of Vulnerability in the 2010 Haiti Earthquake." *Journal of Medical Ethics* 43 (2017): 613–617. http://dx.doi.org/10.1136/medethics-2015-102896.

Ekstrom, Julia A., Susanne C. Moser, and Margaret Torn. "Barriers to Climate Change. Adaptation: A Diagnostic Framework." Lawrence Berkeley National Laboratory, Berkeley, CA, California Energy Commission, 2011.

Eriksen, Siri H., and P. Mick Kelly. "Developing Credible Vulnerability Indicators for Climate Adaptation Policy Assessment." *Mitigation and Adaptation Strategies for Global Change* 12, no. 4 (2007): 495–524. Doi: 10.1007/s11027-006-3460-6.

Fazey, Ioan, Peter Moug, Simon Allen, Kate Beckmann, David Blackwood, Mike Bonaventura, Kathryn Burnett et al. "Transformation in a Changing Climate: A Research Agenda." *Climate and Development* 10, no. 3 (2018): 197–217. Doi: 10.1080/17565529.2017.1301864.

Fekete, Alexander. "Validation of a Social Vulnerability Index in Context to River-floods in Germany." *Natural Hazards and Earth System Science* 9, no. 2 (2009): 393–403.

Fekete, Alexander. "Social Vulnerability (Re-)Assessment in Context to Natural Hazards: Review of the Usefulness of the Spatial Indicator Approach and Investigations of Validation Demands." *International Journal of Disaster Risk Science* 10, no. 1 (2019): 220–232. https://doi.org/10.1007/s13753-019-0213-1.

Few, Roger, Daniel Morchain, Diane Spear, Adelina Mensah, and Ramkumar Bendapudi. "Transformation, Adaptation and Development: Relating Concepts to Practice." *Palgrave Communications* 3, no. 17092 (2017): 1–9. Doi: 10.1057/palcomms.2017.92.

Ford, James D., Tristan Pearce, Graham McDowell, Lea Berrang-Ford, Jesse S. Sayles, and Ella Belfer. "Vulnerability and its Discontents: The Past, Present, and Future of Climate Change Vulnerability Research." *Climatic Change* 151 (2018): 189–203. Doi: 10.1007/s10584-018-2304-1.

Füssel, Hans-Martin. *Review and Quantitative Analysis of Indices of Climate Change Exposure, Adaptive Capacity, Sensitivity, and Impacts.* Washington, DC: World Bank, 2010.

Garrelts, Heiki, Johannes Herbeck, and Michael Flitner. "Leaving the Comfort Zone. Regional Governance in a German Climate Adaptation Project." In *A Critical Approach to Climate Change Adaptation. Discourses, Policies, and Practices,* edited by Silja Klepp and Libertad Chavey-Rodriguez. Oxon and New York: Routledge, 2018.

GCC. "State and Local Adaptation Plans. Overview of State and Municipal Adaptation Progress". Georgetown Climate Center (GCC). www.georgetownclimate.org/adaptation/plans.html, last accessed November 19, 2019.

Giordano, Francesca, Alessio Capriolo, and Rosa Anna Mascolo. "Planning for Adaptation to Climate Change – Guidelines for Municipalities." Life Project No LIFE08 ENV/IT/000436, ACT – Adapting to Climate Change in Time, published online, 2011. https://base-adaptation.eu/sites/default/files/306-guidelinesversionefinale20.pdf.

Greksch, Kevin. "Adaptive Capacity and Regional Water Governance in North-Western Germany." *Water Policy* 15 (2013): 794–815. Doi: 10.2166/wp.2013.124.

Godfrey-Wood, Rachel, and Lars Otto Naess. "Adapting to Climate Change: Transforming Development?" *IDS Bulletin* 47, no. 2 (2016): 49–62. Doi: 10.19088/1968-2016.131.

Grothmann, Thorsten, Kevin Grecksch, Maik Winges, and Bernd Siebenhüner. "Assessing Institutional Capacities to Adapt to Climate Change – Integrating Psychological Dimensions in the Adaptive Capacity Wheel." *Natural Hazards and Earth System Science* 13, no. 12 (2013): 3369–3384. https://doi.org/10.5194/nhess-13-3369-2013.

Grothmann, Torsten, and Anthony G. Patt. "Adaptive Capacity and Human Cognition: The Process of Individual Adaptation to Climate Change." *Global Environmental Change* 15, no. 3 (2005): 199–213. https://doi.org/10.1016/j.gloenvcha.2005.01.002.

Gupta, Joyeeta, Katrien C. J. M. A. Termeer, Judith, Klostermann, Sander Meijerink, Margo van den Brink, Pieter Jong, Sibout Nooteboom et al. "The Adaptive Capacity Wheel: A Method to Assess the Inherent Characteristics of Institutions to Enable the Adaptive Capacity of Society." *Environmental Science and Policy* 13, no. 6 (2010): 459–471. Doi: 10.1016/j.envsci.2010.05.006.

Hang, Shuanzhu, Shan Ping, Bao Lu, Ao Renqi, Wang Jianwu, Zheng Yan, Wei Yurong et al. "Climate Change and Inner Mongolia." In *Climate Risk and Resilience*, edited Rebecca Nadin, Sarah Opitz-Stapleton, and Xu Yinlong. London: Routledge, 2016.

Ignacio, J. Andres F., Grace T. Cruz, Fernando Nardi, and Sabine Henry. "Assessing the Effectiveness of a Social Vulnerability Index in Predicting Heterogeneity in the Impacts of Natural Hazards: Case Study of the Tropical Storm Washi Flood in the Philippines." *Vienna Yearbook of Population Research* 13, no. 1 (2015): 91–130. Doi: 10.1553/populationyearbook2015s091.

IPCC. "Summary for policymakers. In: Climate Change 2014: Impacts, Adaptation, and Vulnerability. Part A: Global and Sectoral Aspects." In *Contribution of Working Group II to the Fifth Assessment Report of the Intergovernmental Panel on Climate Change*, edited by C. B. Field, V. R. Barros, D. J. Dokken et al., 1–32. Cambridge, UK and New York, NY: Cambridge University Press, 2014.

Javeline, Debra. "The Most Important Topic Political Scientists Are Not Studying: Adapting to Climate Change." *Perspectives on Politics* 12, no. 2 (2014): 420–434. https://doi.org/10.1017/S1537592714000784.

Kaswan, Alice. "Environmental Justice and Domestic Climate Change Policy." *Environmental Law Reporter* 38, no. 10287 (2008).

Kaswan, Alice. "Climate Adaptation and Land Use Governance: The Vertical Axis." *Columbia Journal of Environmental Law* 390 (2014).

Kates, Robert W., William R. Travis, and Thomas J. Wilbanks. "Transformational Adaptation." *Proceedings of the National Academy of Sciences* 109, no. 19 (2012): 7156–7161. Doi: 10.1073/pnas.1115521109.

Keskitalo, E. Carina H., and Benjamin L. Preston. *Research Handbook on Climate Change Adaptation Policy*. Cheltenham, UK, and Northampton, MA: Edward Elgar Press, 2019.

Khunwishit, Somporn, and David A. McEntire. "Testing Social Vulnerability Theory: A Quantitative Study of Hurricane Katrina's Perceived Impact on Residents living in FEMA Designated Disaster Areas." *Journal of Homeland Security and Emergency Management* 9, no. 1 (2012): 13. https://doi.org/10.1515/1547-7355.1950.

Livermann, Diana. "23 Reading Climate Change and Climate Governance as Political Ecologies." In *The Routledge Handbook of Political Ecology*, edited by Tom Perreault, Gavin Bridge, and James McCarthy, 303–319. Abingdon: Routledge, 2015.

Liu, Jie, Zhenwu Shi, and Dan Wang. "Measuring and Mapping the Flood Vulnerability Based on Land-Use Patterns: A Case Study of Beijing, China." *Natural Hazards* 83, no. 3 (2016): 1545–1565. Doi: 10.1007/s11069-016-2375-0.

Mikulewicz, Michael. "Politicizing Vulnerability and Adaptation: On the Need to Democratize Local Responses to Climate Impacts in Developing Countries." *Climate and Development* 10, no. 1 (2018): 18–34. https://doi.org/10.1080/17565529.2017.1304887.

Morchain, Daniel. "Rethinking the Framing of Climate Change Adaptation Knowledge, Power, and Politics." In *A Critical Approach to Climate Change Adaptation Discourses, Policies and Practices*, edited by Silja Klepp and Libertad Chavez-Rodriguez, 55–74. London: Routledge, 2018.

Moser, Susanne C. "Whether Our Levers are Long Enough and the Fulcrum Strong?— Exploring the Soft Underbelly of Adaptation Decisions and Actions." In *Adapting to Climate Change, Thresholds, Values, Governance*, edited by W. N. Adger et al., 313–343, Cambridge: Cambridge University Press, 2009.

Moser, Susanne C., and Maxwell Boykoff. *Successful Adaptation to Climate Change: Linking Science and Practice in a Rapidly Changing World*. London: Routledge, 2013.

Moser, Susanne C., Joyce Coffee, and Aleka Seville. "Rising to the Challenge, A Review and Critical Assessment of the State of the US Climate Adaptation Field." Report, Kresge Foundation, December 2017.

Nadin, Rebecca, Sarah Opitz-Stapleton, and Xu Yinlong (eds.). *Climate Risk and Resilience in China*. London and New York: Routledge, 2016.

O'Brien, Karen. "Global Environmental Change II: From Adaptation to Deliberate Transformation." *Progress in Human Geography* 36, no. 5 (2012): 667–676. Doi: 10.1177/0309132511425767.

O'Brien, Karen. "Climate Change and Social Transformations: Is It Time for a Quantum Leap?" *WIREs Climate Change* 7, no. 5 (2016): 618–626. https://doi.org/10.1002/wcc.413.

O'Brien, Karen, Siri Eriksen, Lynn P. Nygaard, and Ane Schjolden. "Why Different Interpretations of Vulnerability Matter in Climate Change Discourses." *Climate Policy* 7, no. 1 (2007): 73–88. Doi: 10.1080/14693062.2007.9685639.

O'Brien, Karen, and Robin M. Leichenko. "Human Security, Vulnerability and Sustainable Adaptation." UNDP Human Development Report 9, 2007b.

Oels, Angela. "Sustainability, Participation and Power – or: Why Participation Does not Necessarily Lead to Sustainable Development." Original: "Nachhaltigkeit, Partizipation und Macht – oder: Warum Partizipation nicht unbedingt zu Nachhaltigkeit führt." In *Partizipation und Nachhaltigkeit: Vom Leitbild zur Umsetzung*, edited by Helga Jonuschat, Elke Baranek, Maria Behrendt, Kristina Dietz, Bianca Schlußmeier, Heike Walk, and Andreas Zehm, 28–43. München: Oekom Verlag, 2007.

Oels, Angela. "From Governance to Governmentality: The Added Value of a Foucaultian Discourse Analysis for International Politics: The Example of Climate Politics." Original: "Von Governance zu Gouvernementalität: Der Mehrwert einer Diskursanalyse nach Foucault für die internationale Politik – das Beispiel Klimapolitik." In *Der Klimawandel – Sozialwissenschaftliche Perspektiven*, edited by Martin Voß, 171–186. Wiesbaden: VS-Verlag für Sozialwissenschaften, 2010.

Oels, Angela. "The Promise and Limits of Participation in Adaptation Governance: Moving Beyond Participation Towards Disruption." In *Research Handbook on Climate Change Adaptation Policy*, edited by E. Carina H. Keskitalo and Benjamin L. Preston, 138–156. Cheltenham: Edward Elgar Press, 2019.

Omodeo, Pietro Daniel. "Towards a Political Epistemology: Positioning Science Studies." In *Political Epistemology*, edited by Pietro Daniel Omodeo, 13–49. Springer, 2019. https://doi.org/10.1007/978-3-030-23120-0_2.

Parry, Luke, Gemma Davies, Oriana Almeida, Gina Frausin, André de Moraés, Sergio Rivero, Naziano Filizola et al. "Social Vulnerability to Climatic Shocks Is Shaped by Urban Accessibility." *Annals of the American Association of Geographers* 108, no. 1 (2018): 125–143. Doi: 10.1080/24694452.2017.1325726.

Pelling, Mark. *Adaptation to Climate Change. From Resilience to Transformation.* New York: Routledge, 2011.

Purdon, Mark, and Philip Thornton. "Research Methodology for Adaptation Policy Analysis: Embracing the Eclectic Messy Centre." In *Research Handbook on Climate Change Adaptation Policy*, edited by E. Carina H. Keskitalo and Billie L. Preston, 157–192. Cheltenham and Northampton: Edward Elgar Publishing Limited, 2019.

Ribot, Jesse. "Editorial. Vulnerability before Adaptation: Toward Transformative Climate Action." *Global Environmental Change* 21, no. 4 (2011): 1160–1162. Doi: 10.1016/j.gloenvcha.2011.07.008.

Ribot, Jesse. "Cause and Response: Vulnerability and Climate in the Anthropocene." *The Journal of Peasant Studies* 41, no. 5 (2014): 667–705. Doi: 10.1080/03066150.2014.894911.

Schlosberg, David, Lisette B. Collins, and Simon Niemeyer. "Adaptation Policy and Community Discourse: Risk, Vulnerability, and Just Transformation." *Environmental Politics* 26, no. 3 (2017): 413–437. https://doi.org/10.1080/09644016.2017.1287628.

Street, Roger, Sarah Opitz-Stapleton, Rebecca Nadin, Cordia Chu, Scott Baum, and Declan Conway. "Climate Change Adaptation Planning to Policy: Critical Considerations and Challenges." In *Climate Risk and Resilience in China*, edited by Rebecca Nadin, Sarah Opitz-Stapleton, and Xu Yinlong, 11–37. New York: Routledge, 2016.

Taylor, Marcus. *The Political Ecology of Climate Change Adaptation. Livelihoods, Agrarian Change and the Conflict of Development.* London and New York: Routledge, 2015.

Voß, Jan-Peter, and Rene Kemp. "Sustainability and Reflexive Governance: Introduction." In *Reflexive Governance for Sustainable Development*, edited by Jan-Peter Voss, Dierk Bauknecht, and René Kemp, 3–28. Cheltenham, UK, and Northampton, MA: Edward Elgar, 2006.

Vogel, Brennan, and Daniel Henstra. "Studying Local Climate Adaptation: A Heuristic Research Framework for Comparative Policy Analysis." *Global Environmental Change* 31 (2015): 110–120. Doi: 10.1016/j.gloenvcha.2015.01.001.

Walsh-Dilley, Marygold, and Wendy Wolford. "(Un)Defining Resilience: Subjective Understandings of 'Resilience' from the Field." *Resilience* 3, no. 3 (2015): 173–182. Doi: 10.1080/21693293.2015.1072310.

Winterfeld, Uta von. "What Is and to What End Do We Conduct Social-Scientific Adaptation Research?" *GAIA* 21, no. 3 (2012): 168–170. https://doi.org/10.14512/gaia.21.3.4.

Zahran, Sammy, Samuel D. Brody, Walter Gillis Peacock, Arnold Vedlitz, and Himanshu Grover. "Social Vulnerability and the Natural and Built Environment: A Model of Flood Casualties in Texas." *Disasters* 32, no. 4 (2008): 537–560. Doi: 10.1111/j.1467-7717.2008.01054.x.

11 Conclusion

What lock-ins can explain adaptation deficits that are particularly apparent in highly uneven vulnerability to climate change? What factors can explain why vulnerability is such a deeply rooted phenomenon across two different political systems and adaptation occurring only incrementally?

By answering the main research question, this book, on the one hand, makes a contribution to the newly emerging research field of adaptation lock-ins, which examines the interrelation of path-dependent factors, which hamper adaptation action. On the other hand, this book makes a contribution to critical adaptation studies by exploring the parameters against which adaptation action operates and taking a closer look at (political) factors that determine, recreate and maintain vulnerability. Critical adaptation studies examine the way adaptation measures are interpreted, transformed and implemented and how these measures interfere with power relations (Taylor 2015; Klepp and Chavez-Rodriguez 2018). The increase of adaptation critical research coincides with years of criticism on the apolitical nature of adaptation research. The politics of adaptation is now slowly becoming an evolving item of research.

In line with this notion, the book argues that adaptation and especially vulnerability conceptualizations are not neutral drivers of action but carry political and ontological implications (also see Bauriedl and Müller-Mahn 2018). This becomes especially visible in the term "social vulnerability", which suggests vulnerability is a social condition. But as much as uneven susceptibility to climate change is a social condition, this book aimed to sensitize for the entanglement of conditions which result in high uneven vulnerability, which has political, economic and cultural origins. The lock-in perspective helps to unveil the different origins (and understandings) of vulnerability and how they interact. Making different understandings transparent is important, as they often provide the rationale for (political) action and thus can result in different courses of action. Because vulnerability and adaptation are mutually interconnected and oftentimes inseparable, they were analyzed in an integrated way and from a non-sectoral perspective. Adaptation has not been mainstreamed in the local cases examined and uneven human vulnerability to climate change is far from being a self-evident concept. The argument of this book is that lock-ins to adaptation in the sense of addressing "social" vulnerability are deeply connected with questions of social justice.

DOI: 10.4324/9781003183259-11

Despite the increasing number of adaptation critical works, and their importance for questioning adaptation as a self-evident concept, this book laid out why it is important to investigate the role governments can play in addressing concerns of uneven vulnerability to climate change. The extended argument however is that "social" vulnerability is also a result of the way political institutions have operated in the past. Simultaneously, it became obvious that problems with knowledge access, production and bias toward vulnerable populations structure the here and now.

Summary of main findings

The findings are grounded upon instrumental qualitative research and an in-depth analysis of two case units, where lock-ins were examined. Atlanta, in the state of Georgia, the United States, and Jinhua, in Zhejiang province, China, were chosen for the exemplary analysis. In both municipalities, adaptation deficits are enormous given the cities' geographic exposure, high climate and especially high unevenly distributed vulnerability paired with continued governmental inaction. What this book aimed to show, is how adaptation deficits do not just manifest in the adaptation policy sector but across different policy domains, which structure the parameters of climate change adaptation. High uneven vulnerability to climate change is a deeply political phenomenon, which is locked into both political systems of China and the United States.

Conflicting interpretations of vulnerability

Central concepts and the evolution of vulnerability conceptualizations related to climate change and human exposure were briefly revisited in Chapter 2. There is an array of academic work with different vulnerability underpinnings, which crucially inform (adaptation) policy responses and (governmental) interventions. A major challenge this book was facing was to deal with the different, but often times overlapping vulnerability understandings. The book derived the main conceptualization of vulnerability from works in the political ecology, viewing vulnerability as a relational concept and embodied experience. However, when interviewing local decision-makers on matters related to vulnerability, it became visible that vulnerability definitions are themselves a relational boundary object. Definitions and understandings of vulnerability were construed in relation to understandings at the global, national and/or regional level. Making different understandings of vulnerability transparent and how they determine rationales for action will continue to be a core task of research and policy practice in the future. The chapter aimed to argue for the importance of reconciling adaptation scholarship with critical perspectives on adaptation but also with opportunities for policymaking, as the state has important roles to play in mediating the uneven distribution of vulnerability.

Lock-ins to examine adaptation deficits and the manifestation of uneven vulnerability

The complex challenge of addressing system imminent adaptation deficits is commonly examined by research on adaptation barriers. Because these explanatory attempts are often symptomatic and do not touch upon the root causes of how barriers emerge and why they persist, the emerging framework of lock-ins was introduced as main analytical framework in Chapter 3. Lock-ins intend to explain the persistence of policy problems by looking at interrelated path-dependencies and factors, which are locked in at deeper levels of our society and political systems. The analysis of lock-ins enables a systemic perspective by analyzing feedback mechanisms, which can hinder effective adaptation policy-making and/or structure its current parameters. They also help to understand, why uneven vulnerability is a protracted phenomenon and why vulnerability patterns are so difficult to break. This analytical perspective helps our understanding of factors, which prevent larger shifts from dominant pathways. The lock-ins were conceptualized in a political ecology lens. As a result, they were viewed as products of how the political economy and nature-society relations operate.

Adaptation planning in both countries

The background chapter (5) presented the status of adaptation planning at the international level and in both countries. Compared to other policy sectors, climate change adaptation is a relatively new policy field that is picking up traction internationally. In the United States, the adaptation policy domain presents a very mixed picture. The U.S. federal government launched institutional adaptation efforts earlier than China in the early 2000s, most of which have come to a standstill during the Trump administration. In light of the partisan divide on the issue, "all adaptation is local" seems to have become the dominating paradigm in the United States, bringing forward an array of non-state and local government action and also broadening the conceptual field of climate change adaptation by encompassing urban resilience. In China, adapting to changing environments has been part of much of the country's history. Here, climate change adaptation as a specific policy domain is a new concept. In the last decade, political climate change adaptation efforts have diversified across the country and policy efforts continue to be mandated in a top-down manner with backlog demands of local expertise. The central government launched several top-down adaptation efforts in the 2010s, such as a macro-policy framework on adaptation (2013) and several adaptation pilots at the municipal and provincial levels. Overall, China has placed a strong focus on food security, urbanization and water. Engineering and technological adaptation measures dominated much of China's earlier adaptation efforts with recent shifts to ecosystem-based adaptation.

Inequality patterns in both countries

Congruent with the theoretical framework, Chapter 6 argued that we also have to look at broader inequality patterns as manifested in the political economy for understanding vulnerability patterns. They structure the parameters of adaptation when intending to address uneven vulnerability to climate change. Factors which were often outlined to lead to higher social vulnerability to climate change are ethnicity, education and socioeconomic backgrounds. However, these categories too, are only incremental explanations, as they are often related to challenges of access to public goods. Atlanta, for instance, is characterized by limited populations' access to public transportation, health, affordable housing and education – a phenomenon, which is strongly segregation-based. In Jinhua, limited access seemed to be restricted to education and health but unfolded to a greater extent culturally. The rural–urban divide and processes of cultural devaluation appear to be powerful components of the way society is organized in Jinhua and in less urbanized, formerly agricultural regions. Atlanta and Jinhua are examples of broader inequality patterns, which have their origin in the political economies of China and the United States. Both countries stand out in their high levels of inequality when compared to other industrialized countries and emerging economies.

Biased problem recognition and politically sensitive populations

Chapter 7 focused on the problem recognition of local decision-makers in Atlanta and Jinhua regarding social vulnerability to climate change. This part was interested in learning about how the social vulnerability perspective, despite its academic reemergence within the adaptation field, is operationalized in local political practice. It was found that some political practices protract social vulnerability even further in terms of stigmatizing certain populations, and assigning vulnerability in a top-down manner. Based upon the initial analysis, lock-ins were analyzed. This chapter compared local connotations of human vulnerability to climate change and examined the political practicality of social vulnerability assessments together with vulnerability lock-ins. The awareness and problem understanding of local officials about uneven human vulnerability coupled with growing climate impacts were rather low. In Atlanta, the awareness about social vulnerability existed in fragments but was politically sensitive in the context of climate change. In Jinhua, there was a broader lack of understanding and considering (differential) human vulnerability among government officials. It seemed that decision-makers in both contexts deliberately deemphasized and neglected social vulnerability due to reasons linked to political legitimacy. Additional research is needed to verify this claim.

Another impression suggests that even in cases where awareness about the uneven impacts on society was deliberate and high, it remained detached from climate change and thus did not result in corresponding policy responses. This was partially explained through epistemological lock-ins coupled with political power relations. The way vulnerability is ascribed externally onto people not just constitutes a problematic scientific practice but has not proven conducive in the political arena.

Accidental adaptation measures

When examining how decision-makers perceive climate change adaptation in Atlanta, it became clear that a politically motivated bias shaped problem recognition. In Atlanta, climate change was considered a politically sensitive issue with a lack of scientific understanding and shared knowledge of local decision-makers. In some cases, awareness was more pronounced, but the political environment was provided as major reason for the current lack of deliberate adaptation planning. As a result, and although planned differently initially, Atlanta's municipal government opted for the adoption of "less sensitive language" in terms of omitting some climate change related terms, as part of a broader resilience strategy. In light of a major change in the municipal administration, however, the resilience strategy appears to be a largely discontinued effort with no follow-up mechanisms put in place, for actually implementing the different visions the plan had in mind.

Aside from political strategies at the municipal level, certain instances of "accidental adaptation" were observed. One example is the Beltline, the reconstruction of a former railway corridor into a multiuse rail that now connects different neighborhoods and is helpful in abating some of the extreme heat the city has been confronted by. This project was based on the master thesis of former student Ryan Gravel. It was retroactively referred to as quasi-adaptation project. Despite the appraisal of the transformative planning of the Beltline, conflicts of interests and a disconnect with the initial vision have overshadowed the project. It has recently been debated as an instance of neoliberal governance that persists with its growth first approach, instead of implementing its initial targets some of which were tailored to creating a higher amount of affordable housing units and enabling equal access to transportation. Since its construction, the Beltline is said to have been confronted by the political reality on the ground, which is characterized by select participation and a focus away from community advocacy to fundraising.

The construction of floodwater parks and creation of green infrastructure parks is yet another example of accidental adaptation. One flagship is the Old Fourth Ward Park (O4W), located in a historical, formerly predominantly African American district in central Atlanta, also known as home district of Martin Luther King. These types of parks intend to cope with the increasing floodwater run-off as a result of more extreme precipitation and are also an effort of revitalizing certain neighborhoods. However, they too have come at a considerable cost by relocating predominantly African-American populations, some of whom could no longer afford the rising housing prices in light of rising property values. The chapter shows that what was initially, though accidentally, designed as adaptation measure stands in conflict with other adaptation targets related to social vulnerability. Though alleviating some of the stress related to extreme precipitation, these accidental adaptation measures present instances of green gentrification. In the sense of avoiding vulnerability reproduction, it can be argued that they can be considered an instance of maladaptation through displacing residents, who can no longer afford the rising housing prices.

Despite the lack of precautionary measures as part of existing policy efforts, some actors within the city have been well aware and active on the nexus of climate change and racial justice. This is exhibited as part of a recent project run by two higher education institutions to map urban heat island with community science, which was also supported by the city of Atlanta.

Jinhua stands in contrast, where local officials had only little awareness about the distributional effects of climate change. Despite building large-scale ecosystem-based forms of adaptation in the city center and becoming China's first sponge city pilot, awareness outside the water sector appears to be rather low. Decision-makers pointed to the need to develop first and argue that instructions from higher political levels were lacking before any further adaptation action could be taken. Overall, the political awareness and integrated policy responses to flooding appeared to be more pronounced. This was especially evident in the city's take on civil engineering approaches and treating water in an ecological and holistic way. The city is trying to set new ecological standards in the context of provincial guidelines of a new urbanization development strategy. The cross-linking of different policy efforts between the provincial and municipal levels stands in contrast to efforts taken in Atlanta, where no support was expected due to the traditional climate skeptic government. The Five Water Treatment presents one of the more recent policy efforts that were driven by the provincial level. It is a cross-sectoral strategy that involves different departments and aspects of water treatment. In addition to the city's sponge city planning and ecological corridor, it marks another example of Jinhua's (and China's) shift between hard engineering solutions and nature-based forms of adaptation.

In both cases, existing adaptation efforts were embedded in a rationale for (re) development. Accidental adaptation measures were strongly signified by green gentrification approaches and little awareness about the distributional effects of climate change and policies alike. At best, existing policy efforts were only incrementally related to social vulnerability and in the case of Atlanta even present an instance of maladaptation.

The findings were explained with lock-ins at the intersection of knowledge and politics. Three facets of these types of lock-ins were presented: 1) The dominance of certain knowledge paths, 2) the lack of shared knowledge and 3) the perception of overcomplexity. In China, the adequate handling of floods is well known as an important trajectory for political legitimacy. Good water treatment is deeply embedded in China's evolving mandatory target system and political career opportunities. This has mattered in terms of the political motivation of local decision-makers. It is also the result of natural-science-focused knowledge paths. Social science studies continue to be largely absent in China and not informing adaptation processes. The engineering knowledge perspective and access to better education, often equated with technology-heavy focuses, is considered an important marker for good job opportunities.

In Atlanta too, the understanding of flood management in terms of addressing traditional infrastructure seemed to be more pronounced. Heat and social vulnerability to climate change were considered nontraditional topics, for which no good practices existed. But to what extent, this actually is a result of path-dependent knowledge paths needs further research. Chapter 7 hinted at the psycho-social components and aspects of experiential knowledge based on limited representation of some groups within decision-making were pointed out. One interviewee was cited who emphasized that the "total lack of knowledge about a black uneducated teenager, makes governments somewhat resistant. They lack experience. The political machine represents one group of the society." This lack of representation was often considered an extension of limited access to knowledge which also results in limited access to better paying jobs and has the interests of certain parts of the population better represented.

Further aspects related to knowledge include the lack of a shared understanding, also as a result of a science-knowledge gap and how knowledge is distributed to decision-makers. The findings illustrate the importance of improved access to institutions that disseminate knowledge, knowledge co-creation in the planning process and learning how to deal with complex and nontraditional challenges, such as heat and uneven vulnerability to climate change.

Political epistemological lock-ins

Because the empirical cases both presented lock-ins at the intersection of knowledge and politics, Chapter 9 discussed the findings in further detail, by picking up the discussion of newly emerging research in the field of political epistemology. In the cases examined, marginality and vulnerability appear to be protracted phenomena that are the result of how knowledge, political actors and institutions operate as a function and intertwined aspects of the political ecology. Initial findings of the last chapter suggest these also impact other factors such as built physical and critical infrastructure and cultural components (or what Taylor (2015) would call: the way we collectively choose to reproduce ourselves). This chapter uncovered some of the factors that are locked in at the political level, such as selected acknowledgment and cultural stigmatizationted to only selective vulnerability recognition. Political epistemological lock-ins were detected in the way knowledge is produced, used and processed, who has access to education and how etymological legacies operate. It became obvious that knowledge is power for those who can access it and that marginalized groups are often in a position of epistemic disadvantage. The examined cases reveal that the production of vulnerability and adaptation knowledge is problematic, as it reflects a form of one-sided epistemic agency. Initial findings suggest that the formation of different classes has become apparent in both political contexts, which fundamentally relates to differential access to resources such as education and being culturally devaluated and stigmatized through not belonging to the "invulnerable."

Adaptation policy and transformation?

Chapter 10 placed the empirical findings in some of the discourses on transformative adaptation. The first chapter section laid out why climate change adaptation is an inherently political concept and why greater attention needs to be devoted to questions of power and conflict. Next, transformation was presented as a contested knowledge process and the need for self-critical adaptation studies pointed out. The next section revisited some of the dominant discourses on transformative adaptation to place the main argument. Many different notions of transformative adaptation exist. On a meta-level, the two dominant normative aims either revolve around reducing the vulnerability that is already there, e.g., ameliorating the social vulnerability and/or retrofitting existing structures, for instance, distributing water bottles in periods of extreme heat, or avoiding the reproduction of vulnerability that does not yet exist. The latter is often understood as addressing the root causes of vulnerability. In light of rapidly intensifying climate change and the growing awareness that vulnerability is distributed very unevenly across society, there has been an ever-growing interest in finding empirical cases that speak to these two ideas of transformative adaptation from the perspective of vulnerable populations. But when looking at ameliorating uneven vulnerability or avoiding the vulnerability reproduction of people, empirical transformative cases are not readily available. Both are highly complex endeavors – also because the persistence of uneven vulnerability is a deeply path-dependent phenomenon that has various roots, some of which are deeply political, as this book aimed to show.

Knowledge was presented as a fundamental aspect that matters for transformative adaptation. Access to shaping dominant knowledge patterns and discourses related to vulnerability and adaptation were discussed as an important element. Power differentials still determine who gets to assign vulnerability onto others and who gets to determine adaption policy-processes. This begs us to investigate opportunities for educational reform and more egalitarian access to education, as this kind of access corresponds with access to decision-making.

This chapter also discussed some of the implications for dominant scientific practices and what science can do differently, in terms of verifying dominant vulnerability categories, and making underlying assumptions visible. At the methodological level, it became visible how assumptions related to the concept of social vulnerability operate, are largely intransparent and contain questionable conceptual elements. There is a tendency to favor the existent status quo and dominant institutional order (also in line with Oels 2019). SVAs appear to be politically unsustainable in their current form. The examination of decision-makers' perception of human vulnerability to climate change uncovers problematic practices of politicizing certain parts of the population. Thus, socially vulnerable populations, which were identified through the SVAs were deemed to be politically sensitive and were not necessarily considered to qualify for protection. The identification of vulnerable populations fails to result in a fundamental change in how vulnerability reproduction operates. The political process of naming and (partially) acknowledging vulnerable parts of society often coincides with legitimizing the lack of action at the structural level. Thereby vulnerability recognition serves the function of diverting attention away from the underlying factors, enabling a further discharge of responsibility. Lastly, SVAs fail to examine why certain parts of the population continue to be more exposed than others. The findings coincide with a growing critique on the broader practice of vulnerability recognition and identification, which calls for disclosing the causal mechanisms and path-dependencies (Fraser 1995; Moser and Ekstrom 2010; Ribot 2011, 2014; Pulido et al. 2016).

Next, the role policymaking and the state can play in addressing transformative adaptation was explored. What framework conditions can the state set? What role can governments play in addressing uneven vulnerability? Aside from educational reform and offering more egalitarian access to education and decision-making structures, an even higher hanging fruit would begin by investigating the nature of our current economic system, the economic values and how they condition interest-driven path-dependency and uneven vulnerability. What is the corresponding political ideology and what are the related political values of how we constitute ourselves as a society?

Looking at the tree of low-hanging policy fruits, the early inclusion of local populations for assessing local vulnerability and adaptation responses was argued for as an important means to meaningfully structure adaptation responses. In this context, it is important to mention that knowledge co-creation also has limitations. When initiating participation as part of adaptation planning, one must be aware of some of the potential procedural limitations by self-critically asking: Who is (un)able to participate and whose interest are (not) being served? At the same time, these procedural elements of adaptation planning also do not speak to the necessity of initiating structural change at higher political levels and across different policy sectors, which are complementary to local efforts on the ground.

Further, the need for constant policy iteration, self-critical evaluations of what has been done and increasing the learning capacity of governmental actors to deal with complex problems was argued for. The last part of this section presented some policy tools that are available to assess the adaptive capacity of state institutions – which is an important prerequisite for future adaptation options.

Future research

Although lock-ins already go a step further by not just describing and categorizing vulnerability but also explaining path dependencies and the combined nature of factors, which act in agglomeration, a more systematic approach of looking at the combination of lock-ins and their effects is needed. This book could only touch upon some of the aspects of lock-ins that were conceptualized as a source of the political ecology. A crucial step for further research would be to follow up on how lock-ins are interconnected, supported by systematic methods for guiding these multicausal lines of research. Chapter 5 partially looked at other lock-in factors, such as built critical and physical infrastructure, cultural patterns as well as psycho-social processes. These research angles could be systematically deepened. A follow-up study could also examine how these factors are interlinked, where they overlap and where critical junctures may be detected.

Identifying mechanisms for unlocking

The perspective of transformation may help to explain policy processes of vulnerability and adaptation but may not necessarily help to identify and trigger such transformations (Keskitalo and Preston 2019). Little research exists on how to operationalize transformative adaptation through what kind of political institutions. The lock-in perspective is helpful to identify the intersections of different systems, which may end up being the critical junctures that need addressing.

Corvellec et al. (2013) examine opportunities to unlock infrastructures in the case of waste incineration in Göteborg. The recommendations are grounded on the understanding that the way urban infrastructures are demanded and planned has an unlocking potential and that path-dependence can be deliberately terminated (Corvellec et al. 2013). To overcome a lock-in in the chosen example, policymakers and city managers have to reorient an array of interrelated factors. According to this understanding, unlocking processes seek to increase the adaptive capacity of political institutions and ultimately reach transformative adaptation.

Hölscher et al. (2019) examine unlocking capacities as the ability of actors to recognize and dismantle structural drivers of unsustainable path dependencies and maladaptation. They define transformative capacity by the abilities of actors to develop and test new ideas and "embed them in structures, practices and discourses" (Hölscher et al. 2019: 796). Conducting a literature overview of existing studies on critical junctures and pairing it with the empirical findings of this study could potentially culminate in detecting mechanisms for unlocking.

Policy formulation tools

Policy formulation tools and earlier stages of the policy process need more attention (Jordan et al. 2015). Future researchers could focus on analyzing other non-pioneers and improving the related research methodology. Further studies could examine more closely how policy expertise is generated for the policy formulation process. It would be interesting to learn about adaptation-specific policy formulation tools and differentiating those with distinct characteristics from the broader policy toolbox. In China, further research is needed on the local specificities of environmental target setting and to what extent climate adaptation has or could become a concern in the evaluation of local political performance. Exploring geographic and local differences will expand the understanding of differences across local policymaking processes.

Cultural habits and social practices

Local practices of (indigenous) self-governance are another important but commonly overlooked and underestimated field aside from community-based adaptation, which often ends up being quasi-governmental. Means of self-governance and cultural practices were not reflected upon but are important (also see, e.g., Scott 1988). In the adaptation context, different elements have been researched, such as cultural forms of knowledge (e.g., Yeh 2016; Heimann 2019) or indigenous health vulnerabilities (e.g., Berrang-Ford et al. 2012). The idea of being socioeconomically deprived and equating this with a lack of adaptive capacity neglects other sources of ideas or cultural habits certain groups might have (also see Wolf et al. 2009). Studying informal adaptation practices outside the governmental realm is important to understand other impact factors, which can foster resilience. Exploring which cultural factors contribute to population's autarchy would enrich the debate and move away from perceiving local populations as merely governable entities. At the same time, governmental interventions can have negative impacts on the adaptive strategies that communities developed in response to climate change, thereby constituting examples of maladaptation. This has for instance been shown in the context of drought management practices that herders in Inner Mongolia developed in response to frequent droughts (Li and Li 2012; Zhang et al. 2013). Here, external interventions and the government-led privatization of land-use rights led to an elimination of the herders' ability to rely on traditional strategies and weakened different forms of capital that provided a high adaptive capability (Zhang et al. 2013). Further research is needed not just regarding the governmental co-creation of vulnerability in terms of broader political economy patterns, but also more specifically looking at how specific governmental interventions impact populations' means of self-governance.

Accepting vulnerability as a human condition

No validation studies were conducted analyzing how people view and conceptualize their own vulnerability. Follow-up surveys could investigate the perceptions and visions of local populations regarding their susceptibility, risk perception and awareness of local populations regarding extreme heat and other climate impacts and pair these with the studies at hand as well as the perceptions of local decision-makers. Being sensitive to community-based self-perception and developing a shared understanding of societal exposure is important when designing policies and developing population-targeted solutions without disempowering or colonizing. Focusing efforts on ameliorating the uneven distribution of human susceptibility to climate change is the task governments should be set out to do. Accepting that vulnerability is an inherent human condition and rather focusing on the uneven distribution of priviliges could be one step into the right direction. The findings support Bergoffen's demand for a deconstruction of vulnerability embodiment:

> So long as vulnerability is conceptualized negatively, experienced as shameful and equated with victimhood, we will attempt to flee it.
>
> (Bergoffen 2016: 137)

References

Bauriedl, Sybille, and Detlef Müller-Mahn. "Conclusion: The Politics in Critical Adaptation Research." In *A Critical Approach to Climate Change Adaptation Discourses, Policies and Practices*, edited by Silja Klepp and Libertad Chavey-Rodriguez. London: Routledge, 2018.

Bergoffen, Debra. "The Flight from Vulnerability". In *Dem Erleben auf der Spur. Feminismus und die Philosophie des Leibes*, edited by Hilge Landweer and Isabella Marcinski, 137–152. Bielefeld: Transcript Verlag/De Gruyter, 2016.

Berrang-Ford, Lea, Kathryn Dingle, James D. Ford, Celine Lee, Shuaib Lwasa, Didas B. Namanya, Jim Henderson et al. "Vulnerability of Indigenous Health to Climate Change: A Case Study of Uganda's Batwa Pygmies." *Social Science & Medicine* 75, no. 6 (2012): 1067–1077. https://doi.org/10.1016/j.socscimed.2012.04.016.

Corvellec, Hervé, María José Zapata Camposa, and Patrik Zapata. "Infrastructures, Lock-In, and Sustainable Urban Development: The Case of Waste Incineration in the Göteborg Metropolitan Area." *Journal of Cleaner Production* 50 (2013): 32–39. https://doi.org/10.1016/j.jclepro.2012.12.009.

Fraser, Nancy. "From Redistribution to Recognition? Dilemmas of Justice in a 'Post-Socialist' Age." *New Left Review* 1, no. 212 (1995): 68–93.

Heimann, Thorsten. *Culture, Space and Climate Change. Vulnerability and Resilience in European Coastal Areas.* Series: Routledge Advances in Climate Change Research. London and New York: Routledge, 2019.

Hölscher, Katharina, Niki Frantzeskaki, and Derk Loorbach. "Steering Transformations under Climate Change: Capacities for Transformative Climate Governance and the Case of Rotterdam, the Netherlands." *Regional Environmental Change* 19 (2019): 791–805.

Jordan, Andrew J., and John R. Turnpenny. *The Tools of Policy Formulation: Actors, Capacities, Venues and Effects.* Cheltenham: Edward Elgar, 2015.

Keskitalo, E. Carina H., and Benjamin L. Preston. *Research Handbook on Climate Change Adaptation Policy.* Cheltenham, UK, and Northampton, MA: Edward Elgar Press, 2019.

Klepp, Silja, and Libertad Chavez-Rodriguez. *A Critical Approach to Climate Change Adaptation Discourses, Policies and Practices.* London: Routledge, 2018.

Li, Wenjun, and Yanbo Li. "Managing Rangeland as a Complex System: How Government Interventions Decouple Social Systems from Ecological Systems." *Ecology and Society* 17, no. 1 (2012): 1–15. http://dx.doi.org/10.5751/ES-04531-170109.

Moser, Susanne C., and Julia A. Ekstrom. "A Framework to Diagnose Barriers to Climate Change Adaptation." *Proceedings in the National Academies of Sciences* 107, no. 51 (2010): 22026–22031. Doi: 10.1073/pnas.1007887107.

Oels, Angela. "The Promise and Limits of Participation in Adaptation Governance: Moving Beyond Participation towards Disruption." In *Research Handbook on Climate Change Adaptation Policy,* edited by E. Carina H. Keskitalo and Benjamin L. Preston. Northampton: Edward Elgar Press, 2019.

Pulido, Laura, Ellen Kohl, and Nicole-Marie Cotton. "State Regulation and Environmental Justice: The Need for Strategy Reassessment." *Capitalism Nature Socialism* (2016). Doi: 10.1080/10455752.2016.1146782

Ribot, Jesse. "Editorial. Vulnerability Before Adaptation: Toward Transformative Climate Action." *Global Environmental Change* 21 (2011): 1160–1162. Doi: 10.1016/j.gloenvcha.2011.07.008.

Ribot, Jesse. "Cause and Response: Vulnerability and Climate in the Anthropocene." *The Journal of Peasant Studies* 41, no. 5 (2014): 667–705. Doi: 10.1080/03066150.2014.894911.

Scott, James C. *The Art of Not Being Governed: An Anarchist History of Upland Southeast Asia.* New Haven, CT: Yale University Press, 1988.

Taylor, Marcus. *The Political Ecology of Climate Change Adaptation. Livelihoods, Agrarian Change and the Conflict of Development.* London and New York: Routledge Explorations in Development Studies, 2015.

Wolf, Johanna, Irene Lorenzoni, Roger Few, Vanessa Abrahamson, and Rosalind Raine. "Conceptual and Practical Barriers to Adaptation: Vulnerability and Response to Heat

Waves in the UK." In *Adapting to Climate Change*, edited by Adger et al., 181–197. New York: Cambridge University Press, 2009.

Yeh, Emily T. "How Can Experience of Local Residents Be "Knowledge"?' Challenges in Interdisciplinary Climate Change Research." *Researching the Hybrid Geographies of Climate Change: Reflections from the Field* 48, no. 1 (2016): 34–40. https://doi.org/10.1111/area.12189.

Zhang, Chengcheng, Wenjun Li, and Mingming Fan. "Adaptation of Herders to Droughts and Privatization of Rangeland-use Rights in the Arid Alxa Left Banner of Inner Mongolia." *Journal of Environmental Management* 126 (2013): 182–190. http://dx.doi.org/10.1016/j.jenvman.2013.04.053.

Appendices

Appendix 1
Biophysical and geographical impacts

Climate, topography and impacts in Georgia

Geographically, the southeast of the United States spans over 11 states and covers a total area of 1,449,455 square kilometers, or about 15 percent of the United States.[1] In line with the "mega-diverse" ecology of Northern America (Sommer 2015), the Southeast is considered a "biodiversity hotspot" characterized by geological complexity (e.g., rock, soil and ecological diversity), high annual rainfall, a wide latitudinal range of seasonal temperature, as well as diverse altitudinal zoning with cooler temperatures in higher altitudes and warmer temperatures in lower spheres. The topography of Georgia is signified by stretches of coastal areas to the East, low-lying areas to the Northwest and diverse mountainous regions in between.

With roughly 160 km (100 miles), Georgia has a relatively short coastline. Approximately one-tenth of Georgia's counties are located within Georgia's Lower Coastal Plain, which is located in the lowest-lying area of the Atlantic coastal plain. Georgia's coast encompasses 14 major barrier islands, which serve important ecosystem functions. The islands do not just act as buffers blocking off ocean waves and winds, but also enable the 988,421 hectares (400,000 acres) of saltwater marshes to flourish, thereby further protecting the mainland. The islands range from Tybee Island in the North to the largest barrier, Cumberland Island, in the South.

North of the coastal plain, the Piedmont plateau is located in the central part of the state, with the Ridge-and-valley Appalachians to the Northwest and the Blue Ridge Mountains as the highest mountain group in the Appalachian highlands to the Northeast. Georgia has a primarily humid subtropical climate and moderate weather with occasional extreme weather. The North Georgia Mountains receive moderate to heavy precipitation with heavier downpour in the northeastern parts of the state. In sum, the climate of the Southeast is considered variable with divergent latitude and elevation, topography and proximity to water resources.

The observed climatic changes in the Southeast include higher exposure to sea-level rise, increasing temperatures and extreme heat events, as well as precipitation and superstorms, such as hurricanes and tornadoes (Ingram et al. 2013; Carter et al. 2014, 2018). The decade of the 2010s is already the historically warmest in all seasons regarding average daily minimum temperature and average daily maximum temperature in the spring and winter periods (Carter et al. 2018). Globally, 2018 was the fourth hottest year on record, with the coastal Southeast being considerably warmer than the rest of the United States (NOAA 2019). The warming trends of the Southeast have been similar to other parts of the country since the 1960s, however, 61 percent of cities in the southeastern region are experiencing worsening heat waves (Carter et al. 2018). Increases in nighttime temperature are significant aside from an observed moderate increase in daily temperatures. Overall, the length of freeze-free seasons across the Southeast has increased between 1950 and 2016 (Carter et al. 2018).

This correlates with earlier findings from Stone (2012), who examines heat vulnerability in the most sprawling U.S. cities. Recently developing and fast-growing cities like Atlanta, Birmingham, Greenville as well as Tampa are all located in the Southeast and experience higher heat-wave frequency than compact cities.

Aside from considerable warming, the region has been confronted by an increase in extreme rainfall since the 1990s with deteriorating trends, particularly since the 1980s. The Southeast is among the regions, where annual precipitation reaches a continental maximum (Carter et al. 2014). Further, the region is confronted by severe weather in the form of thunderstorms, tornadoes and hail more than any other region in the world (IPCC 1997; Carter et al. 2014). The NCA4 reports on the historic 2017 hurricane season during which three major hurricanes made landfall (Harvey, Irma and Maria), with the most significant impacts being felt in the Southeast (Carter et al. 2018). The combined health impacts from vector-borne diseases, heat and flooding determine the particularly high southeastern community vulnerability.

The projected impacts for both lower and higher emission scenarios (RCP 4.5 and RCP 8.5) predict increases in temperature and extreme precipitation across the Southeast (Carter et al. 2018). Concurrently, the number of warm nights will increase, making the region even more suitable for the transmission of certain vector-borne diseases. Under higher emission scenarios, global sea-level rise in the Southeast is projected to exceed 8 feet (2.4 meters) by 2100, with high tide flooding becoming increasingly common, regardless of the scenario (Carter et al. 2018). Tybee Island at the eastern coast of Georgia is one example that has already been affected severely by rising water levels. Local researchers expect the road to Tybee Island to flood up to 60 times per year by the end of 2060 and two-thirds of the island to be under water by the end of the century (Chapman 2016). Aside from impacting coastal regions by intensified degrees, the combined effects of rising sea levels and storm surges will also have greater consequences for the island (Carter et al. 2018).

In their Georgia-specific climate vulnerability assessment, KC et al. (2015) report warming trends in northern Georgia after the mid-1970s and drier conditions with increases in severe drought. For Atlanta, an above-national average increase of heat wave frequency and duration was observed between 1961 and 2010 (Habeeb et al. 2015). Aside from anomalies in temperature and precipitation, extreme climatic hazards such as floods increased in frequency, particularly in Metro Atlanta's Fulton County. Metro Atlanta and coastal counties saw the highest concentration of hydro-climatic events (floods, droughts and heat waves) in the past four decades (KC et al. 2015). With a warming climate, flood frequency as well as rainfall intensity are expected to increase further (also see Carter et al. 2014; Romero-Lankao et al. 2014).

Climate, topography and impacts in Zhejiang

By occupying only 1.06 percent of China's land surface, Zhejiang is one of China's smallest provinces (Shi and Ganne 2009). China has a coastal line of 14,500 km (9,010 miles), out of which eastern Zhejiang province occupies 2,414 km.[2] The province is also known for its water reservoirs and rivers, and over 3,000 islands. Seventy percent of the province's total terrestrial area (101,800 km^2) is covered by mountains and hills (Zhao et al. 2010). The province has a forest coverage of approximately 61 percent and is rich in fertile grasslands and rice paddies.

Zhejiang has a humid subtropical monsoon climate with the rainy season occurring mainly from April to October, reaching its height from May to June. High-intensity rainfall and the province's elevation differences cause frequent landslides and debris flows (Zhao et al. 2010). The crop growing season has profited from the climate and rich natural conditions. With an annual average temperature of 15 to 19 °C (59 to 66 °F) and strong precipitation, the province is considered a sea–land transitional zone in the eastern monsoon region (Wu et al. 2014).

The YRD region is a typical floodplain that was confronted by three major floods in the 1990s (1991, 1998 and 1999) (Ge et al. 2013). A major flood happened in 2010, affecting over 28 provinces with 4,000 people reported dead or missing (Chen 2010). One of the primary objectives of the Three Georges Dam was to minimize the floodwater impact for the downstream regions. Zhejiang, Fujian and Guangdong are the three southeastern coastal provinces most acutely affected by influential tropical cyclones (Wen et al. 2018). Sea-level rise and increased precipitation are serious problems. Zhejiang has experienced the highest rate of sea-level rise in China with an annual increase of 3.3 mm between 2004 and 2006 (Jin and Francisco 2013). Extreme heat events (EHEs) are also likely to worsen across Zhejiang province, due to the combined effects of global warming and rapid urbanization (Hu et al. 2017). Heat health risk greatly varies with particular concentrations in inner city areas, such as the downtowns of Hangzhou, Ningbo and Jinhua (Hu et al. 2017).

Xia et al. (2015) examined the localized climatic changes in the Qiantang River basin and its three sub-basins: Lanjiang, Xin'anjiang and Fuchunjiang, located predominantly in eastern Zhejiang province. The study of Xia et al. (2015) confirms the findings of previous studies, which detect significant warming trends throughout the region in both, mean and extreme, temperatures. The observed changes include an increase in extreme hot events between 1960 and 2006 and extreme cold events of more complex nature. It also finds increased extreme precipitation patterns under all three of the general assessment models it examined (Xia et al.

2015). This is in line with Zhang et al. (2015) who project frequent and severe floods across the Lanjiang catchment in the period 2011–2040 (Zhang et al. 2015). The county-level city of Lanxi and prefectural city of Jinhua are located in the southern part of the Qiantang River basin.

Within the administrative region of Jinhua, the county-level city Lanxi experienced more than 100 floods at the super alert floodwater level from 1949 to 2015. During this period, the next higher stage of floodwater warning, the critical water level was exceeded 41 times (Wang 2016). The region has a historical consciousness about flooding events. The years 1955, 1989, 1992, 1993, 1994, 1997, 1998 and 2011 are known because of their disastrous historical impact, causing severe human and property losses.

The flood of 1955 affected people in about 285 villages. Despite major historical peaks, the displacement of people, the cutoff of entire regions and outfall of critical infrastructure have occurred almost every year. Similar to the flood of 1955 and 1992, 2011 was considered "the flood of the century." A direct economic loss of 1,139 billion RMB was reported for the Jinhua region alone, with 325,000 affected people, and large agricultural losses. Only five years later, 2017 became known as the next "60-year flood." Most of these flooding incidents are marked according to their month of appearance, followed by the day of the event. Thus, the 2017 flood became known as six twenty-five (*liu ershiwu*).

The historical significance of the 2017 flood was reiterated throughout the interviews and considered historically the severest since 1949, as one Jinhua government official remembers:

> Lanxi's flood occurred on June 25th, it is flood six twenty-five, then there was also the six nineteen [in 2011]. Which else? It is just that the six twenty-five was the second largest flood after our country was built and one of the most severe in the past four to nine years.
>
> (I-38: 41)

Besides the United States, China is among the countries most frequently and severely affected by storm surge disasters (Jin et al. 2018). Tropical cyclones (TCs) account for roughly 50 percent of all weather-related economic losses worldwide (Wen et al. 2018). With a significant rise of 1.8 billion RMB per year, TCs caused an estimated direct economic loss of 44.7 billion RMB annually in China between 1984 and 2015 (Wen et al. 2018). Depending on different socioeconomic pathways and emission projections, future losses from TCs are projected to be five times higher at a 1.5 °C global warming level, compared to a sevenfold increase at the 2.0 °C warming level (Wen et al. 2018). Zhejiang and Jinhua are particularly affected by the typhoon season. Local decision-makers had a particularly strong awareness about typhoons and the monsoon season was frequently pointed to in the context of the tourism industry, urbanization as well as ecological replacements. Climate impact studies suggest that awareness is particularly high in the coastal regions due to the direct economic exposure affecting tourism (e.g., Fang et al. 2016).

Notes

1 The states included in the definition of the southeast are based on the categorization of the latest National Climate Assessment (2018). These are Alabama, Arkansas, Florida, Georgia, Kentucky, Louisiana, Mississippi, North Carolina, South Carolina, Tennessee and Virginia. The numbers were retrieved from the last U.S. in Census 2010.
2 For further information about natural resources across Zhejiang province, see Zhejiang Government Official's Web Portal, available at: http://english.zj.gov.cn/col/col1116/index.html, accessed March 2, 2019.

References

Carter, Lynne M., James W. Jones, Leonard Berry, Virginia Burkett, James F. Murley, Jayantha Obeysekera, Paul J. Schramm, and David Wear. "Ch. 17: Southeast and the Caribbean." In *Climate Change Impacts in the United States: The Third National Climate Assessment,* edited by Jerry M. Melillo, Terese (T.C.) Richmond, and Gary W. Yohe, 396–417. U.S. Global Change Research Program, 2014. Doi: 10.7930/J0N- P22CB.

Carter, Lynne M., Adam J. Terando, Kirstin Dow, Kevin Hiers, Kenneth E. Kunkel, Aranzazu R. Lascurain, Doug Marcy, Michael J. Osland, and Paul Schramm "Southeast." In *Impacts, Risks, and Adaptation in the United States: Fourth National Climate Assessment, Volume II,* edited by: David Reidmiller, C. W. Avery, D. R. Easterling, K. E. Kunkel, K. L. M. Lewis, T. K. Maycock, and B. C. Stewart, 743–808. U.S. Global Change Research Program, 2018. https://doi.org/10.7930/NCA4.2018.CH19.

Chapman, Dan. "Tybee Island Acts Against Rising Seas and a Warming Climate." *Atlanta Journal Constitution,* April 15, 2016. https://www.ajc.com/news/state--regional-govt--politics/tybee-island-acts-against-rising-seas-and-warming-climate/9RSm0YO0u0JKog4LzT1OtK/, accessed February 8, 2019.

Chen, Jie. "Transnational Environmental Movement: Impacts on the Green Civil Society in China." *Journal of Contemporary China* 19, no. 65 (2010): 503–523. http://dx.doi.org/10.1080/10670561003666103.

Fang, Yan, Jie Yin, and Bihu Wu. "Flooding Risk Assessment of Coastal Tourist Attractions Affected by Sea Level Rise and Storm Surge: A Case Study in Zhejiang Province, China." *Natural Hazards* 84, no. 1 (2016): 611–624. Doi: 10.1007/s11069-016-2444-4.

Ge, Yi, Wen Dou, Zhihui Gu, Xin Qian, Jinfei Wang, Wei Xu, Peijun Shi, Xiaodong Ming, Xin Zhou, and Yuan Chen. "Assessment of Social Vulnerability to Natural Hazards in the Yangtze River Delta, China." *Stochastic Environmental Research and Risk Assessment* 27 no. 8 (2013): 1899–1908. Doi: 10.1007/s00477-013-0725-y.

Habeeb, Dana, Jason Vargo, and Brian Stone Jr. "Rising Heat Wave Trends in Large US Cities." *Natural Hazards* 76, no. 3 (2015): 1651–1665. Doi: 10.1007/s11069-014-1563-z.

Hu, Kejia, Xuchao Yang, Jieming Zhong, Fangrong Fei, and Jiaguo Qi. "Spatially Explicit Mapping of Heat Health Risk Utilizing Environmental and Socioeconomic Data." *Environmental Science and Technology* 51, no. 3 (2017): 1498–1507. Doi: 10.1021/acs.est.6b04355.

Ingram, Keith T., Kirstin Dow, Lynne Carter, and Julie A. Anderson. "Climate of the Southeast United States: Variability, Change, Impacts, and Vulnerability." *NCA Regional Input Reports book series (NCARIR),* 2013. Doi: 10.5822/978-1-61091-509-0.

IPCC. *The Regional Impacts of Climate Change: An Assessment of Vulnerability.* Edited by Robert T. Watson, Marufu C. Zinyowera, and Richard H. Moss. Cambridge, UK: Cambridge University Press, 1997.

Jin, Jianjun, and Hermi Francisco. "Sea-level Rise Adaptation Measures in Local Communities of Zhejiang Province, China." *Ocean and Coastal Management* 71 (2013): 187–194. Doi: 10.1016/j.ocecoaman.2012.10.020.

Jin, Xue, Xiaoxia Shi, Jintian Gao, Tongbin Xu, and Kedong Yin. "Evaluation of Loss Due to Storm Surge Disasters in China Based on Econometric Model Groups." *International Journal of Environmental Research and Public Health* 15, no. 4(604) (2018). Doi: 10.3390/ijerph15040604.

KC, Binita, Marshall Shepherd, and Cassandra Johnson Gaither. "Climate Change Vulnerability Assessment in Georgia." *Applied Geography* 62 (2015): 62–74. http://dx.doi.org/10.1016/j.apgeog.2015.04.007 0143-6228/.

National Oceanic and Atmospheric Administration (NOAA). *Assessing the Global Climate in 2018. For the Globe, 2018 Becomes Fourth Warmest Year on Record.* National Center for Environmental Information, NOAA, February 6, 2019. www.ncei.noaa.gov/news/global-climate-201812, last accessed February 8, 2019.

Romero-Lankao, P., J. B. Smith, D. J. Davidson, N. S. Diffenbaugh, P. L. Kinney, P. Kirshen, P. Kovacs, and L. Villers Ruiz. "2014: North America." In: *Climate Change 2014: Impacts, Adaptation, and Vulnerability. Part B: Regional Aspects. Contribution of Working Group II to the Fifth Assessment Report of the Intergovernmental Panel on Climate Change* edited by V. R. Barros, C. B. Field, D. J. Dokken et al., 1439–1498. Cambridge, UK and New York, NY: Cambridge University Press, 2014.

Shi, Lu, and Bernhard Ganne. "Understanding the Zhejiang Industrial Clusters: Questions and Re-evaluations." *Asian Industrial Clusters, Global Competitiveness and New Policy Initiatives*, (2009): 239–266. https://www.worldscientific.com/action/showCitFormats?doi=10.1142%2F9789814280136_0009.

Sommer, Bernd (ed.). *Cultural Dynamics of Climate Change and the Environment in Northern America.* Leiden and Boston: Brill, 2015.

Stone, Brian. *The City and the Coming Climate. Climate Change in the Places We Live.* Cambridge and New York: Cambridge University Press, 2012.

Wang, Enkuang. "80 Years Ago, Lanxi Floods Old Photos Were First Discovered." (Original: 八十年前兰溪水灾老照片首次发现.") *Lanxi Official Broadcasting Television Station*, September 12, 2016. http://news.lxzc.net/2016/0912/33493.shtml, last accessed September 14, 2017.

Wen, Shanshan, Yanjun Wang, Buda Su, Chao Gao, Xue Chen, Tong Jiang, Hui Tao, Thomas Fischer, Guojie Wang, and Jianqing Zhai. "Estimation of Economic Losses From Tropical Cyclones in China at 1.5 °C and 2.0 °C Warming Using the Regional Climate Model COSMO-CLM." *International Journal of Climatology* (2018): 1–14. https://doi.org/10.1002/joc.5838.

Wu, Li, Cheng Zhu, Chaogui Zheng, Chunmei Ma, Xinhao Wang, Feng Li, Bing Li, and Kaifeng Li. "Impact of Holocene Climate Change on the Prehistoric Cultures of Zhejiang Region, East China." *Journal of Geographical Science* 24, no. 4 (2014): 669–688. Doi: 10.1007/s11442-014-1112-4.

Xia, Fang, Xingmei Liu, Jianming Xu, Zhonggen Wang, Jingfeng Huang, and Philip C. Brookes. "Trends in the Daily and Extreme Temperatures in the Qiantang River Basin, China." *International Journal of Climatology* 35, no. 1 (2015): 57–68. https://doi.org/10.1002/joc.3962.

Zhang, Xujie, Martijn J. Booij, and Yue-Ping Xu. "Improved Simulation of Peak Flows under Climate Change: Postprocessing or Composite Objective Calibration?" *Journal of Hydrometeorology* 16 (2015): 2187–2208. https://doi.org/10.1175/JHM-D-14-0218.1.

Zhao, Yuelei, Hongjuan Yang, and F. Wei. "Soil Moisture Retrieval with Remote Sensing Images for Debris Flow Forecast in Humid Regions." Conference Paper, Conference: DEBRIS FLOW 67, 2010. Doi: 10.2495/DEB100081.

Appendix 2

Table A2.1 List of interview partners, anonymized

Interview number, code and date	Comments
US-based research phase I: (n = 11)	
I-01 (October 17, 2016)	Adaptation expert on questions of adaptation justice, informing case selection
I-02 (October 18, 2016)	Non-state actor and public health policy advisor, informing case selection
I-03 (October 25, 2016)	Non-state actor, working on matters related to climate equity, informing case selection
I-04G (October 25, 2016)	Group interview with two non-state actors and climate policy advisors, informing case selection
I-05G (October 28, 2016)	Group interview with two non-state actors and climate policy advisors, one of them an adaptation expert on Atlanta
I-06 (October 28, 2016)	Adaptation expert on equity issues
I-07 (October 28, 2016)	Academic and policy advisor as well as adaptation expert at the East Coast, informing selection of case unit
I-08 (October 31, 2016)	First interview related to case unit, government official and non-state actor working for an environmental organization in Atlanta
I-09 (October 31, 2016)	Adaptation academic and policy advisor in Georgia
I-10 (November 2, 2016)	Fulton County Government Emergency Management, policy practitioner
I-11 (November 4, 2016)	Georgia state employee working on matters related to the environment and climate change in Atlanta
US-based research phase II: (n= 15)	
I-19 (April 11, 2017)	Employee at the Mayor's Office of Resilience, Atlanta
I-20G (April 28, 2017)	Two interviewees: academics and policy advisors to the municipal climate change plan of Atlanta

(Continued)

Table A2.1 (Continued)

Interview number, code and date	Comments
I-21 (May 2, 2017)	Atlanta-based academic and policy advisor on urban climate adaptation
I-22 (May 3, 2017)	Employee at the Atlanta's Mayor's Office of Resilience
I-23-US-ATLM-AC-PP-EM (May 3, 2017)	Senior Administrator at the Office of Critical Event Preparedness and Response, Emory University, providing leadership in emergency management and disaster response throughout the Atlanta Metropolitan Area
I-24 (May 10, 2017)	Fulton County government official
I-25 (May 11, 2017)	Fulton County government official
I-26 (May 11, 2017)	U.S.-based Chinese academic working on vulnerable populations
I-27 (May 12, 2017)	State actor and employee working at the environmental protection division at the Department of Natural Resources in Atlanta
I-28 (May 12, 2017)	Fulton County government official
I-29 (May 16, 2017)	Atlanta-based adaptation expert, academic and policy advisor
I-30 (May 16, 2017)	Fulton County government official
I-31 (May 17, 2017)	Employee at an Atlanta Interstate Agency working on climate resilience
I-32 (May 18, 2017)	Atlanta city planner and policy advisor
I-33 (May 19, 2017)	Government official and member of the Atlanta City Council
Pre-China research phase: (05/07/2016) (n=13), informing case selection	
I-00 (June 5, 2016)	China adaptation expert (academic and policy advisor)
GI-01 (July 5, 2016)	Group interview with government officials at the Shanghai political leadership academy and emergency crisis lab, informing case selection
GI-02 (July 5, 2016)	Group interview with academics at the Shanghai Social Sciences think tank
GI-03-CH-SH-NGO-PA/PP (July 8, 2016)	Not used
GI-04-CH-SH- SSC (July 8, 2016)	Group interview Shanghai Institute for Development and Strategic Studies
GI-04-CH-WL-GO July 9, 2016)	Not used, coding error
GI-05-CH-CQ-CCP-AC (July 11, 2016)	Chongqing political leadership academy
GI-06-CH-BJ-CCP-AC-PP/PA (July 13, 2016)	Beijing Party school research center
GI-07-CH-BJ-AC (July 13, 2016)	Not used
GI-08-CH-BJ-CCP-SPMO (July 14, 2016)	Beijing political leadership academy
GI-09-CH-BJ-NGO (July 14, 2016)	Not used
GI-10-CH-BJ-GO (July 15, 2016)	High-ranking government official
GI-11-CH-BJ-GO (July 15, 2016)	High-ranking government official

Interview number, code and date	Comments
China-based research phase I: (n= 7)	
I-12 (March 7, 2017)	Hangzhou-based academic and policy advisor to the provincial government
I-13 (March 7, 2017)	Zhejiang-based academic and policy advisor
I-14 (March 7, 2017)	Zhejiang-based academic and policy advisor
I-15 (March 9, 2017)	Academic researching climate change policies and emergency management throughout Zhejiang
I-16 (March 14, 2017)	Academic and Researcher at the Shanghai Office of Meteorology
I-17 (March 14, 2017)	Resident of Hangzhou and former citizen of Jinhua City
I-18 (March 22, 2017)	Chinese Teacher
China-based research phase II: (n= 7)	
I-34 (July 25, 2017)	Academic and policy advisor working on matters related to local governments
I-35 (August 10, 2017)	Adaptation expert involved in sponge city planning
I-36 (August 29, 2017)	Environmental policy advisor
I-37 (October 20, 2017)	Adaptation expert advising local government
I-38 (November 20, 2017)	Government official at the Jinhua Municipal Government
I-39G (November 21, 2017)	Five Jinhua officials working for a multi-sectoral governmental agency on matters related to flooding, heat and the environment
I-40 (April 27, 2018)	Advisor to China's adaptation pilots

Index

Note: Page numbers in *italics* indicate a figure and page numbers in bold indicate a table on the corresponding page. Page numbers followed by "n" indicate a note.

60-year flood (Jinhua, China) 272
100RC *see* 100 Resilient Cities
100 Resilient Cities (100RC) (United States) 172–176, 202; *also see* World Bank Climate Resilient Cities Program

Academy of Sciences *see* Chinese Academy of Sciences (CAS)
ACCC *see* Adapting to Climate Change in China
ACCCI *see* Adapting to Climate Change in China
ACCCII *see* Adapting to Climate Change in China
accidental adaptation 5, 78; Atlanta 168–181; Jinhua 185–194; lock-ins related to 13, 52, 196–204; Zhejiang 182–185
accidental adaptation measures, China and the United States 259–261
accidental adaptation policy 168–208
Adapting to Climate Change in China (ACCC) 87, 103; Guangzhou as pilot city 199; phase I (ACCCI, 2009 to 2014) 87; phase 2 (ACCCII, 2014 to 2017; 2019) 87, ACCCI and ACCCII 107n3
adaptation: apolitical nature of 10, 14, 255; areas and types of 38–39; nontraditional areas of 51; priority areas of 86; pilots 85–87, 94, 97, 102, 185–186, 188, 257; framework for assessing 37–38; politics of 10; social scientific understanding of 38; governance in China and the United States 82–113; *see also* climate adaptation governance; transformative adaptation
adaptation barriers 5, 7, 11, 36, 39–42, 43, 51, 234, 257; critique of 41–42; Atlanta 174, research in China 48; research in the United States 49–50
Adaptation Clearinghouse database (United States) 177, 250n2
adaptation deficits 6, 7, 10, 11–12, 36; in Atlanta and Jinhua 247; definition of 37; dependent variable of 78; explaining 39, 46, 50, 64; large scale 216, 228; lock-ins as explanation for 255, 257; politics and policies of 216, 218, 242, 247, 256; research on 51; persistence of 61, 74; social vulnerability and 61; studying 52
adaptation justice 10
adaptation limits 42–43
adaptation lock-ins 13, 43–47, 215, 228, 232, 238, 255; theoretical foundations and main analytical framework 36–60, *also see* vulnerability lock-ins
adaptation planning in China and United States 257, *also see* adaptation governance in China and the United States
adaptation policy and transformation 232–254, 261–263
adaptation research 10, 36, 37, 43, 45; foci in China 47–48; in the United States 49, *also see* adaptation studies; apolitical nature of 254
adaptation studies: China 47–48; United States 49–50, 51

Adaptation Task Force *see* Interagency Climate Change Adaptation Task Force (ICCATF) (United States)
Adaptive Capacity Wheel (ACW) 246–247
ADB *see* Asia Development Bank
ACW *see* Adaptive Capacity Wheel
African American districts in Atlanta 179–181, 259
African Americans 28; Hispanics replacing 143
African Americans in Atlanta 117, 120, 138, 143, 149; limited access to education of 224; relocation efforts involving 195; segregation of 156, 207, 226, 249
agenda-setting *see* thematic agenda-setting
Albuquerque, New Mexico 104
Amazons (Brazil) 240
Anhui province, China 48, 86, 198
anthropogenic causes of climate change 20, 29
anthropogenic skeptics 89, 198
ARC *see* Atlanta Regional Commission
Asia Development Bank (ADB) 184–185
Atlanta Beltline 176–178, *see* Beltline
Atlanta Beltline Partnership 177–178, 208n2
Atlanta-Fulton emergency department 146
Atlanta, Georgia 117–125; accidental adaptation 168–181; adaptation policymaking in 52; climate action and resilience plan adapted by 70; composite vulnerability of 139; decentralized political administration of 117–118, 202; decision-making by local government authorities in 63, 114, 165; education in 227; Emory University 73; ethnic minorities in 37; green infrastructure adaptation in 237; Jinhua compared to 7, 67–70, 126–127, 143, 144, **144**, **154**, **196**; housing in 143; inequality in 158–159, 162; interviews conducted in 71; racial segregation of 117, 119, 120, 148, 223, 225; protracted vulnerability of 137; sea-level rise, impact on 100, 104; social and climate vulnerability of 67–68, 136–137; social dimensions of vulnerability of 138, 142–151; social justice in 232; stigmatization of vulnerable populations in 155–156, 162, 164; trends shaping 130; *see also* African Americans in Atlanta; Fulton County;

Metro Atlanta Region; Metropolitan Atlanta Rapid Transit Authority (MARTA)
Atlanta Municipal Government 158
Atlanta Regional Commission (ARC) 119, 146, 171
Austin, Texas 100, 248
authoritarianism: China 7, 64–65, 216; environmental 102, 215, 228; fragmented 7

Baise, China 87
Baltimore, Maryland 248
barrier classification 41
barriers *see* adaptation barriers
barrier thinking 41
Beijing, China 71, 86; *see also* Chinese Academy of Agricultural Sciences; Turenscape
Beltline **176**, 180, 217, 259 *see also* Atlanta Beltline
Bergoffen, Debra 8, 24, 27, 265
bias 78, 145, **154**; importance of unveiling 248–249; inherent 203; political 165, 199, 219, 232, 243, 246, 249; politically motivated 204; structural (US) 163–164; systemic (US) 162
biased perceptions 247
biased problem recognition 258–259
biased vulnerability acknowledgement 154, 156, *157*
Biden administration **85**, 91, **91**, **94**
Bierbaum, Rosina 49
Biesbroek, G. Robbert 37, 39, 50, 243
Birmingham, Alabama 270
blue-green infrastructure 98, 171
Boston, Massachusetts 248
Bottoms, Keisha Lance 175, 208n1
Boykoff, Maxwell 245

CAAS *see* Chinese Academy of Agricultural Sciences
CAPs *see* Climate Action Plans
Castree, Noel 23–24
CCA *see* climate change adaptation
CDC *see* Centers for Disease Control and Prevention (United States)
Centers for Disease Control and Prevention (CDC) (United States) 146, 171, 202
CEQ *see* Council on Environmental Quality (CEQ) (United States) 90
Chai Jiang 30n4

Chambers, Robert 250
Chattahoochee 217; Riverkeepers 171,
Chen, Wenfang 68, 141, 143
Chicago, Illinois 248
China: adaptation efforts, climate-change-
policy-related **85**; adaptation progress
88; adaptation research in 36, 47–48;
agricultural livelihoods in 46; case
study region 67; climate adaptation
governance 84–89; environmental
consciousness in **106**; imperial 182, 192,
198; increasing inequality *115*; National
Climate Assessments **97**; research and
interviews conducted in 71–73, 77–78;
rural-urban transformation of 70; social
vulnerability research in 27–28, 29;
United States compared to 64–67;
water, history of management of 7; *see
also* Adapting to Climate Change in
China (ACCC); Communist Party of
China (CPC); Five Year Plans (FYP);
Hu Jintao; *hukou*; Jinhua; Lanxi; Mao
Zedong; National Adaptation Strategy
(NAS); Xi Jinping; Zhejiang province
Chinese Academy of Agricultural Sciences
(China) (CAAS) 87, 102, 199
Chinese Academy of Sciences (CAS) 87,
96–97, 102
Chongqing Municipality 86
class and class lines 45, 218, 261
class-based society 13.215; reconstitution
of 224–227, 228
Clean Water Act (EPA, United States) 181
Climate Action Plans (CAPs) (United
States) 73, 92; President's 90, 94;
Atlanta 119, 146, 170, 172
climate adaptation governance: current
state of 83–84; in China 84–89; in
the United States 89–92, *also see*
adaptation: governance in China and
the United States
climate adaptation illiteracy 195
climate adaptation plans (United States)
by state 91–92
climate adaptation policies 89; executive
orders related to (United States) **91**
Climate Adaptation Strategies (United
States) 91
climate change adaptation (CCA) 39;
developed nations, focus on 52; field of
(China) 82; field of (United States) 51,
82; planning in the United States 49;
research in China on 47–48

climate change adaptation (CCA)
measures, comparison 94
climate change policy **85**, 91; need to
depoliticize 201
climate change science: Republican
rejection of 89
climate policy research 20–21; political
ecology and 23–24
climate skepticism 67, 83, 89, 145, 200
climate vulnerability assessments (CVAs)
24, 37, 64, 270; China and the United
States 92–94, **95**
climate vulnerability governance, China
and the United States 94
Clinton administration 90, **91**
Clinton, Bill 90
Communist Party of China (CPC) 72, 76
Confucian ideals 192
Corvellec, Hervé 44, 264
Council on Environmental Quality (CEQ)
(United States) 90
CPC *see* Communist Party of China
critical infrastructure lock-ins 116–117; *see
also* lock-ins
cuiruo qunti 147–148 162; *see also qunti*
cultural stigmatization *see* language and
cultural stigmatization
Cutter, Susan 22, 25–26, 30n2, 46, 140

Dabrowski, Marcin 47, 98, 192
data analysis, methodology and approach
73–78
Dayang, China 192
Da Yu (Emperor) 198, 208n6
decentralization 221, 247
decentralized political administration:
Atlanta 117–118, 202
decentralized liberal democracy: United
States as 7, 49, 145
decision-makers 5; administrative 105;
Atlanta 136, 145, 146, 148, 149, 165,
205, 258; climate change skepticism or
partisanship expressed by 200, 201, 208;
interviews of 19, 77; Jinhua 258, 260;
local 29, 30; perceptions of 137; political
143, 144, 169; political bias among 219,
259; social vulnerability assessments
considered by 38, 138, 161; surveyed 47;
vulnerability issues, awareness (or lack
of awareness) of 157, 160, 256
decision-making: education and access to
218, 227, 262; governmental 7, 11, 51;
inaccess to 163–165, 243; incremental

43, 52, 195; Jinhua 184, 189; local 46, 71; personal values' impact on 233–234; political 13, 23, 64, 232, 238
decision-making processes: environmental 102; local 114; public 94; public inclusion in 195, 236; three general phases of 245
decision-making structures 38, 263
DEE *see* Departments of Ecology and Environment (China)
democratic recession 215
democracy 7, 64, 116, 215–216, 228
Deng Xiaoping 222
Denver, Colorado 100, 104
Department for International Development (United Kingdom) 87
Department of Building Energy Efficiency and Technology (China) 184
Department of Natural Resources, Georgia (United States) 163
Department of Planning and Community Development, Atlanta (United States) 171
Department of Watershed Management, Atlanta (United States) 171
Departments of Development and Reform (China) 87
Departments of Ecology and Environment (DEE) (China) 87
dependent variable problem 39
Deutsche Gesellschaft for International Development Cooperation (GIZ) 185
Diekmann, Andreas 61, 75
differentiality 160, 161–162
disenfranchisement *see* political disenfranchisement
displacement: cultural 148; disaster-related 124, 130; non-displacement zone 181
Dow, Kirstin 162
Du, Yaodong 28, 47
Dupuis, Johann 36–37, 41

Eco-agenda (China) 66
Eco-cities (China) 66, 97, 98
Ecological civilization 65, 97–98, 186, 189–190, **191**; eco-civilization perspective (China) 98;
embeddedness in the 192–193; Ecological Civilization Construction (China) 192–193
ecological corridors (China) 188–189, 194
ecological resettlement (China) 130
Eco-provinces (China) 67, 98, 198

Eco-villages (China) 192
EDF *see* Environmental Defense Fund
education 142, 151, 247, 248; access to (China) 158–159, 218; class power and 226–227; discriminatory policies linked to 221; inaccess to (China) 163–164, 221–223; segregation of (United States) 223, 224; transformative adaptation and 235–237
educational attainment and wages 142
educational reform 249, 262, 263
education revolution, China 222
education campaigns 183
environmental actors: China 100–103
environmental authoritarianism *see* authoritarianism
environmental change 20, 21, 24, 100
environmental consciousness **106**
Environmental Defense Fund (EDF) 103
environmental degradation 47, 88
environmental governance 50, **65**, 89, 102, 216
environmental justice 28, 33, **91**, 236; movement 90, **106**, 236
environmental institutions: United States 103–105
environmentalism 189
environmental laws and regulations 105–107; legislation 65, 105, 108n11
environmental migration 98
environmental pollution 193
environmental problem-solving 82
Environmental Protection Bureau, Lanxi 193
Environmental Protection Division *see* Georgia Environmental Protection Division
"Environmental Protection Law of the People's Republic of China" 190
environmental protection targets 88
environmental protests 27
Environmental Protection Agency (EPA) (United States) 90, 94, 103–105, 131n8, 138–139, adaptation plans prepared by 138; *see also* Clean Water Act
EPA *see* Environmental Protection Agency
epistemic agency 220–221, 223, 228, 261
epistemic realism 23
Evans, David 42
Executive Orders (EO) 90–91, 94
expert interviews 11, 62, 71–73
"external populations" in China 150–153

extreme heat and heat waves 3, 36, 40, 65, 76; American Southeast 269; Atlanta's response to 172, 180, 181, 205, 217, 218; China 271; heat islands and 100; increase of 121, 123
extreme heat events (EHE) 271
extreme poverty 115, 116; definition of 130n1
extreme precipitation 171, 180, 181, 218, 270; American Southeast 270; Atlanta's response to 233, 259
extreme temperature 64
extreme weather events 28

federal adaptation framework on climate preparedness and resilience (United States) 90; *see also* Executive Orders
federal agencies (United States) 49, 89, 90, 95, 146 149, 171, 201, 203
federal government 49, 220, 257
federal guidance 83, 96, 195
federal investigation 103, 175
federal climate legislation (United States) 105, 163
federalism 216
federal protocols 204
Federal Transit Authority (FTA) **170**, 171
Fekete, Alexander 239–240
Fengshui 98, 192
Fieldman, Glenn 41
Fineman, Martha 8–9, 27, 161, 242
Five Water Treatment (FWT) (*wushui gongzhi*) 190–192, 203, 260
Five-Year Plans (FYP) (*wunian jihua*) (China) 73, 87–88, 97, 99; 11th 88; 12th 88, 92, 189, **191**; 13th 88, 147; provincial 182
flooding resilience: Atlanta 179–181
floodplain: Atlanta 122; Yangtze River Delta 271
floods and flooding 36, 40, 45, 64, 66; Atlanta's vulnerability and response to 121, 122, 145, 170, 171, 179, 202, 205, 236; Boulder, Colorado 240; China's exposure to and history of 47, 198, 260; coastal 28; Georgia (state of) 144, 270–271; Germany 3, 239; Jinhua's vulnerability and response to 150, 186, 192, 193, 199–200, 218; Lanxi's experience with 194, 272; Qingdao's risk of 88; urban 187; in Southern China 130; urban adaptation to 98; Zhejiang's experience with 130

floodwater and floodwater management: Atlanta 195, 207, 233, 259; China 198; Jinhua *186, 187*, 203, 208, 272; Lanxi 272; Zhejiang 271
"food deserts" 131n9; Atlanta, Georgia 123, 146
food productivity: China 130; United States 149
food security: China 47, 96–98, 99, 194, 257
Ford, James D. 29, 243
Fujian, China 271
Fulton County Board of Commissioners 145
Fulton County, Georgia: Atlanta, location in *118*, 121, 123; composite vulnerability of 138, *139*; income segregation of 225–226; overview of 117; social vulnerability of 67, 68, 142, 144
Fulton County Government 145, 159, 202
Fuchunjiang sub-basin 271

Gansu province 86
GCP *see* Georgia Climate Project
Garrelts, Heiki 236
GDPH *see* Georgia Department of Public Health
GEMA *see* Georgia Emergency Management Agency
geo-engineering 192
geological complexity 269
Georgia Climate Project (GCP) 171
Georgia Department of Public Health (GDPH) 146
Georgia Emergency Management Agency (GEMA) 136, 146
Georgia Environmental Protection Division 122
Georgia (state of), United States 270; *see also* Atlanta, Georgia
Germany: climate adaptation project in 236; energy transition in 43; floods 3, 239; solar power 206
GIZ *see* Deutsche Gesellschaft for International Development Cooperation
Global Change Research Program (United States) *see* USGCRP
Gore, Al 30n4
GPCR *see* Great Proletarian Cultural Revolution
Gravel, Ryan Austin 176–178, 259
Great Proletarian Cultural Revolution (GPCR) (China) 221–222

green agricultural development: China 193
green-blue infrastructure 207 *also see* blue-green infrastructure
green gentrification 233; Atlanta, Georgia 180–181, 195, 259–260
greenhouse gases 25, 102
green households 192
green infrastructure: Atlanta, Georgia 196, 207, 237; *see also* blue-green infrastructure
green infrastructure landscape: Jinhua, China 187
green infrastructure planning: China 185, 193
green infrastructure programs: Atlanta, Georgia 178–179
Green Infrastructure Strategic Action Plan (Atlanta, United States) 172
greening the environment 4, 172, 207
green roofs 188
green schools 192
Green Shield mission (China) 100–101
green spaces 177, 185, 188
Greenville 270
greenwashing 177
Grounded Theory framework 11, 73
Guangdong province 28, 86, 128, 248, 271
Guangxi province, China 86, 87
Gullah Geechee, Georgia 163

Habermas, Jurgen 243
Haihe-Luanhe River basin 105
Hainan province 86
Haiti: earthquake 240
Hangzhou, China 125, 127, 142; extreme heat events in 271; interviews taking place in 71, 150, 151; MOHURD in 185
Harles, John 62, 116, 159
"hazards of place" 140; model 30n2
Hebei province 86
Heilongjiang province 86
herders in Inner Mongolia 248, 265
Hispanics in the United States 28, 117, 138, 143, 226
Huaihe River basin 105
Hölscher, Katharina 6, 264
Hoppe, Robert, 39, 51, 53n1
household registrations *see hukou*
Hui people 28
Hu Jintao 94, 98
hukou (household registrations) 27–28, 156, 160, 226; access to resources and public goods based on 152, 222, 225; implementation of 151, rural 222, 224

Hurricane Harvey 270
Hurricane Irma 270
Hurricane Katrina 124, 239
Hurricane Maria 270
Hurricane Matthew 124
Hurricane Rita 124
hurricanes 65, 124, 205, 239, 269–270

ICCATF *see* Interagency Climate Change Adaptation Task Force
ICLEI *see* Local Governments for Sustainability (former International Council for Local Environmental Initiatives)
ideology 246; capitalist 221; political 263
ideology and knowledge production 239
ideology in science 224, 228
immigrants: United States 152, 164; Zhejiang 128
indigenous knowledge 224
indigenous populations 107, 138; Alaska 100; Atlanta 181
indigenous self-governance 76, 264
inequality patterns in China and the United States 12, 114–117, 130, 226, 258; manifesting as rural-urban divide 127–128; *see also* social inequality
Inner Mongolia 86, 87, 236, 248, 265
Institute of Urban Environment (China) 87
Interagency Climate Change Adaptation Task Force (ICCATF) (United States) 90
Intergovernmental Panel on Climate Change *see* IPCC
interviews, methodology and approach 70–73
IPCC 1–2, 5, 19–25, 29–30, **38**, 42, 80, 84, **85**, 101, 244, 270

Jiangsu province 48, 125, 198
Jiangxi, China 87
Jilin, China 86, 87
Jin, Jianjun 182
Jinhua, China 7, 10, 11; accidental adaptation by 185–194; adaptation deficits in 247; adaptation efforts in 182–197; administrative division of 126; advanced shared knowledge in 203–204; Atlanta compared to 7, 12, 67–70, 114, 126–127, 143, 144, **144**, **154**, **196**, 201, 218; climate-induced weather events, responses to 201; Ecological Corridor

190; education in 227; Five Water
Treatment Department 190–192, 203;
Five Year Plan (12th) 189–190; flooding,
vulnerability and response to 150, 186,
192, 193, 199–200, 218; floodwater and
floodwater management 186, 187, 203,
208, 272; governmental failure in 242;
green infrastructure landscape 187, 237;
heat hazard index 141; housing 143; lack
of knowledge regarding adaptation 197;
lack of provincial pressure in 220; lack
of understanding in 147–148; migrant
communities 36; migrant workers
128–130; overview of 125–127; permissible
conceptions in 149–150; policy efforts **191**;
rural-urban divide 127–128; rural-urban
population density of **129**; sea-level rise 68;
social and special inequality in 226; social
justice in 232; as Sponge City 185–188,
190, 260; urbanization, pressures resulting
from 217; vulnerability recognition in 158;
vulnerability to climate change, perceptions
of 136, 137–138, 140–142, 143–144, 165,
221; water management, approaches to
201, 206, 208; Yanweizhou Flood Park 186;
see also Lanxi, Jinhua prefecture, China;
Pan'an, Jinhua prefecture, China
Jinhua City Master Plan 189
Jinhua Municipal Government (JMPG) 71,
186, 190
JMPG *see* Jinhua Municipal Government
Jones, Lindsey 39–40, 48

Katrina *see* Hurricane Katrina
KC, Binita 28, 67, 68, 122, 138, **140**, 142,
270–271
Keskitalo, E. 25, 43, 234–235, 263
Kibue, Grace W. 48
Klein, Richard J.T. 25
knowledge: and education 157, 158–159,
215, 221, 223, 248; and information 7,
40; past-dependent knowledge patterns
12, 156, 165; paths 13, 197–200, 207,
219, 260–261; and politics 215–216,
220, 228; production 14, 76, 199, 220,
239, 246, 248
Kresge Foundation 49

laboratory findings: social vulnerability
assessments as 160
language and cultural stigmatization
162–163
Lanjiang catchment 271, 272

Lanxi, Jinhua prefecture, China 68, 69,
70, 126–130; flooding risk of 193, 272;
out-migration trends in 217; social
vulnerability of 141–143
Las Vegas, Nevada 104
Liaoning province 86
limits *see* adaptation limits
Lishui, Zhejiang province, China 87, 127,
184, 185
Local Governments for Sustainability
(former International Council for Local
Environmental Initiatives) (ICLEI) 49,
195
lock-ins: accidental adaptation related
to 13, 52, 196–204; across two
different political systems (China and
the United States) 7–8; adaptation
43–44; adaptation deficits and 257;
characteristics of 44–45; contextual
114–130; critical infrastructure
116–117; future research 263–265;
interrelated dimensions of 216–217;
path-dependent 74, 168, 195; political
dimensions of 51–52; political ecology of
1, 11–12, 36, 45, 45–46, 51, 61, 63, 216,
228; political epistemological 218–224,
232, 238; qualities of 6–7; structural 63;
systemic shifts and 9–10; vulnerability
237, 238 *also see* vulnerability and
adaptation lock-ins
lock-ins as related to political institutions
46–47
lock-ins of political epistemology 215–228
lock-ins related to infrastructure 217–218
lock-ins related to knowledge and politics
137, 156–164, 165, 196–204, 207
Los Angeles, California 104, 248

Mandatory Target System (MTS) (China) 88
Mao Zedong 189; educational revolution
under 221–222; geo-engineering
projects for flood control introduced by
192; *hukou* system introduced by 151;
water concerns under 6
MARTA *see* Metropolitan Atlanta Rapid
Transit Authority
Mayor's Office of Sustainability (MOoS)
(Atlanta, Georgia) 171
Mayor's Office of Resilience (MOoR)
(Atlanta, Georgia) 171, 178, 202
McGuire, Chad J. 50
MEE *see* Ministry of Ecology and
Environment (China)

MEM *see* Ministry of Emergency Management

Meng, Meng 98, 192

Metro Atlanta Region 67, **70**; food insecurity of 123; hydro-climatic events experienced by 271; Jinhua, compared to 143, **144**; metropolitan planning organization of 171; racialized segregation of 120; socio-climatic vulnerability of 138, *139*

Metropolitan Atlanta Rapid Transit Authority (MARTA) 119–121, 171, 217

metropolitan planning organization (MPO) (Atlanta, Georgia) 171

Miami, Florida 104

migrant communities: China 189; Jinhua 36

migrant workers (China) 27, 128–130, 150–153, 163; children 158, 249; as "external group" 151, 156; "highly disputed" status of 144, 155, 158; urban hegemony over 226

Ministry of Agriculture (MoA) (China) 86, 187

Ministry of Climate Change and Energy (Switzerland) 87

Ministry of Ecology and Environment (MEE) (China) 87, 100, 102–104, 184

Ministry of Finance (MoF) (China) 86, 187

Ministry of Housing and Urban-Rural Development (MoHURD)(China) 86, 96, 102, 184–185, 187

Ministry of Public Health (China) 96

Ministry of Science and Technology (MOST) (China) 96

Ministry of Transport (MoT) (China) 86

Ministry of Transport, Construction, and Urban Development (Germany) 185

Ministry of Water Resources (MWR) (China) 86, 187

mitigation efforts related to climate change 1, 2, 7, 50; Atlanta 70, 195, 219; China 193; China's Five Year Plan, emphasis on 73, 88, 92; Jihua 190; United States 94; Zhejiang 66, 183

MoA *see* Ministry of Agriculture (China)

MoF *see* Ministry of Finance (China)

MoHURD *see* Ministry of Housing and Urban-Rural Development (China)

monsoon 271–272

Morchain, Daniel 49, 233–234, 235

Moser, Susan 2, 7, 39–40, 42, 49, 51, 83–84, 89, 91, 99–100, 103, 107n2, 235, 243, 245–246, 263

MOST *see* Ministry of Science and Technology (China)

MoT *see* Ministry of Transport (China)

MPO *see* metropolitan planning organization

MTS *see* Mandatory Target System (China)

multicausal explanatory framework 74–75

MWR *see* Ministry of Water Resources (China)

NAACP *see* National Association for the Advancement of Colored People

Nadin, Rebecca 2, 5, 7, 8, 26, 27, 29, 39–40, 83–84, 86–87, 92, 102, 198, 244

NARCC *see* National Assessment Report on Climate Change (China)

NAS *see* National Adaptation Strategy, China

NASA *see* National Aeronautics and Space Administration

National Adaptation Strategy (NAS), China 50, 86, 87, 98, 99, 104, 147, 183

National Aeronautics and Space Administration (NASA) (United States) 95

National Assessment Report on Climate Change (NARCC) (China) 96

National Association for the Advancement of Colored People (NAACP) 236

National Centre on Climate Change Strategy and International Cooperation (China) 87

National Climate Assessments (China) 96, **97**

National Climate Assessments (NCAs) (United States) 95, **96**, 102

National Climate Assessment Three (NCA3) (United States) 65, **96**

National Climate Assessment Four (NCA4) (United States) 94, 95, **96**, 270

National Congress of the Communist Party (China) 99

National Development and Reform Commission (NDRC) (China) 86, 87, 94; "China's Policies . . . on Climate Change" (White Paper) 96, 183; Ministry of Environment and Ecology (MEE) displacing role of 102–103; urban adaptation pilots assessed by 102; "Sponge City" assessment by 188; Urban Adaptation Climate Change Action Plan 184

National Environmental Policy Act (NEPA) (United States) 105

National Oceanic and Atmospheric Administration (NOAA) (United States) 90, 204
National Plan to Address Climate Change (2014–2020) (China) 86
National Science Foundation (United States) 95
Nature Conservancy 171
NOAA *see* National Oceanic and Atmospheric Administration
NDRC *see* National Development and Reform Commission (China)
NEPA *see* National Environmental Policy Act (United States)
New Orleans 104; Hurricane Katrina 124
New York City 248
Ningbo, China 125, 127, 142, 185, 271
Ningxia province, China 86, 87, 236; five-year plan for 99; Hui population of 28; migration 130; poverty 194
non-displacement zone 181
Nussbaumian capabilities approach 10

O4W *see* Old Fourth Ward Park
Obama administration 49, 89–92; Clean Power Plan 103
Obama, Barack 49, 92
O'Brien, Karen 23, 26, 235–236
Oels, Angela 234, 237, 240, 262
Office of Science and Technology Policy (OSTP) (United States) 90
Old Fourth Ward Park (O4W) (Atlanta, Georgia) 179–180, 259
Omodeo, Pietro Daniel 220, 239
ontological monism 23
OSTP *see* Office of Science and Technology Policy
Oxfam 107

Pan'an, Jinhua prefecture, China 68, 69, 126; as cultural offshoot 128, 204; family size, importance of 142; social vulnerability of 141, 142, 143, 189
Partnership for Southern Equity (PSE) (United States) 121, 177
passive adaptation frameworks 242
passive house 185
Pelling, Mark 5, 14n4, *38*, 235
Phoenix, Arizona (United States) 104
Policies and Actions for Addressing Climate Change (China) 183
political bias *see* bias
political disenfranchisement 162, 221, 223

political ecology: climate policy research and 23–24; lock-in dynamics anchored in 1, 11–12, 36, *45*, 45–46, 51, 61, 63, 216, 228; power and agency, insights into 30, 42–43; vulnerability, understanding of grounded in 19, 22
political ecology of adaptation 2
political ecology studies 3, 8, 23, 46
political epistemology 13, 215–228, 228n2, 232, 239, 245–246, 248
poverty: in Jinhua 149–150; suburbanization of 148
power: and agency 43; and knowledge 218, 237–238; political power 101, 137, 140, 158, 178, 226, 258
Preparedness Plans (United States) 92
Preparing the United States for the Impacts of Climate Change (Executive Order 13653) (United States) 90
President's Climate Action Plan of 2013 (United States) 90, **94**
Preston, Benjamin L. 49, 234
PSE *see* Partnership for Southern Equity

Qiantang River basin 271, 272
Qingdao municipality 87
Qingdao System Risk Model 88
Qinghai province 86
Qinghai-Tibetan Plateau 48
qunti 155; *cuiruo* 147–148, 162; *ruoshi* 155, 163

Racial Dot map 120
racial justice 181, 260
racial segregation *see* Atlanta
rainwater 188, 191
Rawls, John 243
RCS *see* River Chief System
Reckwitz, Andreas 226–227
Reid, Kasim 175
Republican party (United States) 89–90, 208
Resilience Strategy (Atlanta, United States) 170, 172–174, 195, 259
Ribot, Jesse 10, 22, 41, 52, 76, 155, 233, 239, 254
River Chief System (RCS), China 198–199
Rockefeller Foundation (RF) 49, 195, 219; 100 Resilient Cities Challenge 172, 175–176, 200
ruoshi qunti 155, 163

San Diego, California 104
Schlosberg, David 10, 235–236, 242

Scott, James C. 76, 264
SCP *see* Sponge City Program
SDGs *see* Sustainable Development Goals
sea-level rise (SLR): Atlanta, Georgia 68, 124; Guangzhou province, China 47; Jinhua, China 68; Southeast United States 65; United States 92; Zhejiang 182, 194
seawall 182
SFGA *see* State Forestry and Grassland Administration
Shandong province, China 87
Shanghai 71, 86, 125, 127; low social vulnerability scores of 142
Shengsi, China 68
Sichuan province 86
Siebenhüner, Bernd 4, 6–7, 43–45, 50, 53n2, 74, 84, 117, 216
SLR *see* sea-level rise
Smithsonian Institution 95
SOA *see* State Oceanic Administration
social capital 142
Social Darwinism 155
social dimensions of vulnerability *see* vulnerability
social drivers of vulnerability *see* vulnerability
social indicators of climate change 2
social inequality 22, 27, 28, 216, 226, 228, 234
socialism 45; post-socialism in China 64, 98
social justice 2, 28, 100
social marginalization 116
social practices 76, 221, 264
social sciences 11, 38, 62; environmental 22; in China 239
social stratification 144
social variables: vulnerability and 10
social vulnerability: bottom-up approach to 22; comparative approach to 9; heightened 61; high 8; lock-ins in the context of 44, 52; manifestation of 5; political drivers of 30; political institutions and 46; politics and 42; pre-existent 21; research questions pertaining to 5–6; root causes of 1; "scientist" nature of 75; socioeconomics of 114; systemic study of, need for 30; term, implications of 255; uneven 12; *see also* Atlanta; Jinhua
social vulnerability assessments (SVAs): China 29; bottom-up approach to 25–26; Fekete's review of 240;

knowledge produced by 160–161; as laboratory findings 160; operationalizing 241; political unsustainability of 262; political utilization of 136–137, 146–147; superficiality of 164
social vulnerability conceptions 29, 30
Social Vulnerability Index (SoVI) 138, 142, 144, 146; Georgia and Fulton County 68; Yangtze River Delta 68, 69, 141
social vulnerability frameworks 137
social vulnerability knowledge: methodological limitations of 160–161
social vulnerability literature 26
social vulnerability research 20; China 27–28; United States 28
social vulnerability variables **140**
social welfare 41, 151, 158
Southeast Atlanta Green Infrastructure Initiative (SAGII) 181
Sponge City program 70, 97, 98, 193, 203–204; Jinhua as pilot 185–188, 190, 260
State Forestry and Grassland Administration (SFGA) (China) 86
"State, Local, and Tribal Leaders Task Force on Climate Preparedness and Resilience" (United States) 90
State Oceanic Administration (SOA) (China) 86
Stone, Brian 2, 100, 117, 122, 224, 270
Stone, Clarence 223, 226
storm surge 100, 270, 272
Suichang, China 68
SVAs *see* social vulnerability assessments

Taizhou, China 185
Tampa, Florida 270
TAR3 *see* Third Assessment Report (China)
Taylor, Marcus 2–4, 8, 10, 19, 23–24, 46, 233, 234, 237, 239, 255, 261
TC *see* tropical cyclones
thematic agenda-setting: political institutions and 163–164
Third Assessment Report (TAR3) (China) 94
Thousand Mile Seawall 182
Three Gorges Area 86
Three Georges Dam 99, 132n18
three-stage model of transformation 14n4
three types of adaptation intent 38
Tianjin Municipality 86
tornadoes 206, 269, 270
tourism 95, 182, 272

transformative actions 5
transformative adaptation 1, 4, 262;
    adaptation policy and 243–248;
    critiques of 237; debates on 9; discourses
    on 248, 262; dominant notions of
    236, 248; knowledge and 232, 235,
    238; knowledge essentialism and 23;
    lock-ins and 45; operationalizing 264;
    policymaking and 263; SVAs and 38
transformative adaptation strategies 242
transformative change: barriers to 40;
    difficulties of 46
transformative learning 246, 247
transformative planning of the Beltline
    177, 259
transformative responses 10
tropical cyclones (TC) 271, 272
Trump administration 82; administrative
    blocking and/or climate-adaptation-
    harming policies of 77, 91, 107, 257;
    federal agency rewriting or removal of
    documents related to climate change
    89, 94
Trump, Donald J. 207; climate skepticism
    of 65
Turnbull, Nick 39, 51n1
Turenscape 186
Tybee Island, Georgia 205
typhoons 88, 150, 218, 272

UHI *see* urban heat island
uneven vulnerability: as adaptation
    challenge 8; ameliorating 262; decision-
    making related to 46; importance of
    institutions/governments in dealing with
    256, 261; inequality and 258; lock-ins
    and 6, 63, 228, 255, 257; politics/policies
    and 241, 242; role of governments in
    addressing 263
UNFCCC *see* United Nations Framework
    Convention on Climate Change
United Nations 116
United Nations Framework Convention
    on Climate Change (UNFCCC) **85**, 98,
    101
United Nations Habitat Cities and Climate
    Change Initiative 49
United Nations Human Rights Council 116
United States: adaptation efforts in
    85; adaptation practices, dominant
    89–96; adaptation progress in 93;
    adaptation research in 49–50; case
    study region in 66; China compared to

64–66; condescending environmental
    institutions in 103; Democratic states
    92; environmental consciousness in
    106, 107; extreme inequality in 116;
    federal agencies 89, 90, 95; field research
    conducted in 11; food production in
    149; increasing inequality in 114, *115*,
    115–116; interviews conducted in 71,
    76, 77; lack of clear policy focus in
    99–100; minorities in 153; partisan
    climate agenda of 201; political
    obliviousness to climate change as
    choice 203; political systems of 7, 145;
    Republican party 89–90, 208; research
    trips to 73; sea-level rise projected in
    124; social vulnerability studies in 28,
    29; thematic priorities in 163; urban-
    rural sphere, shifts in 70; vulnerability
    and adaptation governance in 82–84,
    87, **97**; *see also* 100 Resilient Cities
    program; Atlanta, Georgia; Biden
    administration; Clinton administration;
    Environmental Protection Agency;
    executive orders; Obama administration;
    Trump administration
United States Congress 90–91, 94, 105
United States Global Change Research
    Program (USGCRP) 92, 94–95, 104
Urban Adaptation Climate Change Action
    Plan (NDRC and MOHURD) 184
UrbanHeatATL (Atlanta, Georgia) 181
urban adaptation pilot cities (China)
    184–185
urban heat islands (UHI) 141, 181
urban-rural development *see* Ministry of
    Housing and Urban-Rural Development
    (MoHURD) (China)
USGCRP *see* United States Global Change
    Research Program

VA *see* vulnerability assessments
vulnerable, the 241
vulnerability: adaptation and 255;
    adaptation lock-ins and 36–52;
    conflicting interpretations of 256; as
    constant condition 8; core concepts
    19–30; etymological and cultural
    legacies of 162; high uneven 255, 256;
    main terms used in China to describe
    162–163; non-sectoral perspective on 9;
    political institutions and 27; as relational
    concept 8, 26–27; social dimensions of
    24–25; social drivers of 19, 26; studies

of root causes of 10; *see also* uneven vulnerability
vulnerability angles, difficulty of reconciling 9
vulnerability as human condition 265
vulnerability assessments (VA) 8, 11, 22, 92–94; climate 37, 64, 270; conceptualization of 26; four prototypes of 25, 30; participatory 244, 248, 249; as policy instruments 24–27; political functions of 161; prevailing 164; subjectivity of 241; *see also* climate vulnerability assessments, China and the United States; social vulnerability assessments (SVAs)
vulnerability conceptualization 26, 219, 255; climate adaptation policy and 25; discursive 12; evolution of 11, 256; idealized 233, 241; social 29
vulnerability constructions 142, 165
vulnerability embodiment 265
vulnerability governance *see* climate vulnerability governance, China and the United States
vulnerability research 21–22
vulnerability thinking 20–24

Walsh-Dilley, Marygold 235
Wang, Weijun 48
Washington, DC 71, 248; Georgetown University 73
water: being exposed *to* 192; being safe *from* 192; China's focus on 96–98, 107, 257; policy responses to 206; *see also* flooding; floodwater and floodwater management; rainwater
water access: China 143
water availability: United States 65
water conservancy: Jinhua 192
water drainage: Jinhua 192
water governance: China 98, 198; Zhejiang 198–199
water infrastructure 43
water levels, rising: Georgia 270
water management: Atlanta 201–202; China 47, 193
water policy: China 190; Jinhua 206
"water reservoir migrants": China 129
water reservoirs: China 270
water resources: Atlanta 217; China 29; United States 104
water safety: China 190–192, 194; five different elements of (China) 192

water scarcity: Atlanta, Georgia 122–124, 130, 181; China 185
water sector: China 98, 260
water shortages: China 28, 48
water systems: China 105
Water Wars: United States 217
waterways: Ningbo 185
White House Council on Environmental Quality *see* Council on Environmental Quality (CEQ) (United States)
"wicked problems" 39, 51, 53n1
Wolford, Wendy 235
Worldbank Climate Resilient Cities Programm 185

Xia, Fang 271
Xi administration 92
Xi Jinping 65, 98
Xin'anjiang sub-basin 271

Winterfeld, Uta von 234
World Bank 184
World Bank Climate Resilient Cities Program 185
Wu, Hongmei 183, 189
Wucheng district 126
Wuyi county 126

Xia Baolong 191
Xia Dynasty 198
Xinjiang province 86

Yangtze River, China 68
Yangtze River Delta (YRD) region 68, 69, 125, 141–142; flooding of 271
Yanweizhou Flood Park, Jinhua 186, 187, 188, 208n4
Yellow River basin, China 105
Yin, Robert K. 10, 62, 78
Yiwu, China 126, 127, 185, 190, 192
YRD *see* Yangtze River Delta
Yunhe, China 68
Yunnan province 86
Yu the Great 198

Zhejiang Meteorological Development Plan 189
Zhejiang province (China): adaptation efforts 182–196; adaptation priorities **184**; Eco-province, plans to become 66–67; agricultural insurance pilot program 182; Climate Change plan 183, **184**; climate and topology,

overview of 271–272; floating population **129**; "Forest Zhejiang" 182; Georgia compared to 65–70; green infrastructure planning 185; Hangzhou 71, 125; heat hazard index *141*; in-migration to 217; Lishui 87, 127, 184, 185; local language dialects in 76; major cities *125*; migrant workers 128, 158; Ningbo 125; policy efforts directed at 190; poverty in 194; regional-local connotations of human vulnerability in 140–142; river chiefs 198; sea-level rise, vulnerability to 182; Thousand Mile Seawall 182; wealth, perception of 156; Wenzhou 125; vulnerability to climate extremes 66–67; *see also* Jinhua; Lanxi

Zhejiang Provincial Department of Land and Resources 182

"Zhejiang's Response to Climate Change" 183

"Zhejiang Urbanization Development Program" 188

Zheng, Yan 28, 47